Semiconductor Transport

Semiconductor Transport

David K. Ferry

CRC Press
Taylor & Francis Group
Boca Raton London New York

CRC Press is an imprint of the
Taylor & Francis Group, an **informa** business

First published 2000
by Taylor & Francis
11 New Fetter Lane, London EC4P 4EE

Simultaneously published in the USA and Canada
by Taylor & Francis Inc.
29 West 35th Street, New York, NY 10001

Taylor & Francis is an imprint of the Taylor & Francis Group

Typeset in Sabon by Graphicraft Limited, Hong Kong
Printed and bound in Great Britain by Biddles Ltd, Guildford
and King's Lynn

British Library Cataloguing in Publication Data
A catalogue record for this book is available
from the British Library

Library of Congress Cataloging in Publication Data
Ferry, David K.
Semiconductor transport / David K. Ferry.
p. cm.
Includes bibliographical references and index.
ISBN 0–7484–0865–7 (hb : alk. paper). — ISBN 0–7484–0866–5 (pb : alk. paper)
1. Semiconductors. 2. Electron transport. I. Title.
QC611.6.E45F47 2000
537.6'22—dc21 99–27795
CIP

ISBN 0–7484–0865–7 (hardback)
ISBN 0–7484–0866–5 (paperback)

Contents

Illustrations

Tables

Figures

Preface

The decision to sit down and write a textbook in almost any subject is one that the writer often comes to regret, as the task seems to be never ending. In the present case, this task has been eased as the present work is an outgrowth of an earlier book, *Semiconductors*, which covered far more than the material necessary for two graduate courses. That book was the result of some twenty years of teaching students material I felt they needed to know to begin research in semiconductor materials. Yet, time makes changes to this crucial set of required data, and it becomes more problematic just to maintain a grasp of the ever increasing material in transport theory that is important for modern semiconductors. Since that earlier work, new research areas in mesoscopic physics, where the characteristic length scales are comparable to the various relaxation lengths, have required new approaches to become important. Some of these new areas can be grouped together into *ballistic transport*, the case for which characteristic lengths are smaller than, or comparable to, mean free paths. Consequently, this new book focuses on trying to understand semiconductor transport in a manner that ranges from large bulk sizes, on one hand, to the modern ultrasmall mesoscopic size scales, on the other.

Particularly as one approaches the mesoscopic scale, much of the research has been done by people used to dealing with metallic behavior. We cannot forget, however, that semiconductors are unique materials, and it is erroneous in most cases to consider them as metals. For this reason, we begin early in the book to identify those important properties of semiconductors that differentiate them from metals. Transport is developed then along two somewhat different pathways: one is the traditional Boltzmann transport equation approach, while the second focuses on the kinetic properties of ensembles of carriers and the importance of the correlation functions in transport. As a result, the present book differs significantly from even those parts of *Semiconductors* dealing with transport. The growth of knowledge in this area has blossomed so much in the last decade, that it has been difficult to pull the treatments together, but I felt that some success has been achieved in this regard. Still, there are many areas that remain the subject of intense research and still contain considerable controversy. I have tried to identify these areas within the various chapters and sections. Many times, the material is presented to the student as if it were gospel truth (which it may be in some cases, as it seems to depend more upon faith than science), as opposed to merely being our current understanding, which is usually based upon limited information. Those teaching from the material, and particularly the students using the book, should realize that I am trying to convey the latter philosophy and have tried to at least hint at some of the remaining questions. Along the way, I have tried to list the more important references in each chapter, although this is in no way a complete listing.

At Arizona State University, we have a large program in semiconductors, and the material in this book has grown out of a one-semester (first-year) graduate course on semiconductor transport. It is assumed that the student is already familiar with an introductory course in quantum mechanics and an undergraduate course in semiconductor materials. Thus, this book is probably not suitable for a first course in semiconductors to the uninitiated, although the amount of presupposed information is not that extensive. For those familiar with the earlier *Semiconductors*, I have tried to maintain the level of that earlier work, while avoiding advanced topics such as quantum transport theory.

Appreciation and acknowledgements are due to a great many people. First, of course, are my wife and family, without whose support this would not have been possible. The list of scientists and engineers who have contributed to the understanding presented here is simply too long to tabulate; in the time that I have been working in this field, I have been exposed to an exceptionally large number of very bright people, all of whom have influenced (and continue to influence) my understanding. Fortunately, some of these have been my students and colleagues, and special thanks are due to them. Yet, there are a few people with whom I have interacted (and argued) for a long period of time and who have affected my understanding. From this group, I would like to thank John Barker (Glasgow), Carlo Jacoboni and Lino Reggiani (Modena, although Lino has since moved), Karl Hess (Illinois), and Hal Grubin (Glastonbury, CT) for many enlightening discussions and collaborations. In addition, my colleague Jonathan Bird read the entire manuscript during its preparation, while Dragica Vasileska, Shela Wigger, and Lucian Shifren read major parts of the work. In these cases, their resulting questions helped to illuminate many confusing points and catch a significant number of the typographical and mental errors. Certainly this has resulted in major improvements to the final manuscript. Nevertheless, there certainly remain confusions and errors, for which I must be held accountable.

Dave Ferry
Tempe, August 1999

Chapter 1

Introduction

As we approach the end of the twentieth century, the information revolution has significantly changed our lives. It has been more than two hundred years since the beginnings of the industrial revolution, a period characterized by the introduction of machines as an adjunct to multiply human muscle power. Over the last few decades, microelectronics has been used to multiply our computing and reasoning powers, and we speak of this using the term "information revolution." And yet, the beginnings of this latter revolution occurred relatively early in the last century. To understand this, it is necessary to understand that the modern microelectronics revolution would not have occurred without the materials known as semiconductors. Indeed, the properties of these materials, particularly the ability to create transistors and diodes through doping with impurities and control of the local potential by gates, are essential to microelectronics. These effects could not be achieved with metals!

The first apparent measurements, and recognition of the properties, of semiconductors were made by Michael Faraday (1833, 1834; Martin, 1932). It was the property of the conductivity *increasing* with temperature in AgS, rather than decreasing as in a metal, that identified the material as having new properties. Today, we know that this behavior is only characteristic in the so-called intrinsic regime, and can be modified by the inclusion of impurities. Again, it is these modifications to the properties that have proven to be so useful to modern high technology. The next important breakthrough was the observation of rectification in a metal–semiconductor contact by Braun (1874), who joined an iron pyrite to PbS. The useful property of photoconductivity, also not observed in metals, was found at about the same time by Smith (1873). Only a few years later, the observation of a transverse voltage developed across a semiconductor in which a current was flowing in the presence of a magnetic field was made by Hall (1879). The Hall effect remains one of the principal characterization tools for semiconductors even today. The rectification discovered by Braun proved useful in the new radio electronics, where it could be used for direct detection of the waves (Bose, 1904; Pierce, 1907). Yet, it was not until 1938 that Schottky provided the theory of the Schottky diode to explain the action of the metal–semiconductor contact (Schottky, 1938), and by this time several suggestions for surface-field-controlled semiconductor devices had been proposed (Lilienfeld, 1930; Heil, 1935). While it may seem that little was done between 1833 and, say, the Second World War, much work on the properties of semiconductors was carried out, and this led to the discovery of the germanium transistor by Bardeen, Brattain, and Shockley in 1947 (Bardeen and Brattain, 1948; Shockley, 1949). Finally, it was the discovery of the integrated circuit in the late 1950s by Kilby (1976) and the metal-oxide-semiconductor transistor (Shockley and Pearson, 1948; Moll, 1948; Pfann and Garrett, 1959; Khang and Atalla, 1960) that provided the rapid growth in microelectronics of the past few decades. There is no obvious end to the growth or extent of the inroads that microelectronics will make in our lives. Indeed, even in everyday objects such as our automobiles, we find multiple microprocessors being used to control the engine, emission controls, radio, temperature and climate of the passenger compartment, and even the ride, through an "active" suspension system.

The growth of microelectronics has been driven, and is in turn calibrated, by growth of the density of transistors, or gates, on an individual integrated circuit. Considering that the first transistor was invented in 1947, it is indeed phenomenal that we can now routinely place more than 100 million transistors in a single integrated circuit the size of only a few square centimeters. The cornerstone of this technology is silicon, a simple semiconductor material whose properties can be modified almost at will by proper processing technology, and which has a remarkably stable insulating oxide, SiO_2. But Si is just the dominant material currently (and probably for the foreseeable future). The history of the semiconductor electronics community has seen the importance of a wide variety of materials. Indeed, arguments are currently raging over the role of GaAs circuits, particularly for microwave applications, such as cellular phones, and high-speed data processing. Ge–Si heterostructure circuits, special oxide ceramics that compose high-temperature superconducting materials, and polymeric conductors are all under investigation for possible new application technologies. We use light-emitting diodes and laser diodes made of GaAs, AlGaAs, GaP, and other related compound semiconductors composed of the group III and group V elements. Far-infrared detectors depend on the properties of HgCdTe, a compound semiconductor composed of elements from groups II and VI. We not only use these materials in their bulk and relatively well-known forms, but also create artificial *superlattices* and *heterostructures*, which mix various compounds to produce structures in which the primary property, the band gap, has been engineered to have specific properties. What makes this all possible is that semiconductors generally have very similar properties which behave in like manner across a wide range of possible materials. This follows from the fact that all of the useful materials mentioned above have a single crystal structure, the zinc-blende lattice, or its more common diamond simplification.

Thus, although the wide range of properties is obtained by small changes in the basic properties of the individual atoms, the overriding observation is that these materials are characterized by their similarities.

Today's properties that arise in small semiconductor devices require knowledge gained only recently from the study of far-from-equilibrium systems, knowledge acquired in studying the properties of nonlinear transport at high electric fields and in physically small device geometries. Most notably, the latter has led to the study of so-called mesoscopic devices where the characteristic lengths are smaller than, or comparable to, the mean free path for scattering. In understanding how devices perform under a wide range of bias voltages and at the expected very small device geometries, we require a full understanding of the transport properties of the carriers within the device that provide for the current itself. Until a few years ago, the study of transport could be covered in reasonably complete detail simply by understanding the mobility and the diffusion coefficient for the electrons and holes. Then, simple drift and diffusion processes contributed the details of the current, and these were determined by the aforementioned mobility and diffusion coefficient, respectively. This is no longer the case, and a great deal of effort has been expended in attempting to understand just when these simple concepts begin to fail and what must be done to replace them. This field has been termed the study of *hot carriers in semiconductors*, but is really the study of all non-equilibrium properties that are reflected in the physics of transport in semiconductors. In essence, it is also this non-equilibrium behavior that sets semiconductors apart from metals, as the distribution function of the important carriers can be far different from the Fermi–Dirac function found in metals and semiconductors at equilibrium.

It is apparent from the above that we are now faced with trying both to understand and to predict the transport properties of a great many types of small and/or heterostructure devices, in which the properties of the host semiconductor, or semiconductors in heterostructures, are modified by local configurations that may involve superlattices, strain, high fields and voltages, and illumination, and for which these variations can occur over distances that are small compared to mean free paths or wavelengths of the electrons (or holes). To understand and to predict these properties, it has become essential to understand thoroughly the manner in which the properties of an individual semiconductor depend on the subtle differences between different semiconductors.

1.1 High-fields and non-equilibrium effects in devices

The properties of semiconductor devices that are necessary are those related to the transport of non-equilibrium carriers, and the control of this transport by various biases. Normally, in determining the transport properties, particularly from theoretical considerations, one wants to deal with an "average" electron (or hole), if possible, and write some sort of equivalent single-particle equation for its motion, thereby determining the motion of the ensemble. However, when ensemble equations are formulated, there quite often arise terms that cannot be identified with single-particle effects. For example, the diffusion equation implicitly incorporates effects arising from fluctuating forces as well as the deterministic terms. In the case of a linear or weakly nonlinear system, the fluctuating forces disturb the system to give rise to a component of entropy production that subsequently will decay toward an equilibrium with a characteristic decay time usually greater than the characteristic scattering time of the system. It is therefore apparent that one must deal with a multitude of characteristic times for the semiconductor system.

The importance of the various time scales is obvious in the analysis of non-equilibrium systems, since these systems present considerably more difficulty than that of equilibrium systems, due to the necessity of evaluating the time dependence of the various measurable properties. This will become apparent in later chapters. Indeed, these properties must be determined from equations describing them as evolving ensemble averages. The latter statement follows from the observation that the non-equilibrium distribution function is itself evolving (and is therefore non-stationary) in the nonlinear case over the time scale for variations of the measurable quantity. This implies that the system is non-ergodic (by which is meant simply that time averages do not equate to ensemble, or distribution, averages, the latter of which are the important averages) over this time scale. In fact, it is important to evaluate carefully the various collection of time scales that are important. In a semiconductor, numerous collisions occur and it is these collisions that provide the mechanism of exchange of energy and momentum, and relax these quantities toward their equilibrium values. There are collisions between the carriers which randomize the energy and momentum *within* an ensemble but do not relax either of these quantities for the ensemble as a whole. There are also elastic collisions between the carriers and impurities or acoustic phonons which relax the momentum but do little to relax the energy. Finally, there are inelastic collisions between the carriers and lattice vibrations which relax both the energy and the momentum. In general, four generic time scales can be identified (Mori *et al.*, 1962; Chester, 1963):

$$\tau_c < \tau < \tau_R < \tau_H \tag{1.1}$$

Here the average *duration* of a collision is denoted by τ_c. Generally, this time scale is quite short and not of importance to most considerations. However, on the scale of fast femtosecond laser experiments, this may no longer be true. The collision duration is the time required to establish the energy-conserving delta function (in the Fermi golden rule) used in Chapter 3. In distinction, the average time *between* collisions, the mean free time, is denoted by the simple τ. For time scales such that $t \leq \tau$, the evolution of the system depends strongly upon the details of the initial state. Generally, $\tau \gg \tau_c$, but this is not always the case at high electric fields, and the breakdown of this inequality can lead to new transport effects, which must be treated in a quantum mechanical manner.

The establishment of equilibrium, or a non-equilibrium steady state, can be achieved within a few or a few tens of τ, and the characteristic time associated with this process is the *relaxation time* τ_R. Typical quantities characterized by a relaxation time are the momentum relaxation process, and in high fields the energy relaxation process. If configuration space gradients exist, the situation becomes more complex. Relaxation in momentum space proceeds on the scale of τ_R and establishes a "local" equilibrium over regions smaller than a macroscopic scale, perhaps only a few mean free paths in extent. The achievement of a uniform equilibrium or non-equilibrium steady state requires a longer time, the hydrodynamic time $\tau_H > \tau_R$. Only for times large compared to this hydrodynamic time can the ensemble truly be said to be stationary, and only for times on this scale are the processes even beginning to become ergodic. Examples of the hydrodynamic time scale are diffusion times, arising from recombination processes for excess carriers as well as from local non-homogeneous carrier distributions and sometimes energy relaxation times.

For equilibrium or near-equilibrium linear transport, memory of the initial configuration is essentially destroyed on time scales $t \gg \tau$. In this case, no matter what the initial distribution, the final distribution is essentially a functional of the one-particle distribution. However, in many applications, it is desirable to follow the transient process on a shorter time scale without worrying about the details of the initial state. This can be achieved *if there is a faster scattering process* whose details are not important on the scale of interest. The usual implication is a dominance of carrier–carrier scattering, and it is assumed that there are two time scales of scattering. On the very fast time scale, inter-carrier scattering maintains a local equilibrium distribution, while the one-particle scattering time scale τ is a slower process (Bogoliubov, 1946). Thus, on the longer time scale of interest, a generalized distribution can be utilized for transport studies, whether classical or quantum mechanical. Indeed, for a non-degenerate semiconductor, it is usually assumed that the rapid carrier–carrier scattering produces a generalized Maxwellian distribution (Fröhlich, 1947), and that the Boltzmann equation can be used to determine the distribution function and the transport.

We can illustrate some of these points by considering a hypothetical device, such as that shown in Figure 1.1. Electrons are *injected* into the active region from the left and exit on the right in these panels. If the device is a bipolar transistor, then the left is the emitter and the active region is the base, with the potential barrier being controlled by the base–emitter bias. On the other hand, if the device is a field-effect transistor, then the left is the source and

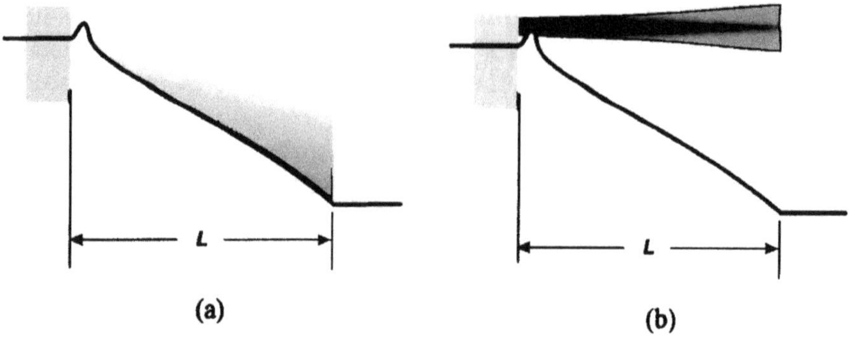

Figure 1.1 A prototypical device structure, and the characteristic distribution function as indicated by the shading, for the case (a) $L \gg l_R$ and (b) $L < l_R$, where l_R is a characteristic relaxation length.

the active region is the channel, with the potential barrier being controlled by the gate–source bias. We have illustrated two cases here. In Figure 1.1(a), it is assumed that the device is large, with $L \gg l_R \sim v_{dr}\tau_R$. That is, the length of the active region is larger than any characteristic relaxation length, whether this is the momentum relaxation length or the energy relaxation length. In this case, a non-equilibrium distribution function exists at each point in the channel, and this is indicated by the decreasing density of shading as the energy increases. While this distribution function is implicitly spatially varying, it can be found from a steady-state solution of the Boltzmann transport equation. If the total bias across the device is 5 V, and the channel is 0.5 μm long, the average field in the device is 10^5 V/cm, which is quite high. This will lead to high-field deviations of the transport, and to the case of ***hot electrons***. It will be noted in Figure 1.1(a) that the spread in the distribution increases as one approaches the right-hand side (collector or drain of the device), and this means that the characteristic temperature of the distribution is much higher than that of the lattice. This is what is meant by hot electrons. Shortly after the invention of the transistor, experiments into the high-field behavior of semiconductors were begun by Ryder and Shockley. These measurements showed that significant ***velocity saturation*** occurred for fields as small as 3 kV/cm in germanium at room temperature (Ryder, 1953). Subsequent work has shown that saturation velocity sets in at fields on the order of 20 kV/cm at room temperature in Si. Thus, in short-channel devices (ca. $L < 0.3$ μm), significant velocity saturation affects the carrier transport, and the device characteristics deviate significantly from the simple gradual-channel approximation usually studied.

The opposite extreme arises when $L < l_R \sim v_{dr}\tau_R$, that is when the length of the device is shorter than the relaxation length. This leads to what is termed ***quasi-ballistic*** transport, and is shown in Figure 1.1(b). Such situations are normally found in mesoscopic devices, and this behavior may be used to actually define a mesoscopic device. Usually, mesoscopic devices are studied at very low temperatures (10 mK–4 K). However, at room temperature, the energy relaxation length in Si may be of the order of 50–100 nm. It is expected that Si integrated circuits will continue to develop to where individual gate lengths may be as short as 20–40 nm within another decade, so that mesoscopic device effects may well come into importance in real devices. From Figure 1.1(b), it may be seen that the distribution function is very strongly peaked around the injection energy of the carriers. This distribution does not relax to anything close to an equilibrium form until well into the right-hand contact region (drain or collector). This distribution now is a very far-from-equilibrium distribution. Use of the Boltzmann equation in this regime is quite suspect, but the situation can be achieved easily at low temperature. All that is really required in the quasi-ballistic transport regime is that the applied bias $V > k_BT$. Device modeling here is further complicated as it is obvious from the figure that the transport depends very little on the detailed shape of the potential in the active region.

In both extremes discussed above, it is possible to analyze the details of the device behavior for these limiting situations. In general, a transport equation must be solved to ascertain the manner in which the applied fields give rise to current responses and to ascertain just how the energy taken up by the carriers from these fields is dissipated to the lattice and on to the heat bath. From the discussions above, it is clear that the details needed to solve this transport problem can be quite complicated. For a proper understanding of the high-field, far-from-equilibrium transport, it is first necessary to understand the general near-equilibrium transport in the semiconductor. Thus one must progress from an understanding of the structure, through near-equilibrium transport, to far-from-equilibrium transport. One cannot jump to the end and correctly ascertain the physics. One must first understand the underlying concepts and how they are being modified. In some sense, this is the goal of the present book. As a further preliminary, we look further at the hot-electron problem.

1.2 Hot carriers

In essence, any condition of current flow (any type of current – electrical, thermal, acoustic, etc.) in a semiconductor constitutes a condition of non-equilibrium behavior. In many cases, it is possible to calculate the proper response of the semiconductor to the driving forces causing the currents through simple approaches extending from linear response theory. In many cases, however, the system is driven far from equilibrium by the driving fields. The latter is the case in most semiconductor devices today. It was pointed out above that individual carriers, be they electrons or holes, may gain significant fractions of an electron volt within the active regions of a semiconductor device. If this energy is distributed throughout the total distribution of carriers, the effective temperature (if one can be defined) is well above the lattice temperature. In fact, it may easily be shown that this is the case. Consider as an example the simple case of two relatively isolated energy levels in the conduction band of a semiconductor, one lying an energy Δ above the second, where Δ is taken to be the energy of the phonon coupling these two levels (a phonon is a quantum unit of energy of the relevant lattice vibrations, as will be discussed later). Then the net rate of loss of energy by electrons sitting in the upper energy level (denoted as 2) is (this equation and the underlying physics will be developed in later chapters)

$$\frac{df(2)}{dt} = \Omega_{12} N_\Delta f(1) - \Omega_{21}(N_\Delta + 1) f(2) \tag{1.2}$$

where $\Omega_{12} = \Omega_{21}$ is a set of basic constants (the equality is set by detailed balance) and $f(1)$ and $f(2)$ are the occupation numbers of the particular levels. The quantity N_Δ is the occupation number for the phonons, given by the Bose–Einstein distribution. In essence, (1.2) is a transport equation and is a much simplified form of the Boltzmann equation. It will be studied in some detail later. The major point here is that the right-hand side constitutes a gain term and a loss term, in that order. These two terms equal each other in equilibrium and lead to the famous Boltzmann factor for the ratio of occupancies of different levels. Here the field upsets this balance between gain and loss. Now the electric field raises the average energy of the carriers, thus raising the density of electrons in level 2, represented by $f(2)$, above its equilibrium value. For this value to be a steady-state quantity, the decay term in (1.2) must become larger than the gain term, or

$$\frac{f(2)}{f(1)} > \frac{N_\Delta}{N_\Delta + 1} = \exp\left(-\frac{\Delta}{k_B T}\right), \tag{1.3}$$

where T is the lattice temperature. The right-hand form is just the famous Boltzmann factor, found in equilibrium. If a simple distribution for $f(1)$ and $f(2)$, characterized by an electron temperature T_e, is assumed, the only way for (1.3) to be satisfied is for T_e to be greater than T. This is termed the case of *hot carriers*. This result also reinforces the assertion that the carrier distribution function itself must change in high fields to signify that the system really is in a state best characterized as being far from equilibrium.

Hot carriers must be dealt with any time the semiconductor is driven far from equilibrium. Consider, for example, one of the devices mentioned above, say, a MOSFET. If the average electric field is on the order of 50 kV/cm, and the device is carrying a current of 10^5 A/cm^2, the average power input is on the order of 5×10^9 W/cm^3. If the local electron density is about 10^{18} cm^{-3}, this corresponds to an energy input per electron of about 3×10^{10} eV/s. The normal net energy loss rate by an electron in equilibrium is zero, as the emission and absorption of phonons by the electron is in balance. Thus, in order to have a net emission of phonons, it is necessary for the distribution to go out of equilibrium. For normal phonon energies, the energy loss above means that each electron must emit a net of approximately 10^{12} phonons per second if a steady-state condition is to be achieved for these parameters. Otherwise, the semiconductor would undergo thermal breakdown, as the carrier velocities would not stabilize at a finite level. For our case of Si, this means that the mean free time for phonon scattering must be of the order (or less than) 1 ps.

The overriding theoretical problem in hot-electron, or high-field, transport, and in understanding the behavior of these energetic carriers, is that of trying to understand the manner in which the distribution function of the carriers is modified by the electric field. This is true whether the subject of interest is a bulk semiconductor material or the performance of a semiconductor device. It is also a formidable experimental problem, as will be discussed below. The most common approach, whether for low fields and near-equilibrium conditions or for high fields and far-from-equilibrium conditions, is to solve the Boltzmann transport equation. However, to do so, it is first necessary to have an excellent knowledge of the band structure, the phonon spectra, the electron–phonon coupling mechanisms that lead to scattering, and so on. Even with experiment, in order to understand what the measurements are telling us, a good understanding of the possible processes that occur in semiconductors is needed.

The main experimental observable of our hot carriers, at least in homogeneous semiconductors, is the observation of velocity saturation. When the carriers are heated by the field, the temperature rises and the distribution spreads with more carriers at higher energy. Since the scattering rate generally increases with energy, there is more carrier scattering (to accommodate the relaxation of the energy input from the field), the mobility is reduced and the velocity does not increase as rapidly. In fact, in Si, the velocity appears to actually saturate at a value near to 10^7 cm/s at a lattice temperature of 300 K. We show this in Figure 1.2, where the results of a Monte Carlo simulation are plotted. In panel (a), the velocity is plotted for a range of electric field, while in panel (b) the distribution function at 40 kV/cm is plotted. The mobility at low field in Figure 1.2(a) is about 1500 cm^2/Vs, which is the value found in bulk, low-doped Si. In addition, the tail of the distribution in Figure 1.2(b), above about 30 meV, has a temperature decay corresponding to about 930 K. The simulation shows that the average energy is about 110 meV, which would correspond to about 850 K. This difference can be seen in the figure as an overall non-thermal shape in the distribution. A significant number of particles at low energy are not in a thermal distribution, even though the tail is quite thermal. This accounts for

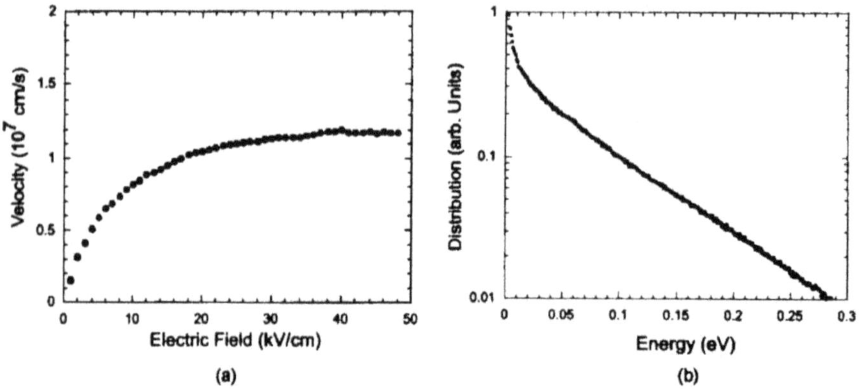

Figure 1.2 (a) The drift velocity for Si at room temperature as calculated with a Monte Carlo procedure. (b) The carrier distribution function at 40 kV/cm.

a difference between the average energy and the characteristic temperature of the tail. This is a quite general property of the non-equilibrium, high-field distribution function in the hot-carrier regime.

1.3 Space and time scales

In the previous sections, we discussed the major new effects that cause the transport to differ significantly from that of equilibrium semiconductor devices. The introduction of hot carriers is quite important in the long device, where the distribution can come to some form of steady-state, in which a balance between the driving field and the relaxation to the lattice is achieved. In the opposite extreme, corresponding to Figure 1.1(b), the transport is quasi-ballistic, and any steady-state transport is much more difficult to compute in that the details of the relaxation within the boundaries become an overwhelming consideration. For example, if the carriers come over the emitter barrier and reach the drain with 0.1 eV excess energy, and the scattering rate is 10^{13} s^{-1} in the drain, the carriers still have a mean free path of some >40 nm. If the active region is only 20–30 nm in extent, it is impossible to avoid calling the boundaries an active part of the overall device. Moreover, the electrons are quite likely to be both coherent and correlated throughout the active region, *and* for 1–2 mean free paths in the boundaries. Thus, it becomes important to understand the relative sizes of important correlation lengths and the device dimensions. The key lengths of the above sections have been taken to be the inelastic and elastic mean free paths, as well as some relaxation lengths. It is worthwhile at this point to make a few estimates of some of the key parameters that will be important in transport, and to assure ourselves that no fundamental limits of understanding are being violated. Some of these key parameters are listed in Table 1.1. In the case for which these numbers

Table 1.1 Some important parameters in typical devices

Parameter	*GaAs (4.2 K)*	*Si (300 K)*	*Units*
Density	0.4	4.0	10^{12} cm^{-2}
Mobility	5×10^5	500	cm^2/Vs
Scattering time	19	0.054	10^{-12} s
Fermi wave vector	1.58	5.0	10^6 cm^{-1}
Fermi velocity	2.73	3.06	10^7 cm/s
Elastic mfp	5.2	0.017	10^{-4} cm
Inelastic mfp	~25	~0.05	10^{-4} cm
Inelastic time	~100	~0.15	10^{-12} s

are tabulated, it is assumed that we have a quasi-two-dimensional electron gas localized at an interface. For the GaAs case, this assumption is predicated upon a modulation-doped AlGaAs/GaAs heterostructure with the dopant atoms in the AlGaAs and the free electrons forming an inversion layer on the GaAs side of the interface, and at low temperature. For the Si case, it is assumed that electrons are introduced at an interface either between Si and an oxide or between Si and, for example, a strained SiGe layer utilizing modulation doping. In the Si case, room temperature is assumed for the lattice. However, we take the lower mobility of the Si–SiO_2 interface for the example. In each case, a modest inversion density (equal for the two examples) is chosen, and the characteristic lengths are worked out from these assumptions.

The density is the sheet density of carriers in the quasi-two-dimensional electron gas at the interface discussed above. The actual density can usually be varied by an order of magnitude on either side of this value; for example, 10^{10}–10^{13} cm^{-2} are possible in Si and somewhat lower at the upper end in GaAs (2×10^{12} cm^{-2}). Again the mobility is a typical value, but in high-mobility structures, as much as 10^7 cm^2/Vs has been achieved in GaAs, and 2×10^5 cm^2/Vs has been achieved in Si (modulation doped with SiGe at the interface) at low temperature. Finding the scattering time from the mobility is straightforward, and masses of $0.067m_0$ and $0.19m_0$ are used for GaAs and Si, respectively. The Fermi wave vector is determined by the density through $k_F = (2\pi n_S)^{\frac{1}{2}}$. (It is often not appreciated that the inversion layer in an Si MOSFET is degenerate even at room temperature at relatively moderate densities.) The Fermi velocity is $v_F = k_F \hbar/m$. The elastic mean free path is then defined by the scattering time and the Fermi velocity as $l_e = v_F \tau_{sc}$. The inelastic mean free path is estimated for the two materials based upon experiments (Ferry and Goodnick, 1997), but it also should be recognized that there will be a range of values for this parameter. The inelastic mean free time, often called the phase-breaking or coherence time, is found through the relationship $l_{in} = v_F \tau_{in}$, and the inelastic, or phase-breaking, time is found from the estimate of the inelastic mean free path.

It should be noted that the *inelastic relaxation time* (such as the energy relaxation time) and the *phase coherence time* are two different quantities, even though these two quantities are often assumed to be interchangeable and indistinguishable. In fact, the inelastic relaxation time is just that, e.g., that to which we have referred as a hydrodynamic time above. That is, not all phase-breaking processes work to generate an inelastic relaxation time equally. Both of these quantities are important in phase interference, and we will try not to confuse the issue by using the two names inappropriately. But they do differ. The inelastic mean free path describes the distance an electron travels *ballistically* in the phase-breaking (or inelastic *scattering*) time τ_{in}. Let us try to explain this a little better. The inelastic scattering time gives us the time between actual inelastic processes, such as phonon emission or plasmon emission (for the electron–electron scattering process). In this sense, this scattering time is very nearly the phase-breaking time. But the inelastic relaxation time depends upon the difference in phonon (or plasmon) emission and absorption, weighted over the distribution function. So, the relaxation time is more of a hydrodynamic time, while the scattering time is more of a ballistic quantity.

Thus, the scattering time is useful in describing those processes, such as tunneling, in which the dominant transport process is one in which the carriers move ballistically with little scattering through the active region of interest. On the other hand, the coherence length l_ϕ is defined as a rate in which phase information is lost by the carrier. In many mesoscopic systems, the latter is associated with the *diffusion* constant D, usually in a disordered region. This means that the carriers are in a region of extensive scattering, so that their transport is describable as a diffusion process in which the coherence length is the equivalent diffusion length defined with the phase-breaking time, $l_\phi = \sqrt{D\tau_\phi}$. In some small nanostructures, the transport is neither ballistic nor diffusive but is somewhere between these two limits. For these structures, the effective phase relaxation length is neither the inelastic mean free path nor the diffusion-defined coherence length. These structures are more difficult to understand, and they are usually still quite sensitive to the boundary conditions. Care must be taken to be sure that the actual length (and the descriptive terminology) is appropriate to the situation under study. In modern devices, the coherence length is more appropriately applied to the energy relaxation length associated with the inelastic scattering time, and that is what is used in the table.

1.4 Modern device modeling

For most of the last fifty years, since semiconductor transistors appeared, these devices were modeled with simple approaches based on the gradual channel approximation and using simple drift mobility and diffusion constants to treat

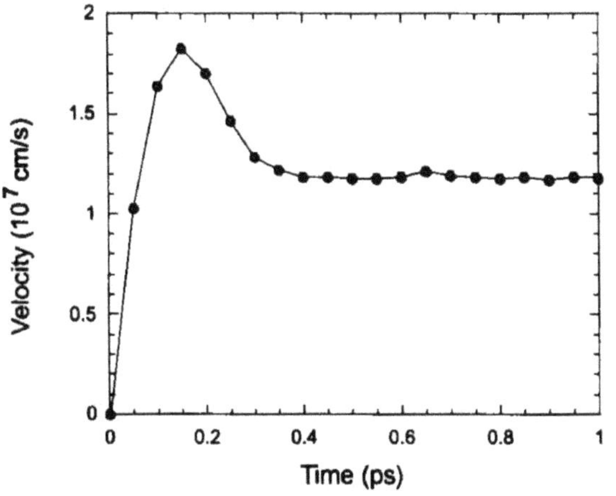

Figure 1.3 The simulated velocity as a function of time for homogeneous Si at 300 K and 40 kV/cm. This illustrates the effect of velocity overshoot.

the transport. As time progressed and field-effect transistors with gate lengths below 1.0 μm appeared, proper short-channel corrections and the inclusion of velocity saturation together with field-dependent mobilities allowed such approaches to still yield relatively good results. However, a turning point in the study of transport in semiconductor devices occurred with the discovery of the Gunn effect in GaAs (Gunn, 1963; Kroemer, 1966), and current instabilities in other bulk semiconductors. This turning point is best summarized by the special issue of the *IEEE Transactions on Electron Devices* in January 1966. The significance of this issue was the large number of papers dealing with the numerical simulation of the space- and time-dependent behavior of the charge distribution within semiconductor devices. These spatio-temporal variations implied – explicitly in some cases, implicitly in others – that these type of calculations could be used to develop the intuition needed to explain much device behavior. Today, simulation is used heavily, together with experiment, in the study of new devices and of scaled-down versions of older devices (Sai-Halasz *et al.*, 1988; Laux and Fischetti, 1988).

The need for more complicated device modeling, which includes specific time and space variations of the distribution functions, was signaled by the increasing importance of *velocity overshoot* in the device. This is illustrated in Figure 1.3 for Si at 300 K and 40 kV/cm. In this, the velocity first rises toward a value characteristic of the low electric field mobility, but as the distribution function expands toward that shown in Figure 1.2(b), the increase in the number of high energy electrons leads to increased scattering and a reduction in the velocity. Thus, the velocity rises initially and then decreases toward its final steady-state value. This only can occur when the energy relaxation time is longer than the momentum relaxation time. Figure 1.3

is for bulk material, with a homogeneous electric field, and the situation is much more complicated in a real device where the fields and densities are quite inhomogeneous. Such behavior is expected in GaAs devices with gate lengths below 0.25 μm and in Si devices with gate lengths below ~0.15 μm. Today, this is just one problem and new effects, such as quasi-ballistic transport, must be considered.

Numerical simulations are generally regarded as part of a device physicist's toolbox and are used routinely in several cases: (1) when the device transport is nonlinear and the device differential equations do not admit to exact solutions, (2) as surrogates for laboratory experiments that are either too costly and/or not feasible for initial investigations, and (3) in computer-aided design. Point 2 brings the study of transport in a semiconductor device environment into the general realm of ***computational science***, a so-called third paradigm of scientific investigation which adds to the previous ones of experiment and theory. In a sense, this new approach amounts to theoretical experimentation, or experimental theory. Suppose that theory suggests an idea, which can be verified by experiments. This new paradigm suggests that the idea can also be tested and verified by simulation *if the simulation contains sufficiently well-known properties of the semiconductor material and its transport properties to be meaningful*. This latter constraint is often not sufficiently appreciated, and it is worth noting that the detailed properties of such a well-known semiconductor as Si are still being debated today!

Simulation, or modeling, of semiconductor devices entails a number of factors. These are an equation describing the manner in which the potential and charge distributions are related in a self-consistent manner – the Poisson equation, and an equation describing the way in which the charge moves in response to the applied potentials. The latter is a *transport* equation. To these two equations there is added the constraint that it is necessary to fully understand the *boundary conditions*, which describe how fields, charge, the movement of charge, and other variables behave at the edges of the device or at interfaces within the device. It may well be the case that the latter constraint is more limiting on good simulation than is suspected. This entire package is then iterated until a self-consistent solution is found for one particular set of conditions, such as bias voltages. These may then be changed and new solutions found.

As pointed out above, the earliest simulations were one-dimensional solutions of the gradual channel approximation. With the reduction in device size, the need for two-dimensional solutions become quickly apparent. These dimensions are usually the direction from source to drain (or emitter to collector in a bipolar transistor), and the direction normal to the gate into the semiconductor. It is normally *assumed* that the transverse direction is sufficiently large that the variation in this direction is minimal. This assumption fails in small devices as the width of the device becomes important as well. As a result, full three-dimensional models have rapidly become quite important. These were first treated with drift and diffusion parameters for computational simplicity, but this is inadequate in small devices, and today full three-dimensional models with quite complicated transport kernals are under development in most laboratories.

A further example of the need for full three-dimensional models can be made by just considering the impurity atoms under the gate. In modern MOSFETs, the channel is not heavily doped to reduce impurity scattering, and a heavier-doped layer lies below the surface as a channel-stop layer. If we consider an example of a 0.15 μm channel length device with channel doping of 10^{17} cm^{-3} for a depth of 0.1 μm, and a channel width of 0.6 μm, then there are only 900 impurities in in the active channel region. The local variation of the charged impurity concentration is $\sqrt{N} = 30$ impurities, which corresponds to a 3 percent fluctuation in the concentration. While this seems to be quite small, it has been estimated that this will lead to a variation in threshold voltage across the chip that may well limit the down-sizing of future designs (Keyes, 1972). Clearly, for smaller devices than this simple example, the treatment of this impurity concentration as a uniform, homogeneous quantity will rapidly fail, and one must treat the individual impurities within the device for accurate simulation results. For our purposes, this greatly complicates the transport problem.

1.5 What this book includes

In the preceding sections, we have talked largely about the understanding of the semiconductor materials in terms of semiconductor devices, for the latter are the driving force for both the microelectronics revolution and further study of physical effects in semiconductors. However, this is not a device book. Rather, it is a book about transport properties and their calculation. Moreover, it is an introduction to those aspects of solid-state physics that distinguish semiconductors from other materials and the properties that are vitally important to understanding the transport in these materials. The impact of devices was used to emphasize a need for understanding the basic physics and chemistry of semiconductors. Although high-field transport and quasi-ballistic transport have been addressed in some detail, it is important to emphasize that full understanding of these can be achieved only when the low-field transport properties are fully understood.

In the next chapter, we begin by considering just how and why a semiconductor is *not* a metal. These differences go beyond just the temperature dependence of the conductivity, but arise from the basic differences of the band structures and the effective masses. Semiconductors are basically narrow-gap materials, and this leads to fundamental properties often referred to as nonparabolic bands and effective masses that are velocity and energy dependent. Then, in the third chapter, an introduction to simple particle transport, correlation functions, and the Boltzmann transport equation will be presented. This is then extended to inelastic processes, nonlinear transport, and high-field behavior in Chapter 4. In both of these chapters, the various scattering processes will be introduced at the appropriate time.

In Chapter 5, the discussion turns to diffusion and optical processes that become important in modern semiconductors. This discussion begins with the near-equilibrium properties and then moves to the dynamic, far-from-equilibrium effects seen in laser illumination of semiconductors. Our discussion turns to multi-band processes in Chapter 6, and these include both impact ionization and transfer between non-equivalent valleys of the conduction band – the Gunn effect (Gunn, 1963; Kroemer, 1966). We conclude with a serious treatment of the electron–electron interaction in Chapter 7, and its role in both high-field and quasi-ballistic transport.

PROBLEMS

1. One view of a surface field-effect transistor – the MOSFET – is that the gate voltage induces an areal charge density due to the oxide capacitance, and that this charge density can be written as

$$Q = -n_s e = C_{ox}[V_G - V_{th} - V(x)],$$

where V_G is the potential on the gate electrode, V_{th} is the threshold voltage where mobile charge begins to accumulate in the channel, and $V(x)$ is the surface potential in the channel itself. Here, C_{ox} is the capacitance per unit area of the oxide. If we write the source-drain current as $I = Qv = Q\mu E$, with the field given by $E = -dV(x)/dx$, then show that:

(a) For the boundary conditions of zero potential at the source $x = 0$ and V_{DS} at the drain $x = L$, the current is given by

$$I_D = \frac{CW\mu}{L}[V_G - V_{th} - \tfrac{1}{2}V_{DS}]V_{DS},$$

where W is the channel width.

(b) If $C_{ox} = 6.7 \times 10^{-4}$ F/m^2, $L = 0.25$ mm, $\mu = 400$ cm^2/Vs, $V_{DS} = 2.4$ V, $V_{th} = 0.5$ V, plot the transconductance $g_m = dI_D/dV_G$ as the gate voltage is varied from threshold to the drain potential.

2. Now consider the situation of Problem 1 for the case in which velocity saturation is found to occur in the channel. If we describe this event by a field-dependent mobility of the form

$$\mu = \frac{\mu_0}{(1 + \mu_0 E/v_{sat})},$$

where μ_0 is the low-field value of the mobility and $v_{sat} = 1 \times 10^7$ cm/s is the saturation velocity, find the new equation for the drain current as a function of the drain and gate voltages. Compute the transconductance as a function of gate voltage, for the same conditions as the previous problem, and compare the two curves. Plot the drift velocity as a function of distance from source to drain for $V_G = V_{DS} = 2.4$ V. How does this compare with the values in the absence of velocity saturation?

3. If we consider the above values for a typical transistor, compute the elastic mean free path from the mobility and the thermal velocity at the source (300 K). If the average energy input per electron is taken to be evE at each point in the channel, use the values of velocity along the channel found in the last problem (for velocity saturation) to plot the energy input per electron along the channel. Now, assume that the electron temperature varies with the field as $3k_B(T_e - T_0)/2 \sim \alpha E^2$ and reaches a value of 1500 K at the drain end of the channel. What is the value of α? Plot the electron temperature along the channel. Using the results you have achieved for the energy input along the channel and the temperature along the channel, plot the energy relaxation τ_{in} time along the channel, if we use the energy balance

$$evE = \frac{3k_B(T_e - T_0)}{2\tau_{in}}.$$

Here, T_0 is the lattice temperature (300 K).

REFERENCES

Bardeen, J., and Brattain, W., 1948, *Phys. Rev.*, **74**, 230.

Bogoliubov, N. N., 1946, *Problemi dinam. teorii. u. stat. fiz.*, Moscow.

Bose, J. C., 1904, US Patent, 755840.

Braun, F., 1874, *Ann. Phys. Pogg.*, **153**, 556.

Chester, G. V., 1963, *Rep. Prog. Phys.*, **26**, 411.

Faraday, M., 1833, *Experimental Researches in Electricity*, Ser. IV, pp. 433–9.

Faraday, M., 1834, *Beibl. Ann. Phys.*, **31**, 75.

Ferry, D. K., and Goodnick, S. M., 1997, *Transport in Nanostructures* (Cambridge, UK: Cambridge University Press).

Fröhlich, H., 1947, *Proc. Roy. Soc.*, **B70**, 124.

Gunn, J. B., 1963, *Solid State Commun.*, **1**, 88.

Hall, E. H., 1879, *Amer. J. Math.*, **2**, 287.

Heil, O., 1935, British Patent, 439, 457.

Keyes, R., 1972, *Science*, 195, 1230.
Khang, D., and Atalla, M. M., 1960, *Proc. Solid-State Devices Res. Conf.*, unpublished.
Kilby, J., 1976, *IEEE Trans. Electron Devices*, 23, 648.
Kroemer, H., 1966, *Proc. IEEE*, 52, 1736.
Laux, S. E., and Fischetti, M., 1988, *IEEE Electron Device Lett.*, 9, 467.
Lilienfeld, J. E., 1930, US Patent, 1,745,175.
Martin, T., Ed., 1932, *Faraday's Diary*, Vol. 2 (London: G. Bell and Sons, Ltd.) pp. 55–6.
Moll, J. L., 1948, *IEEE Wescon Conv. Rec.*, Pt. 3, 32.
Mori, H., Oppenheim, I., and Ross, J., 1962, in *Studies in Statistical Mechanics*, Ed. by de Boer, J., and Uhlenbeck, G. E. (Amsterdam: North Holland).
Pfann, W. G., and Garrett, G. C. B., 1959, *Proc. IRE*, 47, 2011.
Pierce, G. W., 1907, *Phys. Rev.*, 25, 31.
Ryder, E. J., 1953, *Phys. Rev.*, 90, 766.
Sai-Halasz, G. A., Wordeman, M. R., Kern, D. P., Rishton, S., and Ganin, E., 1988, *IEEE Electron Device Lett.*, 9, 464.
Schottky, W., 1938, *Naturwissenschaft.*, 26, 843.
Shockley, W., 1949, *Bell Sys. Tech. J.*, 28, 435.
Shockley, W., and Pearson, G. L., 1948, *Phys. Rev.*, 74, 232.
Smith, W., 1873, *J. Soc. Telegraph Engr.*, 2, 31.

Chapter 2

Semiconductors are *not* metals

When Faraday first measured a semiconductor, the distinguishing feature was the fact that the conductance actually increased with temperature, rather than decreased as expected for the well-known metals (Faraday, 1833, 1834; Martin, 1932). The discovery of photoconductivity (Smith, 1873) and rectification (Braun, 1874) provided still further effects unheard of in metals. All of these effects together suggest that the important major difference between semiconductors and metals is the presence of the energy gap in the spectrum of the semiconductor. The increase in conductance with temperature arises from the thermal excitation of an increasing carrier density across the energy gap. Likewise, photoconductivity arises by photo-excitation of carriers across the energy gap. And the presence of the rectification arises from the ability to excite excess carriers in the semiconductor under one bias direction, but not in the metal under the opposite bias condition. This latter process is viable because of the significantly smaller number of carriers in the semiconductor than in a metal. It is this combination of properties, which were already known by late in the 19th century, that sets the stage for the creation of *p–n* junctions and transistors. It is perhaps remarkable to realize that it took an additional 70+ years for this to occur, although we have to understand that electronics, radio waves and wireless communications were all unknown at the time. As with many discoveries, it required both the technology and the need for a development to provide a timely advance. Still, it is the presence of the energy gap, and a gap of a modest size, that makes semiconductors the useful materials that they are. In this chapter, we want to discuss how this gap occurs, and how it impacts a great many other properties of importance. In this approach, the *one-electron model* is discussed. Here, interactions among the electrons and between the multiple electrons and the lattice are not included, and only the structure for a single electron in the periodic potential of the atoms is discussed.

First, however, it is necessary to discuss how the presence of the atomic lattice introduces periodicity that must be reflected in the electronic energy band structure. This leads to a discussion of the type of wave functions, which are known as Bloch functions, that can occur and the interactions that lead to gaps existing in the spectrum are presented. This discussion is interlaced with a discussion of the effective mass in both metals and semiconductors, and the important distinction in how it must be defined between the two types of materials. This is followed by the chemical bond picture in order to understand how the bonding orbitals interact and lead to the actual conduction and valence bands. This will introduce the LCAO (linear combination of atomic orbitals) method of calculating the band structure, although it will be treated as a semi-empirical theory, which means that the various coupling constants are treated as adjustable parameters to be fit with experimental data. Then, the pseudopotential method of calculating energy bands is briefly introduced primarily because of its use in many modern so-called "full-band" simulations of transport. Finally, the details of the non-parabolicity in semiconductor bands are discussed.

2.1 Nearly-free electrons and periodic potentials

In most crystals, the interaction between the bonding electrons and the nuclei, or lattice atoms, is not negligible. The lattice has certain periodicities, and therefore the potential that arises from the lattice atoms must possess these same periodicities. Thus for a (perfect) one-dimensional crystal, with a lattice constant of a_0, the potential must satisfy

$$V(x + a_0) = V(x), \tag{2.1}$$

for any value of x within the crystal. In three dimensions, this becomes

$$V(\mathbf{r} + \mathbf{L}) = V(\mathbf{r}), \tag{2.2}$$

where

$$\mathbf{L} = n_x\mathbf{a}_x + n_y\mathbf{a}_y + n_z\mathbf{a}_z, \tag{2.3}$$

the n_i are integers and the $\mathbf{a}_i$ define the basic lattice (assumed to be a form of cubic so that these vectors align with the coordinate axes in space). Equation (2.3) defines $\mathbf{L}$ as a vector of the real-space lattice. These properties now must be applied to the Schrödinger equation

$$-\frac{\hbar^2}{2m}\nabla^2\Psi(\mathbf{r}) + V(\mathbf{r})\Psi(n) = E\Psi(\mathbf{r}). \tag{2.4}$$

In this last equation, the interactions between the electrons have been omitted, which leads us to the so-called "one-electron" band structure. The solutions of (2.4) are close to that of the nearly-free electron case if the potential is weak. However, *it is important to note that the basic periodicities in the wave function arise from the periodicities of the potential due to the lattice and not from any quantum mechanics of the system, beyond that of the Schrödinger equation.*

In this section the aim is to examine the general nature of the solutions to (2.4) and the special class of functions – Bloch functions – that arise. The one-electron nearly-free electron bands are computed and it is shown how these are appropriate for the case of metals. The effective mass for the metal will then be introduced. For semiconductors, however, it is important to examine the manner in which the lattice potential creates gaps and bands in the electron dispersion relations. These gaps are the important aspects of the semiconductors, and change the manner of interpreting the effective mass, which is very important in transport. We begin by examining the form of the wave functions obtained as solutions for (2.4).

2.1.1 Bloch functions

Since the potential term in (2.4) is periodic and possesses the symmetries of the real-space lattice through its periodicity in L, it may be expected that the probability of finding an electron on any atom is the same throughout the system. Thus the quantum mechanical probability for this may be expressed as the requirement that

$$|\Psi(\mathbf{r})|^2 = |\Psi(\mathbf{r} + \mathbf{L})|^2. \tag{2.5}$$

This must be satisfied throughout the crystal, that is, for any value of L. For a primitive lattice vector $\mathbf{a}_i$, the wave function will be connected as

$$\Psi(\mathbf{r}) = \lambda\Psi(\mathbf{r} + \mathbf{a}_i), \quad |\lambda|^2 = 1. \tag{2.6}$$

We can go slightly further, by noting that the imposition of periodic boundary conditions upon the entire crystal requires

$$\Psi(x) = \Psi(x + L_{max}), \quad L_{max} = N\mathrm{a}_0, \tag{2.7}$$

where N is the number of atoms in the chain, and a_0 is the unit spacing of the atoms (assumed equal in all three directions). In three dimensions, this will be the maximum value of any of the three n_i. That is, the values in (2.3) must lie in the ranges

$$1 \le n_x \le N_x, \quad 1 \le n_y \le N_y, \quad 1 \le n_y \le N_y. \tag{2.8}$$

With this in mind, we require that

$$\lambda = e^{i2\pi n/N}, \tag{2.9}$$

which, of course, just leads to the ideas of a reciprocal lattice existing for the crystal. The factor λ is a phase factor that accounts for the product of a real space displacement and a reciprocal space vector k. Then, one can easily show that the wave function is required to have the form

$$\Psi(\mathbf{r}) = e^{i\mathbf{k}\cdot\mathbf{r}} u_k(\mathbf{r}), \tag{2.10}$$

where the function $u_k(\mathbf{r})$ has the periodicity of the lattice itself. The wave function (2.10) is termed a Bloch wave function, or a *Bloch function*. It is important to note once again that the Bloch functions arise from the periodicity of the potential and not really from any quantum mechanical result. They are a property of waves in a periodic structure and appear in classical structures as well (Brillouin, 1953).

The foregoing properties of Bloch functions may be seen in another way that is somewhat more formal. This is through the use of Fourier series techniques. Here the crystal potential is expanded in a Fourier series as

$$V(\mathbf{r}) = \sum_{\mathrm{G}} U_{\mathrm{G}} e^{i\mathrm{G}\cdot\mathbf{r}}. \tag{2.11}$$

where G is any vector of the reciprocal lattice (in fact, the sum is over all vectors of the reciprocal lattice). To couple to this, the wave function is also expressed as a Fourier series:

$$\Psi(\mathbf{r}) = \sum_{\mathbf{k}} C(\mathbf{k}) e^{i\mathbf{k}\cdot\mathbf{r}}. \tag{2.12}$$

These two series are then substituted into the Schrödinger equation (2.4), which leads to

$$\sum_{\mathbf{k}} \left[\frac{\hbar^2 k^2}{2m} C(\mathbf{k}) + \sum_{\mathrm{G}} U_{\mathrm{G}} C(\mathbf{k}) e^{i\mathrm{G}\cdot\mathbf{r}} - E C(\mathbf{k}) \right] e^{i\mathbf{k}\cdot\mathbf{r}} = 0. \tag{2.13}$$

It is now necessary to solve (2.13) for the coefficients of the wave function expansion. To proceed, the change of variables $\mathbf{k}' = \mathbf{k} + \mathrm{G}$ is made in the second term, and then the primed set of variables is redefined into an unprimed set of variables to achieve

$$\sum_{\mathbf{k}} \left[\frac{\hbar^2 k^2}{2m} C(\mathbf{k}) + \sum_{\mathrm{G}} U_{\mathrm{G}} C(\mathbf{k} - \mathrm{G}) - E C(\mathbf{k}) \right] e^{i\mathbf{k}\cdot\mathbf{r}} = 0. \tag{2.14}$$

Thus, the final form of the energy can be found by noting that a sufficient condition on the equality of (2.14) is that the bracketed term vanishes, or

$$(E_k - E)C(\mathbf{k}) + \sum_{\mathbf{G}} U_{\mathbf{G}} C(\mathbf{k} - \mathbf{G}) = 0, \quad \text{with} \quad E_k = \frac{\hbar^2 k^2}{2m}. \tag{2.15}$$

The result (2.15) represents an entire set of equations, since one can easily shift the vector **k** by any reciprocal lattice vector **G**′. If the crystal potential is zero (e.g., $U_G = 0$), the free electron parabolic energy bands are recovered. In general, this is not the case and several modifications of the energy structure due to the crystal potential can be expected to occur.

From (2.15), a continuous spectrum of Fourier coefficients is not present. In fact, only a discrete number of values of the vector **k** are allowed by the discretization introduced by the periodic boundary conditions. These are selected by the set of reciprocal lattice vectors and the primitive vectors of the unit cell (of the reciprocal lattice). For a particular momentum state given by

$$\Psi_{\mathbf{k}}(\mathbf{r}) = C(\mathbf{k})e^{i\mathbf{k}\cdot\mathbf{r}}, \tag{2.16}$$

within an arbitrary normalization constant, the wave function is just

$$\Psi_{\mathbf{k}}(\mathbf{r}) \sim \sum_{\mathbf{G}} C(\mathbf{k} - \mathbf{G})e^{i(\mathbf{k}-\mathbf{G})\cdot\mathbf{r}} = \left[\sum_{\mathbf{G}} C(\mathbf{k} - \mathbf{G})e^{-i\mathbf{G}\cdot\mathbf{r}}\right]e^{i\mathbf{k}\cdot\mathbf{r}}. \tag{2.17}$$

The factor in parentheses in (2.17) is just the lattice periodic part of the Bloch function described above and becomes the Fourier series representation of $u_{\mathbf{k}}(\mathbf{r})$. Thus, it is clear that the general solutions of the Schrödinger equation in a periodic potential are the Bloch functions described in (2.12). The Fourier representation will be used below in making estimates of the size of the gap that is opened by the crystal potential (the terms in U_G) and the nature of the bands near this gap.

2.1.2 Free electron bands and the Brillouin zone

The wave functions may now be written as Bloch functions – a plane wave modulated by a function that has the periodicity of the lattice itself. The cell periodic part of the Bloch function, the $u_{\mathbf{k}}(\mathbf{r})$ term, will become quite important in our later treatment of bond orbitals in semiconductors, but for metals is a relatively smooth function. If any interactions between the electrons are ignored, the Bloch functions are the proper one-electron functions. If the Bloch function (2.9) is inserted into the Schrödinger equation (2.4), the result is

$$\nabla^2 u_{\mathbf{k}}(\mathbf{r}) + i\mathbf{k}\cdot\nabla u_{\mathbf{k}}(\mathbf{r}) + \frac{2m}{\hbar^2}\left[E(\mathbf{k}) - \frac{\hbar^2 k^2}{2m} - V(\mathbf{r})\right]u_{\mathbf{k}}(\mathbf{r}) = 0. \tag{2.18}$$

In the case in which $V(\mathbf{r}) = 0$, a valid solution is one in which $u_\mathbf{k}$ is a constant, so that the wave function is just a plane wave, provided that

$$E(\mathbf{k}) = \frac{\hbar^2 k^2}{2m}. \tag{2.19}$$

Thus, in the case in which the lattice potential can be ignored, the free electron case is recovered identically. The first change that may be expected from the presence of the lattice potential is a modification of the basic results of the free electron band structure. These modifications will be pursued in later sub-sections. Here, we want to explore the basic free electron band structure further.

The Bloch wave is not really unique, and it has been sufficient to define **k** only for those values of $n_i < N_i$ according to (2.8). These values correspond to the first Brillouin zone. However, the treatment above tells us that we can shift this *crystal* momentum by any vector **G** of the reciprocal lattice. Suppose that the momentum in the Bloch wave is shifted by a reciprocal lattice vector **G**, so that the wave function (2.10) becomes

$$\Psi_\mathbf{k}(\mathbf{r}) = e^{i(\mathbf{k}+\mathbf{G})\cdot\mathbf{r}} e^{-i\mathbf{G}\cdot\mathbf{r}} u_\mathbf{k}(\mathbf{r}) = e^{i(\mathbf{k}+\mathbf{G})\cdot\mathbf{r}} u_{\mathbf{k}'}(\mathbf{r}), \tag{2.20}$$

where $u_{\mathbf{k}'}$ differs from $u_\mathbf{k}$ only by the phase factor of the reciprocal lattice vector, which is unity. Hence, the wave vector can only be defined modulo a reciprocal lattice vector. The importance of this can be seen by substituting the second form of (2.20) into the Schrödinger equation, so that

$$\nabla^2 u_{\mathbf{k}'}(\mathbf{r}) + i\mathbf{k}' \cdot \nabla u_{\mathbf{k}'}(\mathbf{r}) + \frac{2m}{\hbar^2}\left[E(\mathbf{k}' - \mathbf{G}) - \frac{\hbar^2 k'^2}{2m} - V(\mathbf{r})\right] u_{\mathbf{k}'}(\mathbf{r}) = 0. \tag{2.21}$$

But since the wave functions are identical in (2.20), it is clear that the energy must be a periodic function of the reciprocal lattice vectors, since there is nothing magic about the primed or unprimed wave vectors used in the arguments above. Thus the energy bands in the free electron case consist of an infinite set of parabolas, which for a given value of the reciprocal lattice vector are centered in the extended zone as about different points in the reciprocal lattice, as

$$E_n(\mathbf{k}) = \frac{\hbar^2}{2m}(\mathbf{k} - \mathbf{G}_n)^2. \tag{2.22}$$

If only the values of **k** that lie in the first Brillouin zone are taken, the energy is a multi-valued function of **k**, and different branches are characterized by different lattice periodic parts $u_\mathbf{k}$ of the Bloch function. Each branch, indeed

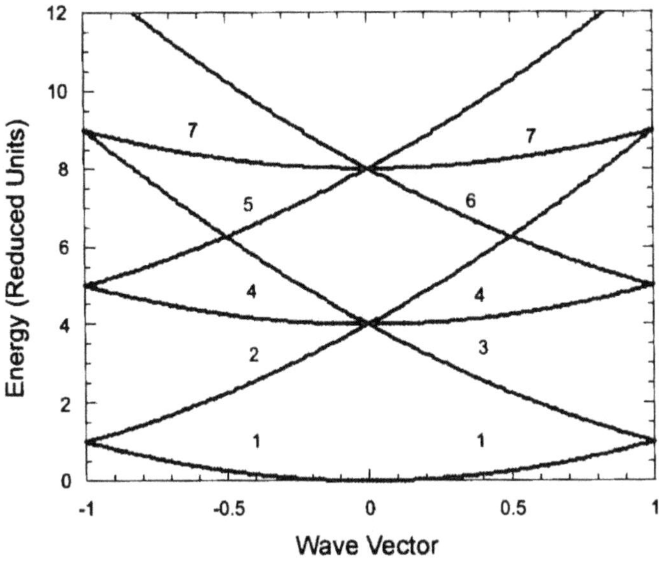

Figure 2.1 The free electron energy bands in the first Brillouin zone. The wave vector is in units of π/a, while the energy is in units of $\hbar^2\pi^2/2ma^2$.

each energy value, in the first Brillouin zone is now labeled with both a momentum index k and a *band index n*. The bands are degenerate (several cross) at the zone edges (at $\pm\pi/a$) in one dimension. There are also degeneracies at $k = 0$ in general, and other accidental degeneracies may occur at other points in the Brillouin zone. It is at these degeneracies that the crystal potential is expected to modify the basic free electron band structure. In Figure 2.1, the set of bands in the first Brillouin zone along the (100) axes is shown. The curve labeled "1" is obvious and corresponds to G = 0. The curves "2" and "3" are also fairly obvious, corresponding to $\mathbf{G} = \pm(2\pi/a)\mathbf{a}_{100}$, where $\mathbf{a}_{100}$ is a unit vector in the (100) direction. The next curve, labeled "4" corresponds to

$$E_4 = \frac{\hbar^2k^2}{2m} + \frac{\hbar^2}{2m}G_{010}^2 = \frac{\hbar^2k^2}{2m} + \frac{\hbar^2}{2m}G_{001}^2 = \frac{\hbar^2k^2}{2m} + \frac{\hbar^2}{2m}G_{0\bar{1}0}^2$$

$$= \frac{\hbar^2k^2}{2m} + \frac{\hbar^2}{2m}G_{00\bar{1}}^2 = \frac{\hbar^2k^2}{2m} + \frac{\hbar^2}{2m}\left(\frac{2\pi}{a}\right)^2.$$

(Here, the line over the number corresponds to a negative direction.) Thus, this set of energy states is four-fold degenerate. Similar considerations lead to the higher lying energy bands, which arise from initial values of G at other, more distant positions.

2.1.3 Metals and the effective mass

Metals like Cu, Ag, and Au are elements from the first column (column IB) of the periodic table. They have a single valence electron which contributes to the bonding, and this electron is in an *s*-state, so that its wave function is relatively isotropic throughout the unit cell of the crystal. This is why we could take the cell-periodic part of the Bloch function to be "constant" and ignore the gradient terms in the Schrödinger equation. This also leads to the principal reason the conductivity is so high in these materials. There are precisely N values of the wave vector $\mathbf{k}$ in the definition that follows from (2.8) and (2.9), in which we recognize that the crystal momentum takes the values

$$k_x = \frac{2\pi n_x}{Na_0}, \quad k_y = \frac{2\pi n_y}{Na_0}, \quad k_z = \frac{2\pi n_z}{Na_0},$$

and N is the number of atoms in the crystal. (Normally, each of the above runs from $-\pi/a_0$ to $+\pi/a_0$, rather than merely the positive values.) But, each state can hold two electrons, if they have opposite spin, as we have ignored the spin angular momentum of the states. Hence, each energy band (such as "1" in Figure 2.1) can hold $2N$ electrons, but the atoms contribute only N electrons. This means that the band is one-half filled, and the Fermi energy lies midway through the possible states. This is shown in Figure 2.2. In one dimension, we would have $k_F = \pi/2a$, but in three dimensions, the factor of 2 is distributed among these three dimensions, so that $k_F = 2^{-\frac{1}{3}}\pi/a \cong 0.79\,\pi/a$, as shown in the figure. As a result, there are N electrons contributing to the conduction through the crystal, and it is quite easy for carriers to gain energy and move above the Fermi energy. While the actual band structure will vary somewhat, we do not introduce much error by the assumption that this is a proper description of the metallic situation.

The key point of the half-filled band is that there is no gap in the energy spectrum. Yet, we can readily take the Fermi energy (at $T = 0$) as the reference energy. If we expand the energy around the Fermi energy, then in one dimension we may write the two terms for the two points in the zone where the energy band crosses the Fermi energy as

$$E(k_F + \delta k) = E(k_F) + \delta k \left.\frac{\partial E}{\partial k}\right|_{k_F} + \frac{(\delta k)^2}{2}\left.\frac{\partial^2 E}{\partial k^2}\right|_{k_F} + \ldots$$

$$E(-k_F - \delta k) = E(k_F) - \delta k \left.\frac{\partial E}{\partial k}\right|_{k_F} + \frac{(\partial k)^2}{2}\left.\frac{\partial^2 E}{\partial k^2}\right|_{k_F} + \ldots \tag{2.23}$$

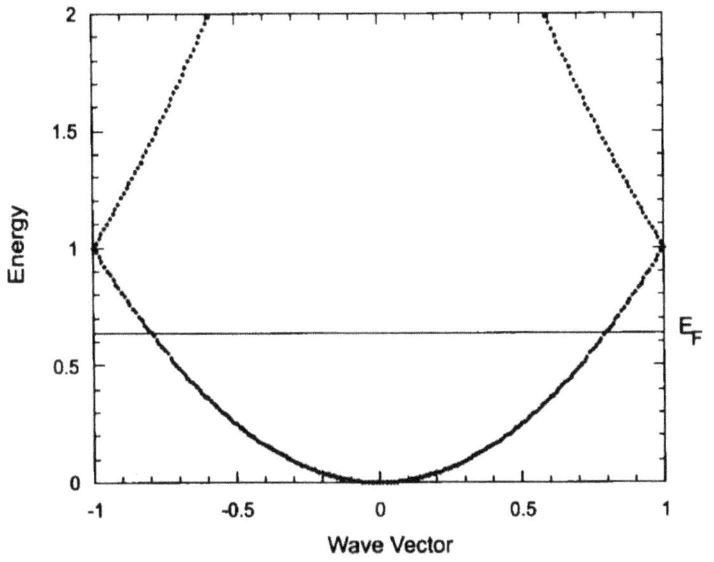

Figure 2.2 The lowest band showing the position of the Fermi energy. We show the three-dimensional case. The Fermi energy does not lie at the mid-point of the energy band, since there are $2N$ states (accounting for spin degeneracy), but only N electrons, and the factor of 2 must be distributed between the three dimensions. The Fermi wave vector k_F is thus about 0.79 in the reduced units (defined in the caption to Figure 2.1). In one dimension, the Fermi wave vector would be at 0.5.

and we may form the expansion by taking one-half of the sum of these two terms (this summing over the number of equivalent points in the zone will reappear in the multi-valley semiconductors like Si), so that

$$E(\delta k) \approx E(k_F) + \frac{\hbar^2 (\delta k)^2}{2m^*}, \tag{2.24}$$

where

$$\frac{1}{m^*} = \frac{1}{\hbar^2}\frac{\partial^2 E}{\partial k^2} \sim \frac{1}{m} \tag{2.25}$$

is the *effective* mass. Because the band structure is nearly that of the free electron approximation, the mass is very nearly the free electron mass $m = 9.1 \times 10^{-31}$ kg.

While we have expanded the energy bands around the Fermi energy, a different approach is normally taken, yet the same result is obtained. Here, we use the Newton equation of motion, with the coupling that the group velocity is obtained from the band structure through the normal energy dispersion relation:

$$v = \frac{\partial \omega}{\partial k} = \frac{1}{\hbar}\frac{\partial E}{\partial k}. \tag{2.26}$$

Now, if we equate the crystal momentum $\hbar k$ with the "particle" momentum m^*v, then we can use the acceleration theorem to yield

$$\frac{dp}{dt} = F = \hbar\frac{dk}{dt} = m^*\frac{dv}{dk}\frac{dk}{dt} = \frac{m^*}{\hbar}\frac{\partial^2 E}{\partial k^2}\frac{dk}{dt}. \tag{2.27}$$

Equating the third and last terms immediately leads to the "definition" of the effective mass that is given in (2.25). This so-called "definition" is quite widely used, even to the point of being construed as the one and only *true* definition of the effective mass. However, there is a flaw in this definition that will prove its non-universal (even wrong) nature. We can summarize the problems with (2.25) and (2.27) as follows. First, these equations say that the mass varies with the value of k, yet it has been brought outside the derivative with respect to k in (2.27). This is mathematically incorrect. In fact, the form (2.27) is only correct for a purely parabolic band structure, not for any real band structure. Nevertheless, it is the common definition. In metals, in fact, the mass is nearly that of the free electron and varies little within the solid crystal structure, regardless of the definition used. A more correct definition of the mass will be developed below, after a more thorough discussion of semiconductor band structure.

2.1.4 Opening energy gaps

At this point we need to ask for just what purpose the crystal potential was included in the Schrödinger equation (2.18). The fact is that the crystal potential is crucially important in those regions where several bands cross in Figure 2.1. There should be no crossings! The crystal potential provides an interaction between the various bands that leads to gaps opening in these regions. This will be very important for semiconductors, as these materials usually have 4 valence electrons per atom, and with two atoms per unit cell, all available states in the valence band are full. While this normally occurs at the $k = 0$ point of the lowest bands (the energy point $E = 4$ in Figure 2.1), the procedure will be demonstrated for the lowest crossing between bands "1" and the "2,3" bands at $k = \pm\pi/a$.

At the points where the various branches of the energy bands cross, it is expected that interaction will occur between the wave functions. However, as will be seen below, it is necessary that the crossing be such that the crossing point (π/a in Figure 2.1) be connected by a reciprocal lattice vector to its image ($-\pi/a$ in Figure 2.1). In essence, this is a Bragg reflection condition, in that the two points are connected by a reciprocal lattice vector. This means that the points where the crystal potential will open major gaps are at the zone edges and at the zone center (the points where the bands cross in Figure 2.1). Consider, for example, the point π/a and its image $-\pi/a$. At these points it suffices to write the plane-wave part of the Bloch wave as a sum of two such waves, in one dimension for example,

$$\Psi \sim e^{ikx} \pm e^{i(k-2\pi/a)x} \to e^{i\pi x/a} \pm e^{-i\pi x/a} \tag{2.28}$$

as the contributions from the two waves representing the two branches that are degenerate at the zone edge. These two wave functions may be combined into two (probability) densities as

$$\rho_+ = |\Psi_+|^2 = 4\cos^2\left(\frac{\pi x}{a}\right), \quad \rho_- = |\Psi_-|^2 = 4\sin^2\left(\frac{\pi x}{a}\right). \tag{2.29}$$

The signs on the wave functions of (2.29) correspond to the appropriate sign taken in (2.28) for the combination wave functions. Now ρ_+ has its peaks at $x = na$, which is at the atomic sites in the real space lattice. Thus, the electron is localized on the atom and is more strongly bound to the atomic potential. This must be the bonding orbital. On the other hand, ρ_- has its peaks at $(n + \frac{1}{2})a$, which lies *between* the atoms. This orbital must correspond to the non-bonding orbital (or anti-bonding, as it is usually called). The crystal potential leads to an interaction that raises the degeneracy of the energy levels for these two different states.

Suppose that only the lowest Fourier coefficient of the potential is considered and written as

$$V(x) = -U\cos\left(\frac{2\pi x}{a}\right), \tag{2.30}$$

which has its most negative values at the atomic positions. The change in energy of the two states may be calculated by normal time-independent perturbation theory (Ferry, 1996), in which the interaction energy is found by integrating the interaction potential (2.30) over the two wave functions, as

$$\delta E_+ = \frac{1}{a}\int_0^a V(x)\rho_+ dx = -\frac{4U}{a}\int_0^a \cos\left(\frac{2\pi x}{a}\right)\cos^2\left(\frac{\pi x}{a}\right)dx = -U. \tag{2.31}$$

The same value is obtained for the anti-bonding wave function. Hence the crystal potential in this simple approach opens a gap between the bonding and the anti-bonding orbitals. The energy of the bonding orbitals is lowered and the energy of the non-bonding orbitals is raised, the consequence of which is an energy gap of $2U$ between the levels.

Let us now examine this in somewhat more detail, using the approach outlined in (2.15). This will allow us to determine the energy band variation in the vicinity of the gap. To be precise, we want to examine the gap that opens at $k = \pi/a$, so that the value of G that is used in the equation is basically $2\pi/a$. We will couple only these two bands, although it is obvious that this also involves the coupling for the negative of G to obtain the other half of the dispersion relation. Thus, we write (2.15) for these two values of k as

$$(E_k - E)C(k) + U_G C(k - G) = 0$$
$$(E_{k-G} - E)C(k - G) + U_G C(k) = 0. \tag{2.32}$$

Obviously, the determinant of the coefficient matrix must vanish if solutions are to be found, and this leads to

$$E = \left(\frac{E_k + E_{k-G}}{2}\right) \pm \left[\left(\frac{E_k - E_{k-G}}{2}\right)^2 + U_G^2\right]^{\frac{1}{2}}. \tag{2.33}$$

At the zone edge, E_k and E_{k-G} are equal, and the energy levels are just

$$E = E_{\pm\pi/a} \pm U_G. \tag{2.34}$$

Thus, the gap is $2U_G$ as discussed above. While we haven't shown it here, the lower energy state is the bonding state and has even symmetry, while the upper state is the anti-bonding state and has odd symmetry.

We can carry this argument a little bit further to point out the problems with the effective mass. Suppose that a small deviation away from the zone edge is considered. What do the bands look like for this variation? To study this, take $k = (G/2) - \delta k$. Then each of the energies may be expanded as

$$E_k = \frac{\hbar^2}{2m}\left(\frac{G^2}{4} - G\delta k + (\delta k)^2\right), \tag{2.35a}$$

$$E_{k-G} = \frac{\hbar^2}{2m}\left(\frac{G^2}{4} + G\delta k + (\delta k)^2\right). \tag{2.35b}$$

Using these values of the free electron energies at the zone edge gives the resultant energy levels as

$$E = E_{G/2} + \frac{\hbar^2(\delta k)^2}{2m} \pm \sqrt{4E_{G/2}\frac{\hbar^2(\delta k)^2}{2m} + U_G^2}\,. \tag{2.36}$$

Here, $E_{G/2}$ is the free electron energy at the zone edge. If we introduce the band edge energies from (2.34) as

$$E_a = E_{G/2} + U_G, \quad E_b = E_{G/2} - U_G, \tag{2.37}$$

the variation of the energy bands for small values of δk may be written as

$$E_a(\delta k) = E_a + \frac{\hbar^2(\delta k)^2}{2m}\left(\frac{2E_{G/2}}{U_G} + 1\right)$$

$$E_b(\delta k) = E_b - \frac{\hbar^2(\delta k)^2}{2m}\left(\frac{2E_{G/2}}{U_G} - 1\right). \tag{2.38}$$

The crystal potential has produced a gap in the spectrum and the bands curve much more strongly than in the free electron model. This gap, and the deviation from the parabolic free electron model introduce an effective mass that is significantly different from the free electron mass m.

2.1.5 A proper definition of the effective mass

In the limit of small deviation from the zone edge, (2.38) leads to a description of the effective masses as

$$\frac{m}{m_a^*} = 1 + \frac{2E_{G/2}}{U_G}, \quad \frac{m}{m_b^*} = \left(1 - \frac{2E_{G/2}}{U_G}\right). \tag{2.39}$$

It is important to note that the various sign conventions that have been introduced lead to a bonding effective mass that causes the energy to decrease. While this may seem to be negative, it is the proper convention here. We show the band variation in Figure 2.3. If it were not for the factor of unity in each term of (2.39), the two bands would be mirror images of each other, when reflected vertically through an energy in mid gap. The two bands are not quite mirror images of each other in (2.36).

For larger values of δk (just how large cannot be said right now), it is not fair to expand the square roots that are in (2.36). The more general case is given by

$$E(\delta k) = E_{G/2} \pm U_G\sqrt{1 + \frac{2\hbar^2(\delta k)^2 E_{G/2}}{mU_G^2}}, \tag{2.40}$$

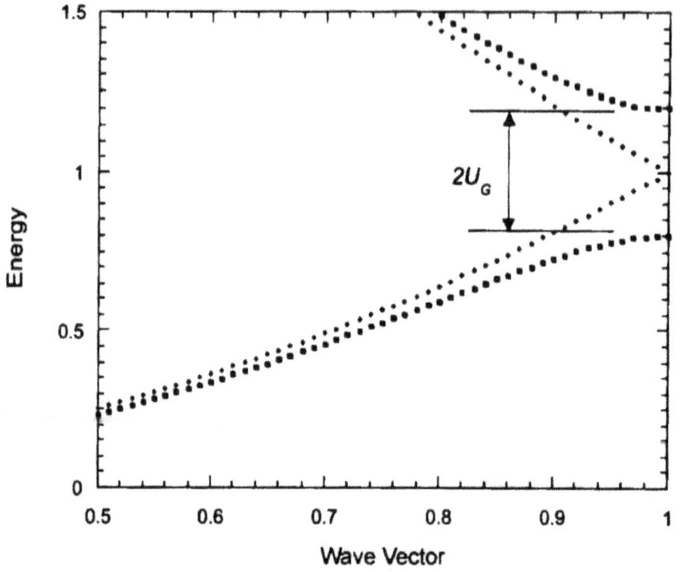

Figure 2.3 The presence of the lattice potentials causes gaps to open in the free electron bands (the light dots here). These gaps lead to semiconductor behavior if the bottom band is fully occupied. The reduced units are the same as those of Figure 2.1.

where the "free electron" term $\hbar^2(\delta k)^2/2m$ has been disregarded as being negligible (in that $m^* \ll m$). The first term is just the bottom of the anti-bonding band shifted downward by one-half of the energy gap (U_G). That is, the first term is just the crossing of the two free electron bands in Figure 2.3. If the bottom of the anti-bonding band is taken as the zero of energy for the present purposes, the two bands can be written as

$$E(\delta k) = -\frac{E_{gap}}{2} \pm \frac{E_{gap}}{2}\sqrt{1 + \frac{2\hbar^2(\delta k)^2}{m_c E_{gap}}}, \qquad (2.41a)$$

where

$$m_c = mU_G/2E_{G/2} = mE_{gap}/4E_{G/2} \qquad (2.41b)$$

is the conduction band effective mass at vanishing δk, given by (2.39) in the absence of the factor of unity, and we recognize that $E_{gap} = 2U_G$. Here the effective mass (for small δk) that has had the free-electron part removed has also been utilized, taken from (2.36). Equation (2.41) is a general *hyperbolic* band variation that is found in nearly all direct-gap compound semiconductors and even in Si and Ge. The method used to calculate the effective mass, and for which one must define the m^*, as in (2.41b), as the value at the band edge, is quite important and the results will differ for different definitions. The parabolic bands are recovered only so long as the kinetic energy (inside the square root) is small compared to the energy gap itself. The cause of the non-parabolicity is quite straightforward. The gap is opened by the repulsion between the bonding and the anti-bonding wave functions generated by the interaction with the crystal potential. As in all perturbation theory, the strength of the interaction is inversely proportional to the gap itself (the energy difference of the two states), so that as δk increases, the gap increases and the repulsive force is decreased. The latter reduces the driving force moving the bands apart, which provides the nonlinearity. Physically, the hyperbolic bands account for the fact that the mass of the electrons must change from the band-edge effective mass at $\delta k = 0$ to the free-electron mass for very large δk in this simple model, which is built up from the nearly free electron bands. As long as one is near the zone edge, or the band extremum, the bands are nearly parabolic with the band-edge value of the effective mass. When we stray from this band-edge position, the complication of an energy-dependent mass must be dealt with.

The important point in this discussion is that the effective mass is now a function of δk. This means that the mass is energy dependent, and velocity dependent as well. And this important point means that the treatment of (2.27) is in error. In this earlier treatment, the derivative of the momentum was translated into a derivative of the velocity. To be correct, we should have written this equation as

$$F = \hbar\frac{dk}{dt} = \frac{d(m^*\mathrm{v})}{dt} = \left(m^* + \mathrm{v}\frac{dm^*}{d\mathrm{v}}\right)\frac{d\mathrm{v}}{dt}$$

$$= \left(m^* + \mathrm{v}\frac{dm^*}{d\mathrm{v}}\right)\frac{1}{\hbar}\frac{\partial^2 E}{\partial k^2}\frac{dk}{dt}. \tag{2.42}$$

This now completely changes the definition of the effective mass. The second derivative of the energy, with respect to k, is not the reciprocal of the mass, but of a more complicated function. How then do we define the effective mass? It turns out that this is rather simple, but requires us to be more careful with our definition of the momentum. In fact, we introduced a connection, valid for a single energy band and within a single Brillouin zone, between the wave momentum $\hbar k$ and the "particle" momentum $m\mathrm{v}$. It is this connection that must be generalized to account for the effective mass in our non-parabolic bands. Now, we must use the energy (velocity) dependent effective mass to make this connection as

$$\hbar k = m^* v, \quad \frac{1}{m^*} = \frac{1}{\hbar^2 k}\frac{\partial E}{\partial k}. \tag{2.43}$$

If this definition is used in (2.42) for Newton's law, we find that

$$m^* + v\frac{dm^*}{dv} = \frac{\hbar k}{v} - v\frac{\hbar k}{v^2} + \hbar\frac{dk}{dv} = \hbar\frac{dk}{dv}, \tag{2.44}$$

which then satisfies the equality (2.42). This means that *we have to recognize (2.43) as the more fundamental definition of the effective mass, a definition based upon establishing a strong relation between crystal momentum and carrier momentum, within a single band and Brillouin zone.*

We can illustrate the more proper definition of the mass with (2.43) even for the case of metals. When the energy gap is opened, we can consider the one-dimensional energy band to be described approximately by

$$E(k) = \frac{E_w}{2} - \frac{E_w}{2}\cos(ka). \tag{2.45}$$

Here, E_w is the *band width*, and the Brillouin zone is described by $-\pi/a < k \le \pi/a$. In one dimension, the metal has sufficient electrons to fill the band up to the half-band point, or $E_F = E_w/2$. The velocity in this cosinusoidal energy band is given by the derivative, as described above,

$$v(k) = \frac{E_w a}{2\hbar}\sin(ka). \tag{2.46}$$

At the half-filled band point, the velocity is a maximum, given by $E_w a/2\hbar$. However, the second derivative of the energy vanishes at the Fermi energy of this model. It is now impossible to use the simple approach of (2.27) to connect an infinite mass times a finite velocity to produce the finite value $k = \pi/2a$. Thus, any attempt to connect the "mass" with the second derivative of the energy band fails as soon as we open a gap and produce a non-parabolic energy band! Our only recourse is to use (2.43) to make a *proper* (and consistent) definition of the effective mass in these non-parabolic energy bands.

In three dimensions, where we may encounter anisotropic bands, such as the ellipsoids that will be found in Si, the mass is traditionally described as a tensor. For our diamond and zinc-blende crystals though, this tensor is a diagonal second-rank tensor, but this is not always the case. For these crystals, we can define the three-dimensional tensor mass in one of two ways:

$$\left(\frac{\overleftrightarrow{1}}{m^*}\right)_{ij} = \frac{1}{\hbar^2 k_i}\frac{\partial E}{\partial k_j}, \quad \text{or} \quad (\overleftrightarrow{m}^*)_{ij} = \hbar^2 k_i \left(\frac{\partial E}{\partial k_j}\right)^{-1}. \tag{2.47}$$

Here, the double-ended arrow indicates a second-rank tensor, and the subscripts refer to a particular element of this tensor.

2.2 The bond-orbital model

At this point it has been shown that the basic structure of the energy bands can be inferred from a knowledge of the atomic lattice and its periodicity. However, the metals normally have a single itinerant electron which is diffusely spread among the atoms and is, consequently, relatively free to move. The semiconductors in which we are interested are quite different, in that they are *tetrahedrally* coordinated. In these semiconductors, there are four electrons (on average) from each of two atoms in the basis of the diamond structure (this is the face-centered cubic structure with a basis of two atoms per lattice site) and these four electrons are in the *s* and *p* levels of the outer shell. For example, the Si bonds are composed of 3*s* and 3*p* levels, while GaAs has bonds composed of 4*s* and 4*p* levels, as does germanium. In each case, the inner shells are not expected to contribute anything at all to the bonding of the solid. This is not strictly true, as the inner *d* levels often lie quite close to the outer *s* and *p* levels when the former are occupied. This would imply that there is some modification of the energy levels in GaAs due to the filled 3*d* levels. This correction is small, and will not be considered further. The *p* wave functions are quite directional in nature, and this leads to a very directional nature of the bonding electrons. Thus, the electrons are not diffusely spread, as in a metal, but are quite localized into a set of hybrids which join nearest neighbor atoms together. This hybrid orbital bonding is shown in Figure 2.4.

The bonds are composed of *hybrids* that are formed by composition of the various possible arrangements of the *s* and *p* wave functions. There are, of course, four hybrids for the tetrahedrally coordinated semiconductors. These four hybrids may be written as

$$\begin{aligned}
|h_1\rangle &= \tfrac{1}{2}[|s\rangle + |p_x\rangle + |p_y\rangle + |p_z\rangle],\\
|h_2\rangle &= \tfrac{1}{2}[|s\rangle + |p_x\rangle - |p_y\rangle - |p_z\rangle],\\
|h_3\rangle &= \tfrac{1}{2}[|s\rangle - |p_x\rangle + |p_y\rangle - |p_z\rangle],\\
|h_4\rangle &= \tfrac{1}{2}[|s\rangle - |p_x\rangle - |p_y\rangle + |p_z\rangle].
\end{aligned} \tag{2.48}$$

The first of these hybrids points in the (111) direction, while the other three point in the $(1\bar{1}\bar{1})$, $(\bar{1}1\bar{1})$, and $(\bar{1}\bar{1}1)$ directions, respectively (the bar

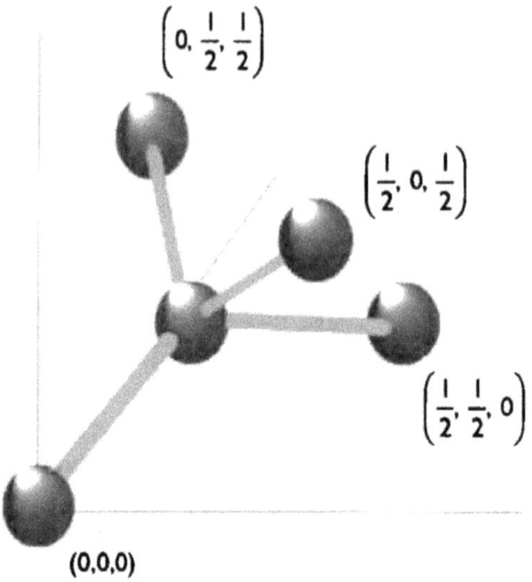

Figure 2.4 The inner atom of the two on the basis located at (0,0,0) is bonded to its four nearest neighbors (on the adjacent faces of the cube) by highly directional *orbitals*. These sp^3 orbitals give highly directional bonds.

over the top indicates a negative coefficient). The factor of $\frac{1}{2}$ is included to normalize the hybrids properly so that $\langle h_i|h_j\rangle = \delta_{ij}$. These hybrids are now directional and point in the proper directions for the tetrahedral bonding coordination of these semiconductors. Thus the bonds are directed at the nearest neighbors in the lattice. For Ga in GaAs, the four hybrids point directly at the four As neighbors, which lie at the points of the tetrahedron. The locations of the various atoms for the tetrahedral bond are shown in Figure 2.4.

Each of the atomic levels possesses a distinct energy level that describes the atomic energy in the isolated atom. Thus the *s* levels possess an energy given by E_s and the *p* levels have the energy E_p. In general, these levels are properties of the atoms, so that the levels are different in the heteropolar compounds like GaAs. The levels will be marked with a superscript *A* or *B*, corresponding to the A–B compound that forms the basis of the lattice. In the following, the compound semiconductors will be treated, as they form a more general case, and the single component semiconductors, such as Si or Ge, are a special case that is easily obtained in a limiting process of setting A = B.

The *s* and *p* energy levels are separated by an energy that has been termed the *metallic* energy (Harrison, 1980). In general, this energy may be defined from the basic atomic energy levels through

$$4V_1^A = E_p^A - E_s^A, \quad 4V_1^B = E_p^B - E_s^B. \tag{2.49}$$

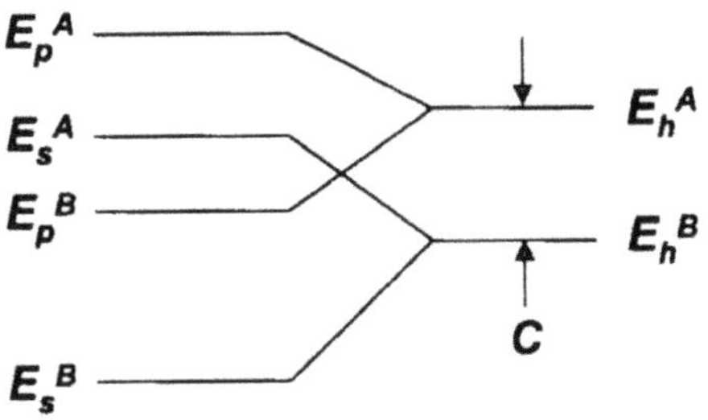

Figure 2.5 The atomic energies form hybrids, as indicated here, for each of the two atoms in the basis. These hybrids are separated by the *heteropolar* energy *C*.

Here the *A* atom is the cation and the *B* atom is the anion in chemical terms. The hybrids themselves possess an energy that arises from the nature of the way in which they are formed. Thus the *hybrid* energy is

$$E_h = \frac{\langle h_i|H|h_i\rangle}{\langle h_i|h_i\rangle} = \frac{1}{4}\frac{\langle s|H|s\rangle + \langle p_x|H|p_x\rangle + \langle p_y|H|p_y\rangle + \langle p_z|H|p_z\rangle}{1}$$
$$= \tfrac{1}{4}(E_s + 3E_p), \tag{2.50}$$

for all values of the index *i*, where *H* is the Hamiltonian operator for the Schrödinger equation representing the crystal but neglecting any interaction between the atoms (without these interaction terms, the cross terms that would arise in (2.50) will vanish by orthonormality). In Figure 2.5, the hybrid energy is derived from the atomic energies in a compound semiconductor.

The two hybrid energies are separated by the *hybrid polar energy* or the *heteropolar energy*, depending on whose definitions one wants to use. The notation C has been used here for this energy. The heteropolar energy is a product of the ionic transfer of charge in the compound semiconductor (since Ga has only 3 electrons and As has 5 electrons, there is a charge transfer in order to get the average 4 electrons of the tetrahedral bond). Of course, this energy vanishes in a pure single compound such as Si or Ge, which are referred to as homopolar materials. That is, they are composed of a homogeneous set of atoms, while the general compound semiconductor is heterogeneous in that it contains two types of atoms. The heteropolar energy C may be easily evaluated using (2.50) as follows:

$$C = E_h^A - E_h^B = \tfrac{1}{4}(E_s^A - E_s^B) + \tfrac{3}{4}(E_p^A - E_p^B). \tag{2.51}$$

It is important to note that the hybrids are not eigen-states of either the isolated atom or of the crystal. Rather, they are constructed under the premise that they are the natural wave function for the tetrahedral bonds. But, we have created them as what seems a natural form (at least to us). In principle,

Table 2.1 Atomic energy levels for the atoms that make up the tetrahedrally coordinated semiconductors

Symbol E_s E_p				
IIB	IIIB	IVB	VB	VIB
		C −17.52 −8.97	N −23.04 −11.47	
	Al −10.11 −4.86	Si −13.55 −6.52	P −17.1 −8.33	
	Ga −11.37 −4.9	Ge −14.38 −6.36	As −17.33 −7.91	
Cd −7.7 −3.38	In −10.12 −4.69	Sn −12.5 −5.94	Sb −14.8 −7.24	Te −17.11 −8.59
Hg −7.68 −3.48				

they will be stable in the crystal once the interactions between the various atoms are included. In fact, they are not orthogonal under action of the Hamiltonian, since

$$\langle h_i|H|h_i\rangle = \tfrac{1}{4}(E_s - E_p) = -V_1, \tag{2.52}$$

so that the metallic energy measures the interaction between the various hybrids. In this sense, the metallic energy describes the contribution to the energy of the itinerant nature of the electrons. In Table 2.1, the values of the atomic energy levels are listed for some representative atoms that contribute to the semiconductors of interest.

Now we have to turn to the interaction between hybrids localized on neighboring atoms. Only standing-wave interactions will be considered here ($\mathbf{k} = 0$) and the propagating wave-function-dependent changes are left to a later section. The interaction energy between two atoms on different sites can arise from e.g. one atom's hybrid pointed in the (111) direction and the nearest neighbor in that direction, displaced $\left(\frac{a}{4}, \frac{a}{4}, \frac{a}{4}\right)$, whose hybrid points in the opposite direction (there is a complete flip of the hybrid directions of the atoms as one moves along the body diagonal direction). Including the angular integration, the interaction energy between these nearest neighbor hybrids is

$$-V_2 = \frac{1}{4}\langle s^A|H|s^B\rangle + \frac{\sqrt{3}}{4}[\langle s^A|H|p^B\rangle + \langle p^A|H|s^B\rangle]$$
$$+\frac{3}{4}\langle p^A|H|p^B\rangle. \tag{2.53}$$

Harrison (1980) has argued that these energies should depend only on the interatomic spacing d ($= \sqrt{3}a/4$), and that they should have the general form

$$V_2 \cong 4.37\frac{\hbar^2}{md^2}. \tag{2.54}$$

Phillips (1973) also argues that V_2 should be a function of the interatomic spacing, but that it should also satisfy another scaling rule. Since the atomic radii are the same for each row of the periodic table, the value of V_2 should be the same in AlP as in Si, the same in GaAs and ZnTe as in Ge, and so on. In other words, the value for this quantity is set by the distance between the atoms, and this really does not change as one moves across a row of the table. Thus this value is the same in heteropolar compounds as in homopolar compounds and Phillips has termed V_2 the homopolar energy, $E_{ho} = 2V_2$ (this should not be confused with the hybrid energy E_h, which differs for the two atoms). The *average energy gap* between bonding and anti-bonding orbitals is thus composed of a contribution from the homopolar energy E_{ho} and a contribution from the heteropolar energy C.

The bonding orbital will be composed of hybrids based on the atoms at each end of the bond, as suggested above. These bonding orbitals can be written as a linear combination of the two hybrids at each atom, as

$$|\psi_{bo}\rangle = u_1|h^A\rangle + u_2|h^B\rangle, \tag{2.55}$$

where the u_i are coefficients to be determined. This is achieved by minimizing the expectation value of the energy determined by the Hamiltonian, which is given by

$$E = \frac{\langle\psi_{bo}|H|\psi_{bo}\rangle}{\langle\psi_{bo}|\psi_{bo}\rangle} = \frac{u_1^2E_h^A - 2u_1u_2V_2 + u_2^2E_h^B}{u_1^2 + u_2^2}, \tag{2.56}$$

with respect to both u_1 and u_2 separately. This leads to two equations (from the partial derivatives)

$$2u_1E = 2u_1E_h^A - 2u_2V_2, \quad 2u_2E = 2u_2E_h^B - 2u_1V_2. \tag{2.57}$$

Introducing the average energy $E_0 = (E_h^A + E_h^B)/2$, the determinant of coefficients for (2.57) may readily be solved to give the energies

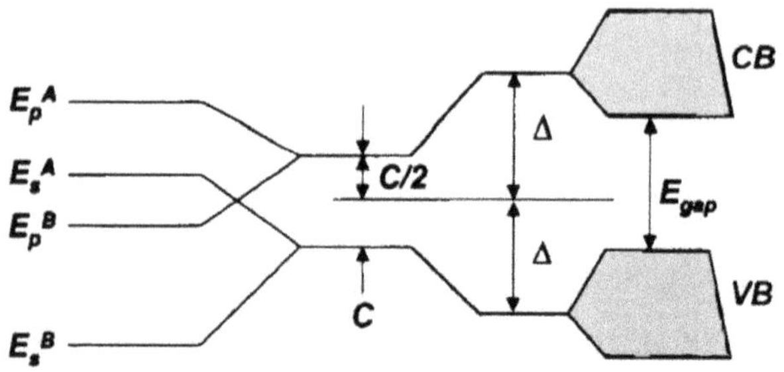

Figure 2.6 The hybridization that forms between adjacent atoms leads to the average energy levels (shown separated by 2Δ) around which the energy bands now form when the periodicity is introduced.

$$E = E_0 \pm \frac{1}{2}\sqrt{(2V_2)^2 + C^2} = E_0 \pm \Delta. \tag{2.58}$$

The bonding (lower sign) and the anti-bonding (upper sign) energy levels are symmetrically spaced about the average hybrid energy of the two atoms. Of course, in a homopolar material, these two hybrid energies are the same and are equal to E_0. The significance of the average energy is unexpected in heteropolar materials but is important for getting the positions of the average bonding and anti-bonding energies correct. The positions of these energy levels are shown in Figure 2.6. Note that the square root appearing in (2.58) is denoted as the quantity Δ in the figure.

We can use the two eigenvalues in either of the two equations of (2.57) to determine the quantities u_i. These lead to the coefficients

$$u_1 = \sqrt{\frac{1+\alpha}{2}}, \quad u_2 = \sqrt{\frac{1-\alpha}{2}}, \quad \alpha = \frac{C}{\sqrt{C^2 + 4V_2^2}}. \tag{2.59}$$

Here, α is the bond polarization fraction. In homopolar materials, $C = 0$, so that the polarization is zero and the two contributions are equal, as one would expect. This polarization is a result of the charge transfer and we will see it discussed again later in connection with the lattice polar optical mode of vibration. We also note that (2.59) relates the polarization to the fraction of heteropolar nature in the bond itself. Thus one can call α a quantity that is related to the fraction of ionicity in the bond.

It is clear that once we know the atomic energy levels, we can immediately determine the hybrid energies, and the heteropolar and homopolar energies. Using V_2 as the constant value given in (2.54), the average energies are now known. However, Phillips (1973) treats the homopolar energy and the heteropolar energy C as adjustable parameters to get better scaling for the average energy gaps, building his results from a need to get the dielectric functions correct. His values are close to the Harrison (1980) values obtained by straightforward application of the atomic energies, but there are differences. In the heteropolar materials, the differences are even greater, and the average gap can differ by more than a volt between the two approaches. To be sure, it is not at all clear that these two methods are comparable or that the two authors are talking about exactly the same quantities, even though it appears to be so. That the numbers are close is perhaps remarkable and points out the basic correctness of the overall picture of the composition of the energy bands in semiconductors.

The values that are obtained, as seen in Figure 2.6 are not the actual energy gaps one associates with the conduction and valence bands, but are average gaps around which the energy bands form. That is, these are levels near mid-band that may be associated with the mean energy in the band. These levels are then broadened by the interaction with other atoms, and the periodicity that is invoked in the structure, and this broadening produces the conduction and valence bands. The number of states in each of these bands is twice the number of lattice sites for the basis of two atoms per lattice site and another factor of 2 for spin. Thus, the four electrons per atom are just sufficient to completely fill the valence band, and it is this full valence band, with a gap to the conduction band, that provides most of the properties of the semiconductors. Before proceeding to the full Brillouin zone, however, we first take a look at a small k expansion around the band extrema in order to further investigate the non-parabolicity in these bands.

2.3 The k · p method for the bands

We have already pointed out above, in the treatment of the nearly-free electron model (Section 2.1), that away from the edges of the bands, the dispersion becomes *nonparabolic*. Indeed, it was shown in (2.40) that the energy bands take a *hyperbolic* form and the effective mass varies with wave vector, and with the velocity of the carriers as a result. The nonparabolicity arises from the interaction between the wave functions of the two bands that would ordinarily (without the gap) be degenerate at the band edge (zone edge or zone center) value of the wave vector. This interaction decreases in amplitude as the gap between the two bands increases, which in turn causes less repulsion and a heavier curvature mass. The limiting case returns to the free-electron bands for a wave vector sufficiently far from the band minimum (or maximum in the case of the valence band).

The interactions between the various bands lead to a mixing of the basic s- and p-derived hybrid wave functions, which alters the admixtures of the components as the wave vector, and hence the energy, vary. Before proceeding to a general treatment of the manner in which the bands are developed across the entire Brillouin zone, we first discuss how the s and p hybrids that form the bonds are modified through the interactions between neighboring atoms to form the bands near the extrema.

In the following, *we will obtain essentially the same results that arose in (2.41)*. The only difference in the present approach is that we began with the localized wave functions of the bond-orbital model, whereas the earlier treatment began with nearly-free electron plane waves. There, the perturbation of the lattice potential *opened* gaps in the nearly-free electron bands. Here, the interactions of the $\mathbf{k} \cdot \mathbf{p}$ model will lead to the bands for values of the crystal momentum, and these bands are based in the bond orbitals described in the previous section.

It was established above that the general solutions of the Schrödinger equation in a periodic potential are Bloch wave functions of the form

$$\Psi(\mathbf{r}) = e^{i\mathbf{k}\cdot\mathbf{r}}u_{nk}(\mathbf{r}), \tag{2.10}$$

where a band index n has been added and $u_{nk}(\mathbf{r})$ has the periodicity of the lattice. It is these general wave functions that will be utilized initially to set up the Hamiltonian for the $\mathbf{k} \cdot \mathbf{p}$ method. Here $\mathbf{k}$ is the wave vector describing the propagation of the wave and is related to the crystal momentum of the electron (in whichever band it is located). On the other hand, $\mathbf{p} = -i\hbar\nabla$ is the momentum *operator*, which is related to motion in real space. To proceed, (2.18) is rewritten as

$$\left[\frac{p^2}{2m} + V(\mathbf{r}) + \frac{\hbar}{m}\mathbf{k}\cdot\mathbf{p}\right]u_{nk}(\mathbf{r}) = E'u_{nk}(\mathbf{r}), \quad E' = E - \frac{\hbar^2k^2}{2m}. \tag{2.60}$$

The third term in (2.60) is the "$\mathbf{k} \cdot \mathbf{p}$" term. In many cases this term is treated by perturbation theory to build up the effective mass from a perturbation summation over all energy bands in the crystal. This is not necessary, and a diagonalization procedure will be utilized below to treat these terms exactly in an approach due to Kane (1957, 1966). First, however, we will add the spin–orbit interaction terms.

In addition, there is a spin–orbit interaction, which splits the triply degenerate (six-fold degenerate when spin is taken into account) valence bands at the maximum, which occurs at the zone center (termed the Γ point). This splitting arises in nearly all tetrahedrally coordinated semiconductors. The interaction mainly splits off one (spin-degenerate) state, leaving a doubly degenerate (four-fold with spin) set of levels that correspond to the light holes and the heavy holes. Thus, to account properly for this band arrangement, it is necessary to include the spin–orbit interaction in (2.60). This interaction leads to two additional terms that arise from coupling of the orbital angular momentum (motion of the electron around the atom) and the spin angular momentum (motion of the electron spinning on its own axis) of the electrons. These are

$$\frac{\hbar}{4m^2}\{[\nabla V(\mathbf{r}) \times \mathbf{p}]\cdot\sigma + [\nabla V(\mathbf{r}) \times \mathbf{k}]\cdot\sigma\}, \tag{2.61}$$

where σ is the spin tensor (a 2 × 2 tensor of the spin up and spin down states; see Ferry, 1996). The form of this tensor is not important for the present treatment, as it will be parameterized below. The second of these terms gives a term that is linear in **k** at the zone center and actually shifts the maxima of the valence band a negligibly small amount away from Γ in most compound semiconductors where there is no inversion symmetry of the crystal. It is usually a very small effect, and this term can be treated in perturbation theory. When coupled with effects from higher-lying bands, this term will be quite important in actually producing a mass different from the free-electron mass for the heavy-hole band, and these results are discussed below. This term, however, is ignored in the present calculations. The first term of (2.61) is called the **k**-independent spin–orbit interaction and is the major term that must be included in (2.60).

2.3.1 Valence and conduction-band interactions

To proceed, it is convenient to adopt a different set of wave functions. While one could use simply the *s*- and *p*-atomic functions, the resulting Hamiltonian is somewhat more complex, and appears in a simpler format with an initial formation of some hybrids, including the spin functions. For this purpose, the basic sp^3 hybridization of the conduction and valence bands suggests the use of *s*, p_x (we will denote this as *X*), p_y (*Y*), and p_z (*Z*) wave functions. These are not proper Bloch functions, but they have the symmetry of the designated band states near the appropriate extrema and are the local cell part of the Bloch states. The four interactions of (2.60) and (2.61) leave the bands doubly (spin) degenerate, which eases the problem and size of the Hamiltonian matrix to be diagonalized. To be sure, we should take *s* and *p* hybrids for the two atoms in the basis separately, as discussed above, but the energies and wave functions discussed here are assumed to be appropriate averages. The manner in which these averages are obtained is discussed in the following section where the full zone energy bands are developed. For the basis states, the wave functions are now taken as the eight functions (the arrow denotes the direction of the electron spin in the particular state)

$$
\begin{aligned}
&|is\downarrow\rangle, && |is\uparrow\rangle, \\
&\left|\frac{X-iY}{\sqrt{2}}\uparrow\right\rangle, && \left|-\frac{X+iY}{\sqrt{2}}\downarrow\right\rangle, \\
&|Z\downarrow\rangle, && |Z\uparrow\rangle, \\
&\left|\frac{X+iY}{\sqrt{2}}\uparrow\right\rangle, && \left|\frac{X-iY}{\sqrt{2}}\downarrow\right\rangle.
\end{aligned}
\tag{2.62}
$$

The left-hand column forms one set of states, which are degenerate with those of the right-hand column. In setting up the Hamiltonian matrix, the wave vector is taken in the z direction. If it is not in this direction, the matrix is more complicated but is easily set up by a simple coordinate rotation of the basis functions.

Since the basis functions lead to doubly degenerate levels, the 8×8 matrix separates into a simpler block diagonal form, containing two 4×4 matrices, one for each spin direction, as

$$\begin{bmatrix} H_d & 0 \\ 0 & H_d \end{bmatrix}. \tag{2.63a}$$

The 4×4 matrices are identical, and given by

$$H_d = \begin{bmatrix} E_s & 0 & kP & 0 \\ 0 & E_p - \Delta/3 & \sqrt{2}\Delta/3 & 0 \\ kP & \sqrt{2}\Delta/3 & E_p & 0 \\ 0 & 0 & 0 & E_p + \Delta/3 \end{bmatrix}. \tag{2.63b}$$

The constant Δ is a parameter, and is a positive quantity. This is the spin–orbit splitting energy that describes the energy difference between the split-off valence band and the top of the valence band. This parameter is given by the matrix elements of (2.61) with the p wave functions (the spin is taken in the p_z direction), as

$$\Delta = \frac{3i\hbar}{4m^2}\left\langle X \left| \left(p_y \frac{\partial V}{\partial x} - p_x \frac{\partial V}{\partial y} \right) \right| Y \right\rangle. \tag{2.43}$$

The momentum matrix element P arises from the extra term in (2.60) and is given by

$$P = -\frac{i\hbar}{m} \langle s | p_z | Z \rangle, \tag{2.44}$$

while E_s and E_p are the averaged atomic energy levels. Within the crystal, there is no true reference level for atomic energies. Normally, they are referenced to a vacuum level, but it is known in quantum mechanics that energy level shifts do not change the properties, and, for convenience, the top of the valence band will be taken as the zero of energy. With this choice, the fourth line in the matrix (2.63b) is an isolated level, and is the heavy-hole band. Since, this isolated level is at the top of the valence band, it is necessary to set $E_p = -\Delta/3$. This heavy-hole band has an energy which is just the free electron curvature of the second term in (2.60). This will be corrected later for interactions with other bands to give the proper variation, as this band has the wrong curvature for holes at present, as well as the wrong mass. Once this energy level choice is made, the characteristic equation for the determinant of the remaining 3×3 matrix is

$$E'(E' - E_{gap})(E' + \Delta) - k^2P^2(E' + 2\Delta/3) = 0, \tag{2.66}$$

where we have set $E_s = E_{gap}$ (this choice will be justified by the results to follow). This suggests, in turn, that the bottom of the conduction band is *s*-like while the top of the valence band is *p*-like. In fact, this is the proper result, and this point will be illustrated more clearly in the next section.

2.3.1.1 *Small* kP

If the size of the *k*-dependent term in (2.66) is quite small (e.g., the energy is near to the band extremum), the solutions will be basically just those arising from the first term, with a slight adjustment for the *kP* term. For this case, the three bands are given by

$$E_c = E_{gap} + \frac{k^2P^2}{3}\left(\frac{2}{E_{gap}} + \frac{1}{E_{gap} + \Delta}\right) + \frac{\hbar^2k^2}{2m}$$

$$E_{lh} = -\frac{2k^2P^2}{3E_{gap}} + \frac{\hbar^2k^2}{2m}$$

$$E_\Delta = -\Delta - \frac{k^2P^2}{3(E_{gap} + \Delta)} + \frac{\hbar^2k^2}{2m}, \tag{2.67}$$

for the conduction band, the light-hole band, and the split-off (spin–orbit) band, respectively. The free-electron contribution has been restored to each energy from (2.60). The bands are all clearly parabolic for small values of *k*. This will not be the case for larger values of *k*, however. The band-edge values of the effective mass, which makes these energies have the normal parabolic form, are all slightly different, but all depend on both the energy gap and the momentum matrix element. Near $k = 0$, the effects of the higher bands are relatively minor, but the heavy-hole band will be treated separately below.

2.3.1.2 $\Delta = 0$; *the two-band model*

If the spin–orbit interaction is neglected, then the split-off band becomes degenerate with the heavy-hole band, and the remaining interaction is just between the mirror image conduction and light-hole bands. In this case, these two energies are given by

$$E_c = \frac{E_{gap}}{2}\left(1 + \sqrt{1 + \frac{4k^2P^2}{E_{gap}^2}}\right) + \frac{\hbar^2k^2}{2m}$$

$$E_v = \frac{E_{gap}}{2}\left(1 - \sqrt{1 + \frac{4k^2P^2}{E_{gap}^2}}\right) + \frac{\hbar^2k^2}{2m}. \tag{2.68}$$

The general form achieved here is just the hyperbolic description that resulted in (2.36) from the nearly-free electron model, except here the interaction energy is defined in terms of the momentum matrix element and the energy gap. This suggests that this matrix element can be used to define an effective mass at the band edge (in the small k limit), and this will be done below.

2.3.1.3 $\Delta \gg E_{gap}, kP$

In the case for which the spin–orbit splitting is large, a somewhat different expansion of the characteristic equation can be obtained. For this case, the spin–orbit energy is taken to be larger than any corresponding energy for which a solution is being sought, and the resulting bands are found to be

$$E_c = \frac{E_{gap}}{2}\left(1 + \sqrt{1 + \frac{8k^2P^2}{3E_{gap}^2}}\right) + \frac{\hbar^2k^2}{2m}$$

$$E_v = \frac{E_{gap}}{2}\left(1 - \sqrt{1 + \frac{8k^2P^2}{3E_{gap}^2}}\right) + \frac{\hbar^2k^2}{2m}$$

$$E_\Delta = -\Delta - \frac{k^2P^2}{3(E_{gap} + \Delta)} + \frac{\hbar^2k^2}{2m}. \tag{2.69}$$

The conduction band and the light-hole band remain mirror images (except for the free-electron contribution), while the split-off band is parabolic in shape.

2.3.1.4 *Hyperbolic bands*

It may be seen from (2.68) and (2.69) that the conduction band and the light-hole band are essentially hyperbolic in shape. The relation between the momentum matrix element and the band-edge effective mass changes slightly with the size of the spin–orbit energy, but this change is relatively small, as the numerical coefficient of the k^2 term changes by only a factor of $\frac{2}{3}$ as Δ goes from zero to quite large. The major effect is the hyperbolic band shape introduced by the $\mathbf{k} \cdot \mathbf{p}$ interaction, and the variation introduced by the spin–orbit interaction is quite small other than the motion of the maximum of the split-off band away from the heavy-hole band. For this reason, it is usually decided to introduce the band-edge masses directly into the hyperbolic relationship without worrying about whether or not the size of the spin–orbit energy is significant. With this decision, and using the large Δ expressions, the band-edge masses for the mirror bands are

$$\frac{1}{m_c} = \frac{1}{m} + \frac{4P^2}{3\hbar^2 E_{gap}}, \quad \frac{1}{m_{lh}} = -\frac{1}{m} + \frac{4P^2}{3\hbar^2 E_{gap}}. \tag{2.70}$$

For most direct-gap group III–V compound semiconductors, the conduction band and light-hole band effective masses are small and on the order of 0.01 to 0.1, so that the free-electron term is essentially negligible. Since the momentum matrix element arises from the sp^3 hybrids, the results of (2.70) are that the masses scale with the band gap in an almost linear fashion, with modest variations from the momentum matrix element. Materials with narrow band gaps usually have very small values of the effective masses.

It is often useful to rearrange the mirror-imaged bands, which are hyperbolic in nature, to express the momentum wave vector k directly in terms of the energy. This is easily done, using (2.69) and (2.70), with the results that

$$\frac{\hbar^2 k^2}{2m_c} = (E - E_{gap})\left[1 + \frac{(E - E_{gap})}{E_{gap}}\right]$$

$$\frac{\hbar^2 k^2}{2m_{lh}} = -E\left[1 - \frac{E}{E_{gap}}\right], \tag{2.71}$$

where the energy is measured from the valence band maxima, and it is assumed that the free electron contribution can be ignored. The equations above have not been corrected for interactions with the higher-lying bands, and this will modify the non-parabolicity only slightly for energies below the band gap. However, an often-used approach is to replace the factor $1/E_{gap}$ with a parameter α in (2.71) for each of the bands, with this parameter being determined by a best fit to the mass enhancement with energy in the appropriate band. In Table 2.2, several parameters are listed for some representative semiconductors.

It is clear that the non-parabolic nature of the energy bands will lead to a mass that varies with the energy. For this variation, it is important to distinguish the particular mass of interest, as it is likely to have its own

Table 2.2 Nonparabolic band parameters for some representative semiconductors

	Gap	m_c/m	α_c	m_{lh}/m	Δ	m_Δ/m
Si	1.1	0.19, 0.91[a]	0.5	0.15	0.044	0.24
Ge	0.7	0.22[b]	1.4	0.049	0.29	0.092
GaAs	1.41	0.066	0.61	0.087	0.34	0.2
InP	1.35	0.072	0.72	0.12	0.21	0.15
InAs	0.43	0.022	2.44	0.025	0.43	0.08
InSb	0.22	0.013	4.5	0.16	0.81	0.11

Notes:
[a] Transverse and longitudinal masses for the equivalent ellipsoids
[b] Average of the transverse and longitudinal masses for the ellipsoids

variation. For example, the variation of the momentum mass described earlier may be determined for the non-parabolic conduction band, using the simple forms of (2.71) to be

$$\frac{1}{m^*} = \frac{1}{\hbar^2 k}\frac{\partial E}{\partial k} = \frac{1}{m_c(1 + 2E/E_{gap})}. \tag{2.72}$$

2.3.2 *Wave functions*

The basic wave functions that form the basis set for the k · p method were given in (2.62). Once the energies are found, the appropriate sums of the basis vectors give the new orthonormal basis vectors. The doubly degenerate wave functions that result from the diagonalization of the Hamiltonian, by the methods described above, may be written in the form (Kane, 1957, 1966)

$$\psi_{i\alpha} = a_i|is\downarrow\rangle + b_i\left|\frac{X - iY}{\sqrt{2}}\uparrow\right\rangle + c_i|Z\downarrow\rangle$$

$$\psi_{i\beta} = a_i|is\uparrow\rangle + b_i\left|-\frac{X + iY}{\sqrt{2}}\downarrow\right\rangle + c_i|Z\uparrow\rangle \tag{2.73}$$

for the k-vector in the z direction, where the subscript i takes on the values c, lh, or Δ. The heavy hole is represented by the doubly degenerate wave functions

$$\psi_{hh\alpha} = \left|\frac{X + iY}{\sqrt{2}}\uparrow\right\rangle, \quad \psi_{hh\alpha} = \left|\frac{X - iY}{\sqrt{2}}\downarrow\right\rangle. \tag{2.74}$$

Finally, the general expressions for the coefficients in (2.73) are

$$a_i = \frac{kP}{N}\left(E_i' + \frac{2\Delta}{3}\right), \quad b_i = \frac{\sqrt{2}\Delta}{3N}(E_i' - E_{gap}),$$

$$c_i = \frac{1}{N}(E_i' - E_{gap})\left(E_i' + \frac{2\Delta}{3}\right), \quad N^2 = a_i^2 + b_i^2 + c_i^2. \tag{2.75}$$

For small kP, the conduction band remains s-like, and the light-hole and split-off valence bands are composed of admixtures of the p-symmetry wave functions. In the hyperbolic band models, however, the conduction band also contains an admixture of p-symmetry wave functions. This admixture is energy dependent, and it is this energy dependence that introduces a similar (energy dependent) effect into the overlap integral calculated for the scattering processes in later chapters. As an example, the coefficients for the conduction-band wave functions, in the limit of large spin–orbit splitting, are just

$$a_c = \sqrt{1 + \frac{E}{E_{gap}}}, \quad b_c = \frac{c_c}{\sqrt{2}} = \sqrt{\frac{E}{3(2E + E_{gap})}}, \tag{2.76}$$

and the energy is now measured from the conduction-band edge. It may be noticed that the wave functions are normalized, so that any overlap integral corrections will arise only from inelastic scattering events for which the initial and final states have different admixtures of p-symmetry wave functions. As with the details of the hyperbolic bands, the size of the spin–orbit splitting makes only slight changes in the numerical factors, but these can be ignored as being of second order in importance.

2.3.3 *Interactions with other bands: heavy holes*

Interactions with conduction and valence bands, other than the four that have been treated here, give rise to corrections to the energy variations above. In general, these lead to the linear k terms mentioned above, which are important only at very small values of k in the valence band. There are also quadratic terms in k that lead to variations of the masses beyond those of the hyperbolic bands, and quadratic cross terms that lead to warping of the band away from a spherically symmetric shape (at $k = 0$). These extra terms are traditionally handled by perturbation theory (Dresselhaus *et al.*, 1955). The dominant effect of these terms, however, is on the light- and heavy-hole bands, and these bands may be written as

$$E_{hh} = -\frac{\hbar^2}{2m}\left[Ak^2 \pm \sqrt{B^2k^4 + C^2(k_x^2k_y^2 + k_x^2k_z^2 + k_z^2k_y^2)}\right], \tag{2.77}$$

Table 2.3 Heavy-hole parameters of some representative semiconductors

	A	B	C	m_{hh}/m
Si	4.0	1.1	4.1	0.49
Ge	13.4	8.5	13.1	0.16
GaAs	7.3	4.5	6.2	0.45
InP	4.8	3.54	–	0.8
InAs	21.2	18.8	–	0.41
InSb	24.6	21	16	0.39

where the upper sign is for the light holes and the lower sign is for the heavy holes, which are of interest here. More complicated forms can be developed, but this brings into focus the major effects that can be expected to occur. In Table 2.3, the heavy-hole mass and the three parameters A, B, and C are listed for some representative semiconductors.

Another effect that occurs, for example, in GaAs, is the crossing of the split-off band and the light-hole band. Since Δ is rather small in GaAs, it is possible for light holes to reach energies for which the band becomes degenerate with the spin–orbit split band. When this occurs, interaction between the bands can be treated by perturbation theory to develop a mathematical form for the anti-crossing behavior of these two bands.

2.4 Linear combination of atomic orbitals

So far, only the manner in which the crystal potential interacts with the nearly free electrons to open gaps has been discussed, together with the manner in which the periodicity creates bands in the nearly free electron case. In the latter, this approach leads to the actual momentum variations across the Brillouin zone that lead to the band structure one would calculate. We have also examined the manner in which the bands are initially formed away from the principal extrema of the conduction and valence band. In this section, we turn our attention to a formulation in which the atomic wave functions are used to develop the proper crystalline energy bands through creation of the Bloch functions and imposition of specific periodicities arising from the crystal structure. This technique is termed the linear combination of atomic orbitals (LCAO) method, or in simpler parlance, the tight-binding method. The approach is rather straightforward, and the fact that we already know most of the eigenvalues for the atomic wave functions is important to its success. Moreover, in its normal applications, known experimental values for gaps and masses are used to create the *empirical* tight-binding method (ETBM).

The LCAO (or ETBM) method follows from a very few simple assumptions. First, one can take the hybrid that is located on one atom, say the central atom of the basic tetrahedral structure, and translate *its wave functions* through the use of the Bloch theorem to the nearest-neighbor atoms. Once the wave function is translated, the *overlap integrals* $\langle h_A|H|h_B\rangle$ are constructed. These integrals now involve interaction energies between the s and p orbitals on the A atom and the s and p orbitals on the B atom. To ensure the symmetry of the basic crystal, this displacement/integration procedure is repeated for each of the atoms in the crystal. However, it is sufficient to limit the procedure to some set of neighbors, say the nearest neighbors, which entails four sets of interactions for the four nearest neighbors, or 12 interactions for the second neighbors. In the nearest-neighbor set, the interactions are always A–B (or B–A). In the second neighbor sets, though, the interactions are A–A or B–B.

As a general rule, the actual interaction energy will be the same, within a sign, for each of the atoms in the neighbor set, and the difference will be the sign and the plane wave factor for the shift of the central wave function to the four neighbors. The summation over the various plane wave phase factors is termed the *Bloch sum*. For example, where there is no preferred angle in the atomic wave function, such as in s–s interaction, we expect that the result for the nearest-neighbor sum will look like

$$\begin{aligned}\langle s_A|H|s_B\rangle &= E_{ss}\sum_{i=1}^{4} e^{i\mathbf{k}\cdot\mathbf{r}_i} \\ &= E_{ss}(e^{i\mathbf{k}\cdot\mathbf{r}_1} + e^{i\mathbf{k}\cdot\mathbf{r}_2} + e^{i\mathbf{k}\cdot\mathbf{r}_3} + e^{i\mathbf{k}\cdot\mathbf{r}_4}),\end{aligned} \tag{2.78}$$

where the four displacement vectors are defined by the positions of the nearest neighbors, as

$$\begin{aligned}\mathbf{r}_1 &= \frac{a}{2}(\mathbf{a}_x + \mathbf{a}_y + \mathbf{a}_z), \quad \mathbf{r}_2 = \frac{a}{2}(\mathbf{a}_x - \mathbf{a}_y - \mathbf{a}_z), \\ \mathbf{r}_3 &= \frac{a}{2}(-\mathbf{a}_x + \mathbf{a}_y - \mathbf{a}_z), \quad \mathbf{r}_4 = \frac{a}{2}(-\mathbf{a}_x - \mathbf{a}_y + \mathbf{a}_z).\end{aligned} \tag{2.79}$$

As discussed above, the factor E_{ss} is the magnitude of the overlap integral between the s functions on two neighboring atoms, and the exponentials provide the phase shift in moving the central wave function to the four neighbors. The a in (2.79) is not the edge of the face-centered cubic structure, as one might have selected in looking at Figure 2.4, as the *primitive* cell has an edge that is one-half of the value of the face-centered cell. To use the proper lengths, so that the reciprocal lattice vectors have the right units, the a values are replaced accordingly, and each vector in (2.79) has been given the length corresponding to the proper primitive cell. The Bloch sum is defined as the sum of the four exponential factors in (2.79) and this particular one is denoted as $B_0(k)$. This may be rewritten as

$$\langle s_A|H|s_B\rangle = E_{ss}B_0(k)$$

$$B_0(k) = 4\cos\left(\frac{k_x a}{2}\right)\cos\left(\frac{k_y a}{2}\right)\cos\left(\frac{k_z a}{2}\right) - 4i\sin\left(\frac{k_x a}{2}\right)\sin\left(\frac{k_y a}{2}\right)\sin\left(\frac{k_z a}{2}\right). \tag{2.80}$$

Similarly, the p orbitals will also interact in such a manner that those directed in the same direction will also produce a Bloch sum that is given by $B_0(k)$. This will be the case for the interaction between the p_x orbital on the A atom and the neighboring p_x orbital on the B atom. This is because they all point in the same direction as that on the neighbor. This holds true also for the p_y–p_y and p_z–p_z interactions.

Consider now the interaction between an s orbital on the A atom and the p_x orbitals on the neighboring B atoms. On the atoms at $\mathbf{r}_1$ and $\mathbf{r}_2$, the p_x orbital points *away* from the A atom, but at the other two sites, it points *toward* the A atom. This means it will change the sign of the contributions to the Bloch sum as

$$\langle s_A|H|p_{x,B}\rangle = E_{sp}B_1(k) = E_{sp}(e^{i\mathbf{k}\cdot\mathbf{r}_1} + e^{i\mathbf{k}\cdot\mathbf{r}_2} - e^{i\mathbf{k}\cdot\mathbf{r}_3} - e^{i\mathbf{k}\cdot\mathbf{r}_4})$$

$$B_1(k) = -4\cos\left(\frac{k_x a}{2}\right)\sin\left(\frac{k_y a}{2}\right)\sin\left(\frac{k_z a}{2}\right) + 4i\sin\left(\frac{k_x a}{2}\right)\cos\left(\frac{k_y a}{2}\right)\cos\left(\frac{k_z a}{2}\right). \tag{2.81}$$

Similarly, the other two Bloch sums for the overlap between an s orbital and a p orbital are given by

$$\langle s_A|H|p_{y,B}\rangle = E_{sp}B_2(k) = E_{sp}(e^{i\mathbf{k}\cdot\mathbf{r}_1} - e^{i\mathbf{k}\cdot\mathbf{r}_2} + e^{i\mathbf{k}\cdot\mathbf{r}_3} - e^{i\mathbf{k}\cdot\mathbf{r}_4})$$

$$B_1(k) = -4\sin\left(\frac{k_x a}{2}\right)\cos\left(\frac{k_y a}{2}\right)\sin\left(\frac{k_z a}{2}\right) + 4i\cos\left(\frac{k_x a}{2}\right)\sin\left(\frac{k_y a}{2}\right)\cos\left(\frac{k_z a}{2}\right) \tag{2.82}$$

and

$$\langle s_A|H|p_{z,B}\rangle = E_{sp}B_3(k) = E_{sp}(e^{ik\cdot r_1} - e^{ik\cdot r_2} - e^{ik\cdot r_3} + e^{ik\cdot r_4})$$

$$B_1(k) = -4\sin\left(\frac{k_x a}{2}\right)\sin\left(\frac{k_y a}{2}\right)\cos\left(\frac{k_z a}{2}\right)$$

$$+ 4i\cos\left(\frac{k_x a}{2}\right)\cos\left(\frac{k_y a}{2}\right)\sin\left(\frac{k_z a}{2}\right). \tag{2.83}$$

It may be noted that one moves through the last three of these sums by a cyclic permutation of the coordinates $x \to y \to z \to x$, which displays the cubic symmetry of the crystal.

For the interactions between the p orbitals, it turns out that the same Bloch sums will result. The "diagonal" interactions between the same p orbitals on each atom yield the result using the symmetric sum, while the "off-diagonal" interactions yield the asymmetric sums, and

$$\langle p_{x,A}|H|p_{x,B}\rangle = E_{xx}B_0(k), \quad \langle p_{x,A}|H|p_{y,B}\rangle = E_{xy}B_3(k). \tag{2.84}$$

In the latter term, we should interpret the subscript on $B(k)$ to represent the "3" (or z) direction, and then the other terms can be easily obtained by a cyclic permutation of the coordinate axes.

In the heteropolar semiconductors, the actual matrix elements for the crossed interaction energy E_{sp} will depend on which atom, the cation or the anion, donates the s orbital, so that there are two possible values for this term. The other interaction elements between these first neighbors do not have this complication, since the interactions always involve one orbital of the same type from each of the two atoms in the crystal. Of course, the atomic s and p energies will also be different. We are working here with the non-hybridized energies, since it is only necessary to determine the interaction energies of the atomic wave functions. This is a general practice in this approach, although it need not be. The actual values of the interaction energies can be calculated by difficult but straightforward means provided that one does not include three-center integrals (where one wave function comes from each of two different atoms and the potential is centered on a third atom). In most instances, however, a different approach is used, which we termed above as the *semi-empirical tight-binding method* (SETBM), in which the energies are used as fitting constants, either to fit to the measured values of band extreme or to fit to more exact pseudopotential calculations (Slater and Koster, 1954).

Because there are two atoms in the primitive unit cell (or in the tetrahedron shown in Figure 2.4), and each atom has four orbitals, the resulting Hamiltonian matrix is an 8×8 matrix. To calculate the energy levels at an arbitrary point **k** in the zone, it is necessary to fill out the entire 8×8 matrix of energies, and then diagonalize the matrix. Symmetry is introduced into the matrix by the Bloch sums that are included. The basic 8 × 8 matrix is decomposable into a 2 × 2 set of sub-matrices, in which the diagonal pair are themselves diagonal with the atomic energy levels of the A and B atoms on the diagonal. The off-diagonal blocks represent the interactions, at the first-neighbor level, between the atomic orbitals on the two individual atoms. The rows and columns are ordered as s_A, p_{xA}, p_{yA}, p_{zA}, s_B, p_{xB}, p_{yB}, and p_{zB}. In addition, since the energy is a measurable real quantity, the matrix must be Hermitian so that the eigenvalues are real quantities. The energy matrix may then be simply

$$[H] = \begin{bmatrix} [A] & [AB] \\ [BA] & [B] \end{bmatrix}, \tag{2.85}$$

where

$$[A] = \begin{bmatrix} E_s^A & 0 & 0 & 0 \\ 0 & E_p^A & 0 & 0 \\ 0 & 0 & E_p^A & 0 \\ 0 & 0 & 0 & E_p^A \end{bmatrix}, \tag{2.86}$$

and similarly for $[B]$. The off-diagonal matrices are defined through

$$[AB] = \begin{bmatrix} E_{ss}B_0 & E_{sp}^{AB}B_1 & E_{sp}^{AB}B_2 & E_{sp}^{AB}B_3 \\ -E_{sp}^{BA}B_1 & E_{xx}B_0 & E_{xy}B_3 & E_{xy}B_2 \\ -E_{sp}^{BA}B_2 & E_{xy}B_3 & E_{xx}B_0 & E_{xy}B_1 \\ -E_{sp}^{BA}B_3 & E_{xy}B_2 & E_{xy}B_1 & E_{xx}B_0 \end{bmatrix}, \tag{2.87}$$

and $[BA] = [AB]^\dagger$, the Hermitian adjoint, where the general element $(BA)_{ij} = (AB)_{ji}^*$. The minus sign that arises in the first column comes from the inversion of the cell when we start with the B atom rather than the A atom. Now, the energy levels at any momentum are found by solving the determinant

$$|E[\mathrm{I}] - [H]| = 0. \tag{2.88}$$

Here, [I] is the 8 × 8 unit matrix.

2.4.1 The k = 0 points

Before proceeding to the full-zone band structure, it is worthwhile to calculate some important energies at the center of the Brillouin zone (the Γ point). For $\mathbf{k} = 0$, all Bloch sums, except B_0, vanish as each has a sine term, and the latter has the value 4. Thus the only interactions are between the s levels on the two atoms, and between the individual p levels on the two atoms. It is necessary only to solve the simple interaction determinants

$$\begin{vmatrix} E - E_s^A & 4E_{ss} \\ 4E_{ss} & E - E_s^B \end{vmatrix} = 0, \quad \begin{vmatrix} E - E_p^A & 4E_{xx} \\ 4E_{xx} & E - E_p^B \end{vmatrix} = 0. \tag{2.89}$$

The second of these equations is triply degenerate for the three different p levels. The solutions of these two determinants are termed the E_{Γ_1} and $E_{\Gamma_{15}}$ symmetry levels. The values are found to be

$$E_{\Gamma_1} = \frac{E_s^A + E_s^B}{2} \pm \sqrt{\left(\frac{E_s^A - E_s^B}{2}\right)^2 + 16E_{ss}^2}\,, \tag{2.90a}$$

$$E_{\Gamma_{15}} = \frac{E_p^A + E_p^B}{2} \pm \sqrt{\left(\frac{E_p^A - E_p^B}{2}\right)^2 + 16E_{xx}^2}\,. \tag{2.90b}$$

The lower Γ_{15} level corresponds to the top of the valence band, while the upper Γ_1 level corresponds to the bottom of the conduction band at Γ. Indeed, we find that the valence band is triply degenerate at $\mathbf{k} = 0$, as we have not included the spin–orbit interaction of the last section, and it was this interaction that split off one of the p symmetry bands. Another point to notice is that the top of the valence band arises from the lower of the two Γ_{15} energies, which occurs with the negative sign in (2.90b). This means that the anion (B atom) predominantly contributes to this level, which is the source of the statement used earlier that the wave functions at the top of the valence band were primarily derived from anion p-functions. This is not completely true, of course, but these are the primary component in these wave functions.

If we now know the experimental data for the energy gap at Γ, and for the width of the valence band, the two unknown parameters in (2.90) are determined. In reality though, one also often uses the s and p orbital energies as additional parameters to be adjusted. In fact, the energy gap at Γ is easily measured with optical absorption, and the width of the valence band is closely related to the valence plasma frequency, itself given by the density of valence electrons in the solid. So, these parameters are quite easily adjusted from empirical data available for the various materials. Another change in the energies is the often used convention in which the top of the valence band is taken to be the reference energy, and assigned the value $E = 0$. In general, this entails a shift of the s and p atomic energies. So long as the differences between s and p are preserved and the differences

Table 2.4 Energy matrix elements for SETBM (in units of eV)

Material	$E^c_{\Gamma_1}$	$E^v_{\Gamma_1}$	$E^c_{\Gamma_{15}}$	E^A_s	E^B_s	E^A_p	E^B_p	E_{ss}	E_{xx}
GaAs	1.55	−12.55	4.71	−2.7	−8.3	3.86	0.85	−1.62	0.45
GaP	2.88	−13.19	5.24	−2.29	−8.02	4.33	0.91	−1.88	0.50
AlAs	3.04	−11.73	4.57	−1.74	−6.96	3.81	0.75	−1.73	0.42
InAs	0.43	−12.69	4.63	−3.53	−8.74	3.93	0.71	−1.51	0.42
InP	1.41	−11.42	4.92	−1.52	−8.50	4.28	0.28	−1.35	0.36
InSb	0.23	−11.71	3.59	−3.4	−8.08	3.07	0.52	−1.37	0.32
Si	4.1	−12.5	3.43	−4.2		1.72		−2.08	0.43
Ge	0.9	−12.66	3.22	−5.88		1.61		−1.70	0.40

between p energies on the two different atoms are preserved, then this is just a shift of the overall energy scale and does not entail any different physics. However, it often occurs that a single shift of the energy levels for one compound is different than that needed for another. Nevertheless, it is better to keep the differences as indicated in Table 2.1. In Table 2.4, we list the critical energy points for a set of III–V compounds and Ge and Si, and the atomic energies and coupling energies needed to fix the gap properly at the zone center are also listed. Here, the A atom is the cation (group III) and the B atom is the anion (group V). These values are slightly different than those given, for example, by Vogl *et al.* (1983) in their discussion of this approach (the values for the experimental energies are taken from this work).

Now, it should be noted for consistency with other workers, that the notation we have used is that for the III–V compounds. In the case of Si and Ge, which have the diamond structure, the top of the valence band has really been assigned the symmetry notation Γ'_{25} and the bottom of the valence band has been assigned the Γ'_2 symbol. The conduction band notations remain as has been shown above. This is a minor point, and unless one is really interested in the specific details of the symmetry differences, the notation is not all that important.

2.4.2 *The X-points at (100)*

By fitting to experimental data for the central gaps at G, one can determine a great number of the parameters that are in the Hamiltonian matrix (2.85). In fact, there are only 3 parameters that are left to be determined. Not unsurprisingly, the next highest symmetry point, that at the zone edge in the (100) direction – the X point – can be used to determine some of these without any difficult computation of the entire matrix. At the point (100), the Bloch sums can be written as

$$B_0 = 0, \quad B_1 = 4i, \quad B_2 = 0, \quad B_3 = 0. \tag{2.91}$$

This allows us to separate the 8 × 8 matrix into four 2 × 2 determinants, as

$$\begin{vmatrix} E - E_s^A & -4iE_{sp}^{AB} \\ 4iE_{sp}^{AB} & E - E_p^B \end{vmatrix} = 0, \quad \begin{vmatrix} E - E_p^A & 4iE_{sp}^{BA} \\ -4iE_{sp}^{BA} & E - E_s^B \end{vmatrix} = 0, \tag{2.92a}$$

and the doubly degenerate

$$\begin{vmatrix} E - E_p^A & -4iE_{xy} \\ 4iE_{xy} & E - E_p^B \end{vmatrix} = 0. \tag{2.92b}$$

These are normally evaluated from the three energy levels in the valence band at the X point. The two terms in (2.92a) lead to the X_3 and the X_1 levels, respectively, recognized as having significant *s* contributions, but with the former connecting to the anion *p* levels at the zone center, while the latter connects with the anion *s* function at Γ_1 in the valence band. The doubly degenerate levels in (2.92b) are the top of the valence band at X, the X_5 level. These determinants can be solved to yield

$$E_{X_3} = \left(\frac{E_s^A + E_p^B}{2}\right) \pm \sqrt{\left(\frac{E_s^A - E_p^B}{2}\right)^2 + 16(E_{sp}^{AB})^2}$$

$$E_{X_1} = \left(\frac{E_s^B + E_p^A}{2}\right) \pm \sqrt{\left(\frac{E_s^B - E_p^A}{2}\right)^2 + 16(E_{sp}^{BA})^2}$$

$$E_{X_5} = \left(\frac{E_p^A + E_p^B}{2}\right) \pm \sqrt{\left(\frac{E_p^A - E_p^B}{2}\right)^2 + 16(E_{xy})^2}. \tag{2.93}$$

Since we have already found a consistent set of atomic energy levels, it remains a simple task to find the last three parameters from the empirical measurements of the energy levels in the valence bands. In Table 2.5, we list the empirical values of the energies found at these three symmetry points and the values for the remaining parameters.

2.4.3 *The band structure*

With all the parameters now determined, it is possible to compute the energy structure for the entire Brillouin zone. The normal method of plotting this is to choose a set of convenient axes, and the normal path is to plot from the L point (the zone edge in the (111) direction) along the (111) axis to

Table 2.5 Energy matrix elements for SETBM at the X point

Material	X_5	X_3	X_1	E_{xy}	E_{sp}^{AB}	E_{sp}^{BA}
GaAs	−2.89	−6.88	−9.83	1.26	1.42	1.14
GaP	−2.73	−7.07	−9.46	1.27	1.84	1.11
AlAs	−2.2	−5.69	−9.52	1.05	1.26	1.46
InAs	−2.37	−8.64	−10.2	1.1	1.73	1.14
InP	−2.06	−6.01	−8.91	0.96	1.33	1.35
InSb	−2.24	−6.43	−9.20	0.96	1.15	0.93
Si	−2.86	−7.69	−7.69	1.14	1.43	
Ge	−3.29	−8.76	−8.76	1.23	1.37	

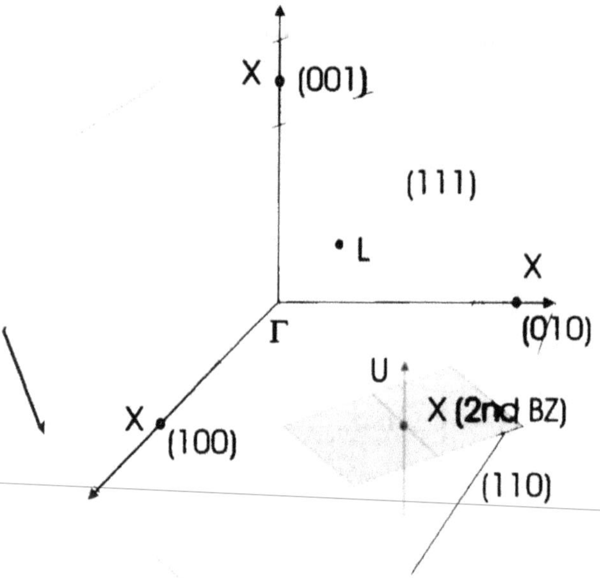

Figure 2.7 The Brillouin zone of the reciprocal lattice for the diamond (or zinc-blende) structure. The dark grey area around the point marked X (2nd BZ) is the displacement of the top square and is part of the second Brillouin zone. It is placed here to indicate how the (110) axis passes from the first zone to another X point in the 2nd zone.

the Γ point at the zone center, then along the (100) axis out to the X point. Now, if one proceeds along the (110) axis to the point $(1,1,0)\pi/a$ we have arrived back at another X point, that in the second zone along its k_z axis, as we show in Figure 2.7. The reciprocal lattice of the diamond (or zinc-blende) structure is a truncated cube, in which the corners are cut by planes normal to the 8 (111) directions at the point $(0.5,0.5,0.5)\pi/a$. This leaves

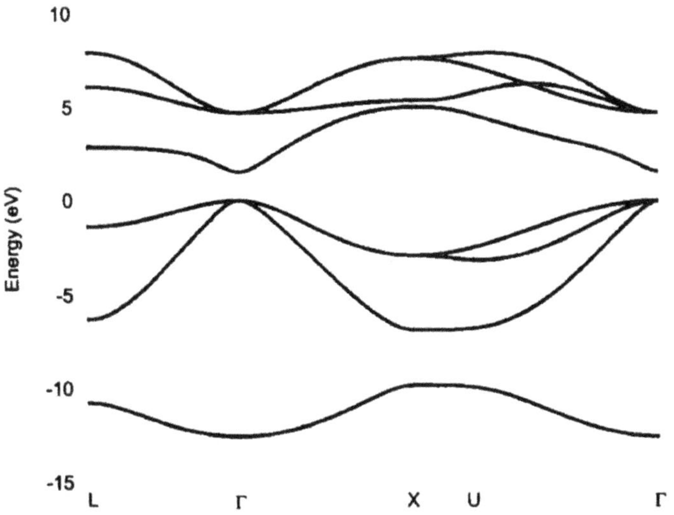

Figure 2.8 The energy bands for GaAs, computed in the SETBM with only nearest neighbor interactions.

square areas normal to the (100) axes at the X points, with the second truncated structures rotated so these squares adjoin as shown in the figure. Hence, the logical plot for the band structure is to return to the Γ point along the (110) line from the X point of the second zone. This means a traverse across the square face as indicated. Such a band structure plot is shown in Figure 2.8 for GaAs. There are four conduction bands and four valence bands arising from the 8 hybrids that are formed. Of course some of these are degenerate in many areas of the zone, a situation caused by the symmetry.

While the energy bands appear quite pleasing, there are several problems with these bands, even for the case of GaAs. First, it is known that the minima of the conduction band at X and L are approximately 0.5 and 0.3 eV above the minimum at Γ. This is not the case at all in Figure 2.8. Instead of 2.05 eV (1.55 + 0.5), the minimum at X is at 5.0 eV, and instead of 1.85 eV the minimum at L is at 2.87 eV. One could adjust the constants to fix part of this problem, but at the expense of no longer being able to fit to the valence band energy levels. Thus, we are faced with the fact that the experiment can measure far more energy levels than can be fit with the limited number of parameters available in this first-neighbor interaction model. This becomes even more significant in the case of Si, as the lowest minimum in the conduction band is not at one of the major symmetry points, but at a point along Δ (from Γ in the direction of X) approximately 0.85X. It is not possible to induce a minimum away from one of the main symmetry points without introducing either higher Fourier coefficients or a more complex interaction in the band structure, which means we have to consider other interactions with more parameters and/or more complicated Bloch functions. There are two major approaches that have been pushed to achieve this goal. The more recent approach is to add the excited *s* states (termed the s^* states) to the basis set, which increases the Hamiltonian to 10×10 when an additional s^* level is added for each atom (Vogl *et al.*, 1983). The more traditional approach is to add the second neighbor interactions, the *A–A* and *B–B* interactions to the 8×8 Hamiltonian matrix that we already have, and this has been used effectively in treating, e.g., Si surfaces (Pandey and Phillips, 1976). For our purposes, it turns out that the addition of the s^* levels give us both desired results for the lowest additional effort. We consider this approach in the following paragraph.

2.4.4 The sp^3s^* SETBM

To remedy the above deficiencies of the sp^3 SETBM, Vogl *et al.* (1983) have added the excited *s* states to the calculation. In addition to the atomic energies $E_{s^*}^A$ and $E_{s^*}^B$, the s^* states are allowed to interact only with the *p* states on the nearest-neighbor atom. This allows these excited states to work to repel (lower) the conduction band energies at the X and L points. As these authors put it, this s^* state is actually an *ad hoc* device which solely is there to permit adjustment of these band points. Since these levels are at higher energy, they repel the lower conduction band states to push the appropriate levels downward. The interaction brings in two additional parameters, the coupling energies $E_{s^*p}^{AB}$ and $E_{s^*p}^{BA}$, which couple with the same Bloch functions as the normal *s–p* coupling terms. All other sp^3 coupling parameters are retained without modification of the values already determined, and these latter two added parameters would be naturally zero in the case where there were no s^* levels. In Figure 2.9, we show the band structure for GaAs using an sp^3s^* model. The additional parameters are given in Table 2.6, and the values for other semiconductors can be found in Vogl *et al.* (1983).

It is clear that the minima of the conduction band at X and L are now closer to those of experiment. However, the second conduction bands at these points are not in the correct position. Corrections can be continued almost *ad naseam*, but the present calculation illustrates the point that given sufficient interactions and coupling constants, a very good agreement with the experimental determination of various energy levels can be achieved. Here, the sp^3s^* gives us good agreement for the III–V compounds. In fact, the minimum at L is 0.29 eV above the minimum at Γ, and the minimum at X is 0.5 eV above that at Γ, which are the accepted values. Yet, we have not introduced the possibility of the minimum being away from a high symmetry point, and this requires the higher Fourier components of the second neighbor interactions.

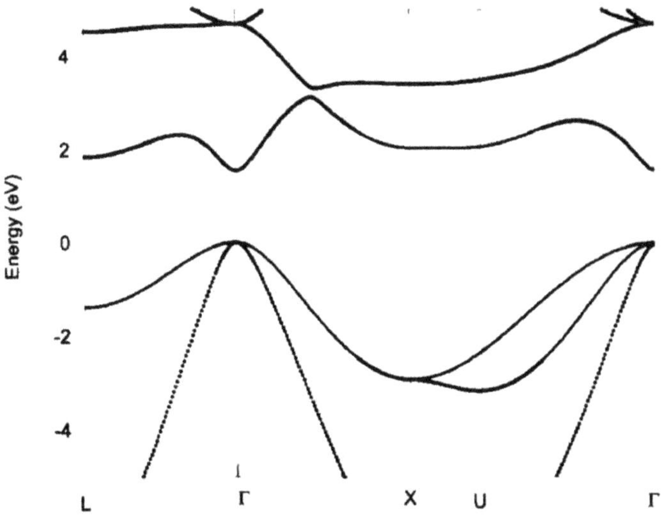

Figure 2.9 The band structure of GaAs as calculated by an sp^3s^* approach, as discussed in the text.

Table 2.6 The s* parameters for GaAs and Si

	GaAs	*Si*
$E^{A}_{s^*}$	6.74	6.2
$E^{B}_{s^*}$	8.6	6.2
$E^{AB}_{s^*p}$	0.75	1.3
$E^{BA}_{s^*p}$	1.25	1.3

In Figure 2.10, we show the energy bands for Si in the region around the principal gap. Again, the addition of the s^* levels, and their interactions with the p functions has produced an acceptable set of conduction bands, including having the minima away from the symmetry point at X. The band gap is essentially the experimental value, but the minima are further from the X point than the accepted positions. Nevertheless, the quality of the band structure is quite good. The parameters for the excited state, and its interactions, are also shown in Table 2.6.

It is obvious that a good set of energy bands can be obtained by choosing a computational scheme with a sufficient number of parameters, and then adjusting these parameters to fit the experimental data. This *empirical* approach has been extensively used. Nevertheless, it is important to point out that the emphasis here has been on fitting the band extrema, and no attention has been paid to getting the effective masses of the various extrema to agree

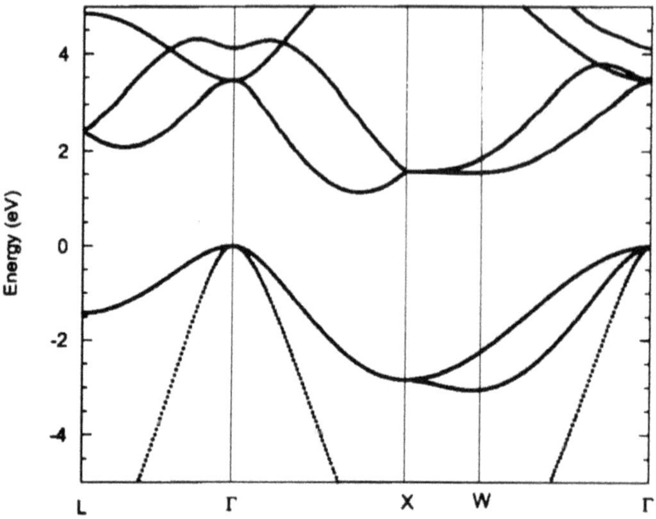

Figure 2.10 The band structure of Si in the vicinity of the band gap. The calculation is done with the sp^3s^* technique.

with the experimental data (where it exists). So long as we understand that these empirical bands are just that, then they are valuable guides to understanding the band structure and its variation among different compounds.

2.5 Pseudopotential band structure

The SETBM approach was discussed in the previous section in the spirit that one could adjust the parameters to fit either experiment or more accurate theoretical approaches, such as the pseudopotential method. Although the latter is really beyond the treatment of this introductory approach, it is informative to briefly review the the method here for completeness. The approach follows closely that of Chelikowsky and Cohen (1976). The wave functions for the valence and conduction electrons in the tetrahedral semiconductors resemble nearly free electrons in many of their properties despite rather strong potentials at the atomic sites. This may be seen from the relative success found earlier in our discussion of the nearly-free electron band structure. Although it is tempting to use free-electron wave functions, this approach is restricted since the true wave functions deviate significantly from true plane waves in the region near the atomic cores. Yet the effect of this region on the energy bands is quite limited in extent. The fundamental concept of the pseudopotential method is that the ion core region can be omitted from the effective potential if a proper renormalized approach to the wave functions is utilized. Computationally, this is crucial, for it implies that the deep core potential is removed and that a simple basis set of plane waves can be utilized and will actually yield rapid convergence. In principle, then, the strength of the pseudopotential method is one of replacing the exact atomic potentials with approximate potentials (the pseudopotentials) that give the right behavior for the conduction and valence electrons. The physics is in the choosing of the pseudopotentials.

Two assumptions are required in formulating the pseudopotential, both of which are crucial to the energy-band calculations. First, the many-electron problem is replaced by a self-consistent potential $V_T(\mathbf{r})$, which includes the effects of the electron–electron interaction as part of the screening of the self-consistent potential. (Since the many-electron problem is not truly of interest here, this is not significant, except that a proper solution of the band structure by such first-principles theory must include at least the exchange and correlation terms in the energy Hamiltonian in the self-consistent loop. This is achieved by solving for a given potential, then calculating the wave functions and using these to calculate the local charge density. The local density is then used within some approximation for the local charge as an exchange and/or correlation term within the Poisson equation for updating the potential.) The second approximation is to divide the various electronic states into core levels and valence/conduction band states and to assume that the core levels are the same as in the isolated atom. With the effective potential, the Schrödinger equation can be written as

$$-\frac{\hbar^2}{2m}\nabla^2\Psi(\mathbf{r}) + V_T(\mathbf{r})\Psi(\mathbf{r}) = E\Psi(\mathbf{r}). \tag{2.94}$$

The core states will satisfy this equation with the same effective potential. For an atomic site i, the core levels provide their own solutions as

$$-\frac{\hbar^2}{2m}\nabla^2\Psi_{c,i}(\mathbf{r}) + V_T(\mathbf{r})\Psi_{c,i}(\mathbf{r}) = E_{c,i}\,\Psi_{c,i}(\mathbf{r}). \tag{2.95}$$

The energy levels of the core states are not those of the isolated atom since these levels are shifted due to the change in the effective potential that arises from the bonding of the atoms into the crystal. To proceed, the wave functions are assumed to be essentially plane waves, but are modified by making them orthogonal to the core-level states. This construction is (with $|\mathbf{k}\rangle = e^{i\mathbf{k}\cdot\mathbf{r}}$)

$$\Psi_{\mathbf{k}}(\mathbf{r}) = |\mathbf{k}\rangle - \sum_{c,i} |c,i\rangle\langle c,i|\mathbf{k}\rangle, \tag{2.96}$$

where we use Dirac notation. The summation term is essentially a Fourier series of the plane wave onto the basis set of core levels, which is then subtracted out. As it stands, (2.96) is an *orthogonalized plane wave* and has been used on its own for different types of band structure calculations (usually called OPW calculations). The sum in (2.96) actually contributes only near the atomic cores, and we build up the total wave function from these modified plane waves as

$$\Psi(\mathbf{r}) = \sum_{\mathbf{k}} C(\mathbf{k})\Psi_{\mathbf{k}}(\mathbf{r}) = \sum_{\mathbf{k}} C(\mathbf{k})|\mathbf{k}\rangle, \tag{2.97}$$

which is similar to that form already used earlier. If the wave function is chosen with the orthogonalized form and used in the Schrödinger equation (2.94), a new potential, which includes the energy levels of the atomic core states, may be introduced as

$$V_P(\mathbf{r}) = V_T - \sum_{c,i} (E_{\mathbf{k}} - E_{c,i})|c,i\rangle\langle c,i|. \tag{2.98}$$

This new potential has the same eigenvalues $E_{\mathbf{k}}$ as the plane waves in (2.94), but the resulting wave functions will be smoothly varying in the core region, due to cancellation of the deep core potential by the second term in (2.98). Thus the pseudo-wave functions have been expressed as plane waves, as in the second part of (2.97). The pseudopotential in (2.98) is sensitive to the angular momentum of the deep core states as well as to the plane wave eigenvalues $E_{\mathbf{k}}$. In general, the number of plane wave states that are required varies from one problem to the next, and the selection of an adequate number is something of an art.

The pseudopotential defined in (2.98) is inherently non-local in nature. Despite this, and its energy dependence, a local simplification yields quite good results for a number of cases. If the pseudopotential is taken to be a simple function of position, it can be written as a Fourier series in the reciprocal lattice

$$V_P(\mathbf{r}) = \sum_{\mathbf{G}} U_{\mathbf{G}} e^{i\mathbf{G}\cdot\mathbf{r}}, \tag{2.99}$$

and the inverse transform is just

$$U_{\mathbf{G}} = \sum_{\alpha} S_{\alpha}(\mathbf{G})V_{\alpha}(\mathbf{G}), \quad S_{\alpha}(\mathbf{G}) = \frac{1}{N_{basis}}\sum_{s=1}^{N_{basis}} e^{i\mathbf{G}\cdot\mathbf{r}_s}, \tag{2.100a}$$

$$V_{\alpha}(\mathbf{G}) = \frac{1}{\Omega}\int_{\Omega} e^{-i\mathbf{G}\cdot\mathbf{r}} V_P(\mathbf{r}) d^3\mathbf{r}. \tag{2.100b}$$

Here, N_{basis} is the number of atoms in the primitive unit cell, $\mathbf{r}_s$ are their positions, and Ω is the cell volume. These terms are not new. S_{α} is no more than the contribution to the form factor used in the scattering of X-rays from the basis of atoms within the unit cell, while V_α is the atomic form factor in this same theory. Here, however, the latter are evaluated using the pseudopotential rather than the true atomic core potential. These equations may be specialized to the case of the tetrahedral semiconductors as

$$U_G = V_S(\mathbf{G})\cos\left(\frac{\mathbf{G}\cdot\mathbf{r}_1}{2}\right) + iV_A(\mathbf{G})\sin\left(\frac{\mathbf{G}\cdot\mathbf{r}_1}{2}\right)$$
$$V_S(\mathbf{G}) = \tfrac{1}{2}[V^A(\mathbf{G}) + V^B(\mathbf{G})]$$
$$V_A(\mathbf{G}) = \tfrac{1}{2}[V^A(\mathbf{G}) - V^B(\mathbf{G})] \tag{2.101}$$

The last two terms are the symmetric and anti-symmetric combinations of the two atomic form factors in the compound. Of course, in the diamond lattice, the anti-symmetric term vanishes since the two atoms are the same. Here $\mathbf{r}_1$ is given by the first term in (2.101). The local pseudopotential method is based on this simplification. If, in addition, the atomic pseudopotentials appearing in (2.100) are taken to be spherically symmetric, the form factors depend only on the magnitude of G. The form factors are often empirically determined from fitting to optical transitions. The general validity of this approach rests upon two additional assumptions. First, it is assumed that the plane wave energy is much larger than the core-level energy, so that the energy bracket in (2.98) can be replaced by a mean energy. Second, it is also assumed that the cancellation theorem for the deep core states holds for all angular momentum states in the core. These assumptions and the local pseudopotential approach have been generally satisfactory until quite recently.

The above approach to the pseudopotential theory has been extended by first using a non-local set of potentials, which include some angular momentum effects in the core functions, and in the pseudopotential itself (through the process of making it orthogonal to the core states). This often produces improved agreement with experiment. Normally, the *ab initio* calculations with pseudopotentials proceed as indicated, with the self-consistent calculation of the pseudopotential and the various Fourier coefficients. On the other hand, as with the SETBM, an *empirical* approach has also appeared which treats the various terms in (2.100) as adjustable parameters, which are varied to fit the experimental data. This latter has come to be called the empirical pseudopotential method (EPM). In general, the *ab initio* pseudopotential approach (which is also termed the *first principles* approach) gives good results for the valence bands, but the band gap is usually too small. More recently, a Green's function self-energy correction, which computes the interaction energy of the electrons in the valence band, provides a correction to the band gap that brings it much closer to that observed in experiment. However, this procedure usually gives a single number for the self-energy, whereas one expects the self-energy to be **k**-dependent as a general rule. Thus, there remains some doubt about the ability to achieve good results for the excited states – the conduction band. Nevertheless, the various types of pseudopotential approach are considered to yield the most accurate band structure for both the valence and conduction band.

2.6 Semiconductor alloys

2.6.1 The virtual crystal approximation

The reader should now be comfortable with the idea that the zinc-blende structure is composed of two interpenetrating face-centered cubic structures, one for each atom in the basis. Thus one face-centered cubic structure is made up of the Ga atoms while the second is composed of the As atoms. We can extend this concept to the case of pseudo-binary alloys, such as GaInAs or GaAlAs, which are supposed to be formulated by a smooth mixing of the two constituents (e.g., GaAs and AlAs in GaAlAs or GaAs and InAs in GaInAs). In such $A_xB_{1-x}C$ alloys, all of the sites of one face-centered cubic sublattice are occupied by type C atoms, but the sites of the second sublattice are shared by the atoms of type A and type B in a random fashion subject to the conditions

$$N_A + N_B = N_C = N, \quad x = \frac{N_A}{N} = c_A, \quad 1 - x = \frac{N_B}{N} = c_B. \tag{2.102}$$

In this arrangement, a type C atom may have all type A neighbors or all type B neighbors, but *on the average* has a fraction x of type A neighbors and a fraction $1 - x$ of type B neighbors. In effect, the structure is a face-centered cubic structure of mixed A–C and B–C molecules, complete with interpenetrating molecular bonding. This structure composes what is called a pseudobinary alloy with the properties determined by the relative concentrations of A and B atoms. In true pseudobinary alloys, it should be possible to scale the properties by a smooth extrapolation between the two end-point compounds.

In recent years, quaternary alloys have also appeared as $A_xB_{1-x}C_yD_{1-y}$ (the most common example is InGaAsP, used in infrared light emitters). Here, C and D atoms now share the sites on the one sublattice. This new compound is still considered a pseudobinary compound composed of a random mixture of two ternaries $A_xB_{1-x}C$ and $A_xB_{1-x}D$, which are only somewhat more complicated than the simple ones discussed in the preceding paragraph. Still, it is assumed that a true random mixture occurs so that the properties can easily be interpolated from those of the constituent compounds. Then any general theory of pseudobinary compound alloys can be applied equally well to the quaternaries as to the ternaries. If these compounds are truly smooth mixtures, the alloy theory will hold, but if there is any ordering in the distribution of the two constituents, deviations from the alloy theory should be expected. For example, $In_xGa_{1-x}As$ may be a smooth alloy composed of a random mixture of InAs and GaAs. However, if perfect ordering were to occur, particularly near $x = 0.5$, the crystal structure would not be a zinc-blende one, but would be a chalcopyrite – a superlattice on the zinc-blende structure with significant distortion of the unit cell along one of the principal axes. In the latter case, changes are expected to occur in the band structure due to Brillouin zone folding about the elongated axis (the lattice period is now twice as long, which places the edge of the Brillouin zone only one-half as far from the origin in the orthogonal reciprocal space direction). For many years, it has been assumed that the ternary and quaternary compounds formed of the group III–V compounds were true random alloys. In recent times, however, it has become quite clear that this is not the case in many situations. We return to this below, as it gives some insight into deviations expected from the random alloy theory, but first the random alloy theory will be presented.

Consider a pseudobinary alloy in which the A–C and B–C molecules are randomly placed on the crystal lattice. Attention will be focused on ternaries, but the approach is readily extended to quaternaries. The general crystal potential for the A and B atoms may be written as

$$U(\mathbf{r}) = \sum_A U_A(\mathbf{r} - \mathbf{r}_0) + \sum_B U_B(\mathbf{r} - \mathbf{r}_0), \tag{2.103}$$

where $\mathbf{r}_0$ defines the lattice site of the particular sublattice on which the A and B atoms are randomly sited. The total crystal potential may now be decomposed into its symmetric and anti-symmetric parts. The former is the "virtual-crystal" potential, and the latter is a random potential, whose average is presumed to be sufficiently small that it can be neglected. This decomposition is just

$$U_s(\mathbf{r}) = \sum_{\mathbf{r}} [c_A U_A(\mathbf{r} - \mathbf{r}_0) + c_B U_B(\mathbf{r} - \mathbf{r}_0)]$$

$$U_a(\mathbf{r}) = \sum_{\mathbf{r}} [U_A(\mathbf{r} - \mathbf{r}_0) - U_B(\mathbf{r} - \mathbf{r}_0)] c_{\mathbf{r}_0}, \tag{2.104}$$

where $c_{\mathbf{r}_0} = 1 - c_A$ for a lattice site containing an A atom and $-c_A$ for a lattice site containing a B atom. The virtual-crystal potential, which is the symmetric potential, is a smooth interpolation between the potential for the A–C crystal and that for the B–C crystal. The random part can contribute either to scattering of the carriers (alloy scattering) or to *bowing* of the energy levels in the mixed crystal. Bowing of a band gap means a deviation from the

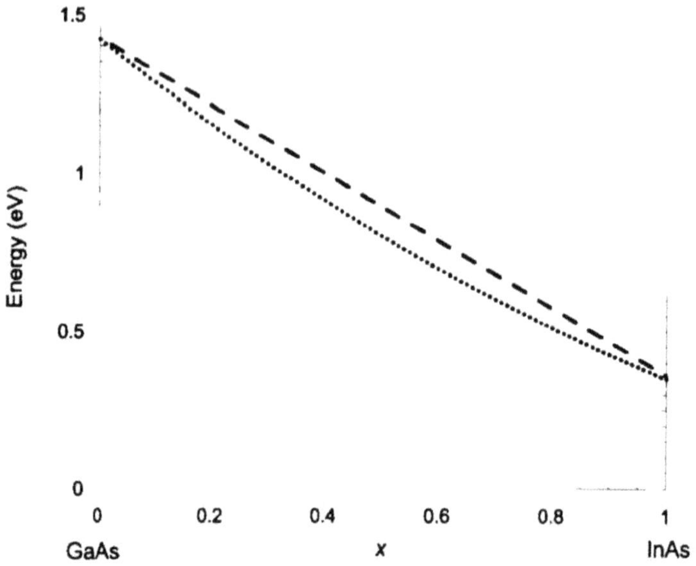

Figure 2.11 The variation of the energy gap in the alloy InGaAs as a function of InAs composition *x*. The virtual crystal approximation is the dashed straight line, while the experimental bowing produces the dotted curve.

linear extrapolation. This is shown in Figure 2.11 for the case of InGaAs, and it can be seen that this bowing is toward a gap that is narrower than predicted by the virtual-crystal approximation. If there is a regularity to U_A, so that it possesses a significant amplitude in one of the Fourier components, it will make a significant impact on the Bloch functions and on the band structure. Thus one definition of a random alloy is that it is one in which the anti-symmetric potential is sufficiently random that none of the Fourier components are excited to any great degree. This means that the anti-symmetric potential must be aperiodic in nature.

The function $c_{\mathbf{r}_0}$ has certain symmetry properties. For example, if this quantity is summed over all lattice sites of the *A–B* sublattice, the result is

$$\sum_{\mathbf{r}_0} c_{\mathbf{r}_0} = N_A(1 - c_A) + N_B(c_A) = [c_A(1 - c_A) - c_A(1 - c_A)]N = 0. \quad (2.105)$$

As expected for a true anti-symmetric potential, the average value is zero. Similarly, a correlation relation may be defined through

$$\sum_{\mathbf{r}_0} c_{\mathbf{r}_0} c_{\mathbf{r}_0+\mathbf{r}_0'} = Nc_A(1 - c_A)\alpha_{\mathbf{r}_0'}, \quad (2.106)$$

where the $\alpha_{\mathbf{r}_0'}$ are the so-called Cowley–Warren order parameters. In a truly random alloy, the correlation coefficient is a delta function: $\alpha_{\mathbf{r}_0'} = \delta_{\mathbf{r}_0'}$. For

Table 2.7 Bowing energy for some alloys

Material	E_{bow}
GaP–GaAs	0.19
GaP–InP	0.63
GaAs–InAs	0.37
GaAs–AlAs	0.4
GaAs–GaSb	0.8
GaSb–InSb	0.51
InP–InAs	0.1
InAs–InSb	0.6
HgTe–CdTe	0.4

small $\mathbf{r}'_0$, clusters of like atoms result in a positive correlation coefficient, while anti-clustering can result in a negative value for this quantity. Thus the only term of interest is that for the on-site summation, and it may readily be determined that (2.106) is the right result, and can be normally defined with the correlation function equal to unity. The factor c_A is indeed just the x term above, and (2.106) is the source of an $x(1-x)$ correction term in the variation of the band gap from one composition of the alloy to the other. This correction to the virtual crystal approximation arises from the anti-symmetric part of the crystal potential, but is present even in true random alloys, as it is the uncorrelated part of the anti-symmetric potential. If there is any correlation at all between the positions of the A atoms or the B atoms, the simple picture above becomes significantly changed. This will be discussed further below.

The experimental measurements of the band gap variation for a typical alloy can be expressed quite generally as

$$E_{gap} = xE_{gap}^{AC} + (1-x)E_{gap}^{BC} - E_{bow}x(1-x). \tag{2.107}$$

The general form of (2.107) is found in nearly all alloys; for example, there is a linear term interpolating between the two endpoint compounds that represents the virtual crystal approximation (the first two terms of this equation, and the dashed line in Figure 2.11), and a *negative* bowing energy [coefficient of the $x(1-x)$ term] that represents the contribution from the uncorrelated anti-symmetric potential, and leads to the actual band gap variation shown by the dotted curve in Figure 2.11. In Table 2.7, the bowing energy found for the direct gap at Γ for some alloys of the group III–V and the group II–VI compounds is listed. It is worth noting that the Sb compounds, which lie low in the periodic table and have significant d-level interactions with the valence electrons, all show large bowing of the gap. In fact, the alloy between InAs and InSb has a smaller gap at the midpoint composition ($x = 0.5$) than at either of the two endpoints.

In the quaternary compounds, it is necessary to extrapolate the band gap and lattice constants from those of the ternaries. There are many possible ternary materials; their number is roughly the number of binaries raised to the $\frac{3}{2}$ power. Usually, however, they are grown lattice matched to some binary substrate. In alloys, a rule known as Vegard's law stipulates that the lattice constant will vary linearly between the values of the two endpoints. For example, the nearest-neighbor distances in GaAs, InAs, and InP are found to be 2.45 Å, 2.63 Å, and 2.533 Å, respectively. From the properties of the face-centered cubic structure, we can then find the lattice constant of the cube edges to be 5.66 Å, 6.07 Å, and 5.85 Å, respectively. Thus the lattice constant of $Ga_xIn_{1-x}As$ is

$$a_{GaInAs} = 6.07 - 0.41x, \tag{2.108}$$

and this is lattice matched to InP for $x = 0.53$. Thus this composition of the alloy may be grown on InP without introducing any significant strain in the layer. The calculation for the quaternary is done in the same manner, but is slightly more complicated. The nearest-neighbor distance for GaP is 2.35 Å, which leads to a lattice constant of 5.43 Å. One wants to have the quaternary InGaAsP lattice matched to InP for this comparison. For the alloy $Ga_xIn_{1-x}P$, the lattice constant is

$$a_{GaInP} = 5.85 - 0.42x, \tag{2.109}$$

and the quaternary value can be obtained by using this value and the value from (2.108), weighted by $(1 - y)$ and y, respectively. This gives

$$a_{GaInAsP} = 5.85 - 0.42x + 0.22y + 0.01xy, \quad x = \frac{0.52y}{1-0.024y}, \tag{2.110}$$

where the last equation is required to maintain the lattice match condition.

2.6.2 Alloy ordering

As indicated above, it is quite possible that these alloy compounds are not perfectly random alloys, but in fact possess some ordering in their structure. The basis of ordering in otherwise random alloys lies in the fact that the ordered lattice, whether it has short-range order or long-range order, may be in a lower-energy state than the perfectly random alloy. In a random alloy $A_xB_{1-x}C$, the average of the cohesive energy will change by

$$E_{coh} = E_{coh}^{BC} + x(E_{coh}^{AC} - E_{coh}^{BC}), \tag{2.111}$$

within the virtual crystal approximation. While the *A–C* compound is losing energy, the *B–C* compound is gaining energy, and this energy comes from the expansion or contraction of the lattice of the two end compounds (and the variation this produces in the energy structure). For example, in $In_xGa_{1-x}As$, the cohesive energy is the average of those of GaAs and InAs, but the gain of energy in the expansion of the GaAs lattice is exactly offset by that absorbed in the compression of the InAs lattice, at least within the linear approximation of the virtual-crystal approximation.

If any short-range order exists, however, this argument no longer holds. Rather, the ordered GaAs regions undergo a loss of energy as their bonds are stretched in the alloy, while the ordered InAs regions gain energy as the bonds are compressed (here gain of energy is to be interpreted in the sense that the crystal is compressed and the equilibrium state now has a lower-energy state). Since one may assume that the cohesive energy varies as $1/d^2$ in the simplest theory, just as any other interaction energy in the crystal, a net increase of cohesive energy in the semiconductor compounds is a very simple calculation. However, the lattice constant, and hence d, varies linearly from one compound to the other due to Vegard's law. Yet, the cohesive energy varies with the inverse square of the change in the lattice constant. Thus, it is not guaranteed that the change in cohesive energy will follow the simple linear law given by (2.111).

The valence band actually contains just the $4N$ (where N is the number of atoms in the structure) electrons in equilibrium. As one alloys two compounds, the absolute position of this band can move, yielding a change in the average energy of the bonding electrons. A decrease in the average energy of the valence band, or an increase of the cohesive energy, both indicate that ordering in the alloy is energetically favored. It is apparent that there are some alloys in which ordering is energetically favored. The data on GaAlAs are mixed, but even if it occurs, it would be only at low temperatures. In this case, only realistic total energy calculations (which are discussed below) can shed much light on the stability of the random alloy. The experimental situation has not been effectively investigated except for a few special cases.

In the case of InGaAs, InGaSb, and InAsP, all indications suggest that the alloy will favor phase separation and ordering at room temperature. Indeed, this tendency to order in the InGaAs and InAsP compounds may produce the well-known miscibility gap in the InGaAsP quaternary alloy that is found in the range $0.7 < y < 0.9$. The actual nature of any ordering that occurs can be quite subtle, however. For example, in pioneering experiments with X-ray absorption fine-structure (XAFS) measurements on InGaAs, Mikkelsen and Boyce (1983) found that apparently the GaAs and InAs nearest-neighbor bond lengths remain nearly constant at the binary values (the covalent radii) for all alloy compositions. The average cation–anion distance follows Vegard's law and increases by 0.174 Å. In addition, the cation sublattice strongly resembles a virtual crystal (this is the sublattice in which alloying occurs), but the anion sublattice is very distorted due to the foregoing tendency. The distortion leads to two As–As (second-neighbor) distances which differ by as much as 0.24 Å, and the distribution of the observed second-neighbor distances has a Gaussian profile about the two distinct values. The distortion of the anion sublattice is clearly beyond the virtual-crystal approximation, and such a structure can be accommodated in a model crystal that resembles closely a chalcopyrite distortion, and can in fact explain the observable bowing of the band structure. If these observations are carried to other semiconductor alloys, it is likely that in alloys in which the alloying is occurring between two atoms of greatly differing size, the nearest-neighbor distance will probably prefer to adopt the binary value.

Zunger and his co-workers (Zunger and Jaffe, 1984; Srivastava *et al.*, 1985; Martins and Zunger, 1986; Mbaye *et al.*, 1986) have carried these theoretical ideas much further to investigate the alloying of group III–V semiconductors. In most of the arguments above, we have only looked at the average compression/expansion of the overall crystal lattice of the two binary constituents and have not included the tendency for the average nearest-neighbor distance to remain at the binary value. For this to occur, there must be a relaxation of the common constituent sublattice within the unit cell as well as a possible charge transfer between the various common atoms on the non-alloyed sublattice. In fact, they find that the latter factors are the dominant ones in alloy ordering. They have investigated the tendency to order by adopting a total energy calculation using the non-local pseudopotential method. They calculate the total energy for a given composition of alloy and then vary the atomic positions to ascertain the lowest energy state. This has proven to be a very powerful approach.

Let us consider the above arguments of Zunger and his co-workers for the manner in which ordering may occur. For an $A_xB_{1-x}C$ alloy, the four cations of type A and B per face-centered cubic cell can assume five different *ordered* nearest-neighbor arrangements around the C atom: A_4C_4, A_3BC_4, $A_2B_2C_4$, AB_3C_4, and B_4C_4. These are denoted as $n = 0, 1, 2, 3, 4$ arrangements. Obviously, n indicates the number of B atoms in the cluster. If the solid is perfectly ordered with these arrangements (which correspond directly to $x = 0$, 0.25, 0.5, 0.75, and 1.0, respectively), the lattice structures are zinc-blende only for $n = 0$ or 4. For the other compositions, the ordered crystal structure is known as either luzonite or famatinite for $n = 1$ or 3, and either CuAu–I or chalcopyrite for $n = 2$. The choice of the particular crystal structure is dominated by whichever is the lowest-energy configuration for the crystal. In any case, it is now thought that a disordered or random alloy must be a statistical mixture of these various crystal structures. This suggests that highly-ordered alloys can generate new types of superlattices with very short periods. Indeed, experiments have observed the highly-ordered $x = 0.5$ structure in both GaAsSb (Jen *et al.*, 1986) and GaAlAs (Kuan *et al.*, 1985). In the former case, both the CuAu–I structure and the chalcopyrite structure are observed, while only the CuAu–I structure seems to be found in the latter case. In addition, the famatinite structure has been observed in the InGaAs alloy for $x = 0.25$ and 0.75 (Nakayama and Fujita, 1986). It is clear that alloys of the binary semiconductors can be anything but random in nature and may be quite different from the simple virtual crystals they are made out to be. Mbaye *et al.* (1987) have calculated the phase diagrams for several alloys utilizing the total energy method discussed above, and find that random alloys, ordered alloys, and miscibility gaps can all occur and that the strain actually stabilizes the ordered stochiometric compounds.

2.7 Heterostructures

One of the principal reasons for developing alloy semiconductors is the desire to tailor the various band gaps to desired levels appropriate to create *heterojunctions* with desired properties. By a heterojunction is meant two dissimilar (in detail) semiconductors that are joined by an atomically abrupt interface. Consider, for example, the growth of AlAs on top of GaAs, where the two semiconductors have substantially different conduction band structure, although the lattice constants are essentially equal. Thus it might be expected that no lattice distortion occurs when AlAs is grown on GaAs. The question then is just how these different bands align themselves at the interface between the two materials.

The earliest theories of heterojunction interfaces generally suggested that the electron affinities, which are the differences between the bottom of the conduction band and the vacuum level, should be matched at the interface and this would give the so-called *band offset* between the two materials. There is no real rationale as to why this should work. The conduction bands have subtle differences between semiconductors due to the variations in the parameters that go into calculating the bands, and there is no reason or rationale in this approach to explain differences between direct- and indirect-gap semiconductors. Where, then, does one choose to evaluate the electron affinity – always at the zone center or at the minimum gap? This is complicated by the fact that the electron affinity is really measured as a surface property and not as a bulk property. As we will see in the next section, the surface is not always representative of the bulk semiconductor. At any rate, it is fairly well known that the electron affinity has no real claim to any credence in defining how various semiconductor bands should be aligned at the heterojunction interface. It might be better to measure from the valence band, since this is the actual *ionization energy* required to remove a valence electron to the vacuum state. However, this has its own problems, as will be discussed below.

This may seem to be concentrating on an unimportant point, but it turns out that one cannot effectively design different heterostructures, which may contain hundreds or thousands of layers of alternating semiconductor materials, without knowing just what physics determines how the bands

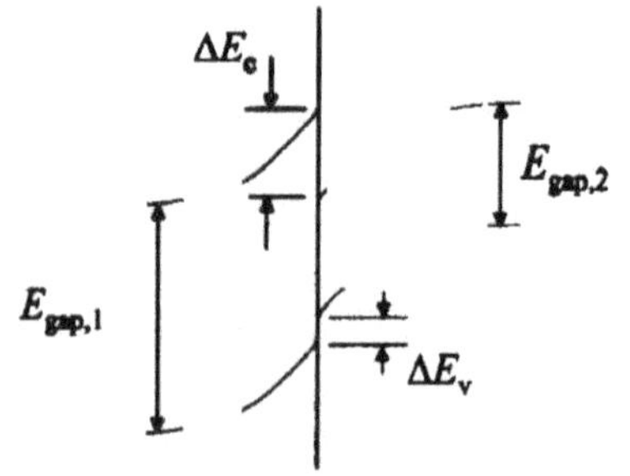

Figure 2.12 The semiconductor bands are bent at the interface, depending upon the doping of the two materials. Here, it is assumed that the left material is much more *n*-type than that on the right. This bending is also required by the *offset* between the two bands, valence or conduction, of the two materials.

are aligned. Only if this alignment is known can one determine what the nature of the energy barriers for electrons (or holes) between different layers will be. Consequently, a great deal of effort has been expended on this problem, but by and large the answer is given mainly by experimental measurements of the actual band offset. We will illustrate the problems by considering the most suggestive, but also probably wrong, approach first – the common-anion rule.

2.7.1 *The common-anion rule*

The problem of interest is indicated in Figure 2.12, where two dissimilar semiconductors are joined at the heterojunction interface. The problem is to ascertain how the difference in the two band gaps is distributed between the conduction-band offset ΔE_c and the valence-band offset ΔE_v. In semiconductor heterojunctions between materials like AlAs and GaAs, the anion (in this case, the As constituent) is common between the two semiconductors that make up the heterojunction. In the absence of the interaction energy E_{xx}, the top of the valence band is given by the atomic *p* level of the anion (see equation (2.90b)). Since the top of the valence band is predominantly composed of anion derived *p*-functions, the common anion rule suggests therefore that the top of the valence band should be continuous across the heterojunction for these two materials (McCaldin *et al.*, 1976). Thus, if the anion is common between the two compounds on either side of the heterojunction, the top of the valence band should be continuous. Even in the presence of E_{xx}, which varies as the reciprocal of the square of the nearest-neighbor distance in the simplest models, one expects the top of the valence band to be nearly the same in the two compounds since E_{xx} will be almost the same in these two compounds.

This reasoning fails, though, first for the obvious reason from (2.90b) that when E_{xx} is not zero, the top of the valence band includes contributions from the p levels of the cation, and this contribution differs for the two constituents of the heterojunction. Thus there is no real reason to expect that the common-anion rule will work. The offsets of interest are shifts in the valence-band edge of fractions of an electron volt in the AlAs–GaAs heterojunction, so the shift, as a fraction of the absolute p-state energy levels, is small and probably subtle.

It has also been suggested that one could modify the common-anion rule by actually using (2.90b) to calculate the shift in the valence-band energy of the atomic p-state energy level of the anion, and this would then give the valence-band offset (that is, in core level shifts of inner energy levels of the atoms). This would incorporate the contributions from the cation differences. The problem lies in the nature of the interface that breaks the normal symmetry of the zinc-blende structure. As pointed out below, the As atoms at the interfacial layer have bonding configurations that are different than the bulk. In addition, the bond-charge contribution (which leads to the core level shifts) is different.

Another modification of the common-anion rule is to suggest that one should use the measured differences in ionization energies for the two constituents (which is a variation of the electron affinity rule as suggested above). This approach replaces the use of atomic energy levels, modified by the interaction energy, with measured energy levels at the top of the valence band, but still depends on the assumption that the character of the wave functions of the valence-band maximum does not change in going from one material to the other. This also does not work very well and there may be an intrinsic problem in using core level shifts.

In GaAs for example, the bonding electrons are the $4s$ and $4p$ electrons. The As $3p$ level lies some 40.2 eV below the vacuum level in AlAs, and an additional 0.6 eV deeper in GaAs. This difference must arise from the difference in the overall potential involved in the bonding electrons. When one goes from AlAs to GaAs, the details of the bond charge, the fractional charge residing in the bond orbitals, will change since the detailed nature of the bonding will change. This is the cause of the core-level shift between the two materials. At the interface between the two materials, As atoms will exist that have (on the average) two Ga neighbors and two Al neighbors. Thus the core-level shift for this layer of As atoms will be somewhere between the values for the two separate compounds. This will lead to an interface dipole charge since the interfacial region will not have the same bonding properties as the bulk material on either side of the heterojunction. To date, there is no real method of calculating the size of this interface dipole, nor is there an effective method of estimating the degree to which it is screened by the valence electrons. As a result, the common-anion rule and its derivatives are a pleasing concept, and provide a zero-order idea for evaluating the expected valence-band offset. However, in reality, the first-order corrections are of the same magnitude as the valence-band offset, and the result is not a theory that can be used to design reliable heterostructures.

There are other problems in setting the alignment in that the results may well depend upon just how the heterostructure is grown. The use of tilted substrates, or of graded interfaces, can change the apparent results. Nevertheless, one should be able to define a real discontinuity for well-prepared heterojunctions, which is related to bulk properties of the two materials.

2.7.2 *Alignment to deep-level impurities*

The problem of determining the band offset is really the problem of determining an appropriate bulk level which can serve as a proper reference level for the band alignments. In the preceding section, this level was the common anion levels at the top of the valence band. Another suggestion is that the reference level can be determined from the deep levels associated with transition metal impurities (Langer and Heinrich, 1985). The transition metals are known to form deep, localized impurity levels in semiconductors. One of the features of these impurities is that they can possess several different charge states separated by only a fraction of the band gap of the host semiconductor, which is different than e.g. the difference in ionization energies of the two materials in the heterojunction. Yet, the evidence is that the impurity levels formed are not pinned to either of the nearby band edges (conduction band or valence band), which is quite different from shallow impurities. On the other hand, it seems that these levels provide a bulk reference level associated with the particular transition metal, and not with its host. This, then, suggests that the valence band discontinuity across the heterojunction interface is given merely by the difference in the transition metal impurity levels in the two materials. Evidence suggests that this holds fairly well, at least for GaAlAs and some II–VI materials, and the prediction of band offsets for other semiconductors is better than that provided by the common anion rule.

2.7.3 *Other approaches*

Frensley and Kroemer (1977) recognized that the dipole calculation would be the most important part of estimating the valence-band offset between the two constituent semiconductors, and suggested that one evaluate the potential in the open interstitial areas that exist in the zinc-blende structure. These points should offer very good values of the *average potential* in the semiconductor, relative to the vacuum level. Then one could use the differences between these average interstitial potentials to estimate the dipole potential existing between the two sides of the heterojunction. Although this potential is a good average for the crystal, there is no reason to expect that it occupies such an exalted position. Yet the values obtained for the band offset are as good as any other approach, and better than most.

Tersoff (1986) assumed that there exists a *charge neutral* energy level in the gap of the semiconductor. If the energy lies below this level, the valence-band holes contribute more to the wave function, while if the energy lies above the level, the electron wave function contributes more. Although quite controversial when it was first presented, the approach has gained a grudging respect, since the predictions seem to be in good agreement with experiment. In this approach, Tersoff used extensive pseudopotential calculations to evaluate the imaginary band structure (the wave vector k is imaginary in the energy gap) to determine the position of this level, and then argued that aligning this level between the two sides of the heterojunction would minimize the dipole charge. The valence-band offset is then envisaged as the difference in the two charge neutral energies, each of which is essentially near the midgap point. The problem in accepting this theory directly lies in trying to estimate why the dipole charge should be minimized. In fact, the dipole charge should arise from the differences in the nature of the bond charge on the two sides of the heterojunction and is thus a difference between bulk properties. It really is not a quantity that can be minimized or maximized; it just is what it is. Further, if the charge neutral energies are to select how the bands line up between the two materials, one would expect that there would be some mechanism to pin the Fermi level near these energies to keep the bands from shifting. Yet the argument for the charge-neutral energy is the same as that used to determine the Fermi energy in an intrinsic semiconductor. Although the intrinsic Fermi energy has a well-defined value, it has no special place in determining just where the Fermi energy will ultimately reside with respect to the valence band maximum. Similarly, while Tersoff's charge neutral energy has a well-defined value, it has no special place in determining just where the Fermi energy will ultimately reside with respect to the valence-band maximum. Without this special ability to balance the Fermi level on the two sides of the heterojunction, particularly the necessity to be insensitive to doping in the two semiconductors, it cannot define the valence-band offset.

The band lineup is even more complicated in the case where one of the two constituents is an alloy. As we discussed in the previous section, it is not at all clear that the alloy is a random mixture of its constituents. This is even more complicated at the heterojunction interface, since it is even more likely that ordering can occur here. Since the tendency to order depends to a large degree on the nature of the substrate, the ordering for growing an AlGaAs layer on top of GaAs may be much different from that for growing this same layer on an AlGaAs layer of the same composition. Because of this, the calculation of the band offset is more difficult in such heterojunctions. This is also the probable cause of differences observed in band offset for a GaAs–AlGaAs heterojunction between the case in which the GaAs is

Table 2.8 Some heterojunction valence band offsets

A *grown on* B	ΔE_v
GaAs–AlAs	0.5±0.1
InAs–GaAs	0.17
InAs–GaSb	0.51
AlSb–GaSb	0.42
CdTe–HgTe	0.12–0.35

grown on top of the AlGaAs and the case in which the AlGaAs is grown on top of the GaAs. Such effects are also at the root of failures in the so-called transitivity relation, which asserts that the sum of the offsets in going around a semiconductor ring A–B–C–A should be zero. This concept is worthy but lacks theoretical rigor, especially with the possibility that the alloys might show ordering at the interface.

It is pretty obvious by now that there does not exist a good theory that explains exactly the band offset observed in heterojunctions. On the other hand, there are several theories that can predict the band offset to within a few tenths of an electron volt. Unfortunately, the band offset itself is usually just a few tenths of an electron volt. In Table 2.8, the valence-band offsets for several heterojunctions (experimental results) are listed, where the first material is grown on the second. Since this area still needs considerable research to provide a universally acceptable theory for the band offsets, readers will probably have good company within the scientific community if they adopt any of the foregoing theories (or several others that have not been discussed here).

2.7.4 *Strained-layer heterostructures*

In all of the discussion above, it was assumed that the alloy can be grown lattice matched to a suitable substrate. Is this what one wants to do? Certainly, it has been argued that one should try to lattice match the heterojunction interface so as not to introduce defects and dislocations arising from release of local strain. On the other hand, it has been found in non-lattice-matched heterostructures that there is a critical thickness of the overgrown layer, below which the strain is not released. Rather, the grown layer is distorted so that its lattice constant along the interface matches the substrate. This results in a distortion of the basic cubic cell in that the cell is compressed (extended) along the interface and therefore is extended (compressed) in the direction normal to the interface (or vice versa depending upon which lattice constant is larger). Yet these layers can be grown quite easily with modern growth techniques such as molecular-beam epitaxy or metal-organic vapor-phase epitaxy. The resulting "strained-layer heterojunction" is a high-quality interface in which the lattice of the grown layer is purposely mismatched to that of the substrate. The layer can be grown as long as it is sufficiently thin (here, this is usually thought to be of the order of 20 nm or less) (Matthews and Blakeslee, 1974). The reason for doing this lies in the dependence of the band structure on the lattice constant. The built-in strain modifies the band structure to produce desirable properties as part of the overall concept of band-gap engineering.

Calculating the band structure of the distorted strained lattice is not simple. The lattice constants of the two materials on either side of the heterojunction are equal in the two directions parallel to the interface. However, the lattice constant in the direction perpendicular to the interface differs for the two materials. Thus there can be straight channels through the material in some directions, but other crystalline directions undergo an effective "tilt" at the interface. This tends to introduce significant tetragonal distortion in the lattice, which complicates the band calculations. In thicker layers, it is the relaxation of this distortion that leads to the development of dislocations and defects. The most extensive data have been taken on InGaAs/GaAs strained-layer heterojunctions, and this suggests that the strain in the layer may be written as

$$e = \frac{w(1-\sigma/4)}{2\pi(1+\sigma)}(1+\ln w), \tag{2.112}$$

where

$$w = \frac{a}{\sqrt{2}h}, \tag{2.113}$$

h is the critical thickness, a is the lattice constant, and σ is the Poisson ratio. (The Poisson ratio measures the difference in expansion/compression of the lattice and is defined as $\sigma = (2G/W) - 1$, where W is Young's modulus and G is the rigidity modulus of the lattice (Nye, 1957).) The experiments on InGaAs–GaAs suggest that $h = 25$ nm.

The strain also creates modifications that are generally outside a dependence on the lattice constant. The valence band of most group III–V semiconductors is threefold degenerate in the absence of spin–orbit splitting (the latter of which splits off one of the three bands by spin–orbit energy). In the presence of spin–orbit coupling, the two remaining degenerate bands at $k = 0$ are the light- and heavy-hole bands. The strain energy splits this degeneracy further. For expansion of the lattice in the plane of the interface, the strain contributes to a lowering of the light-hole band relative to the heavy-hole band. Compression in this plane reverses this trend, so that it is the heavy-hole band that is lowered. The two bands can therefore cross at some point away from $k = 0$ creating a complex energy dispersion relationship with the possibility of new minigaps in the valence band.

2.8 Surfaces

The surface is the ultimate limit of the heterojunction, in which the semiconductor is matched to the true electron vacuum. Between the heterojunction limit and the free-surface limit is the interface between a semiconductor and an amorphous material such as SiO_2. The nature of the surface itself has been an object of study almost as long as there have been semiconductors to study. The aim here is to ask how the surface differs from the bulk. We begin with the idea of a clean surface.

Generally, the perfect surface is considered to be one that has been cleaved in high vacuum so that no "pollutants" are absorbed onto the cleaved surface. The cleavage plane in diamond semiconductors is generally the (111) plane, while in zinc-blende semiconductors it is generally the (110) plane. These are charge neutral planes for the two lattices. Early studies of the surface atomic structure were concentrated on what could be learned with the tools of the day – optical and electron-beam scattering. In 1982, however, the field was revolutionized by scientists at IBM's Zurich laboratory when they demonstrated that a scanning tunneling microscope could be used to achieve resolution of real surfaces on the subnanometer scale (Bennig and Rohrer, 1982). This phenomenal result was rewarded by their being awarded the Nobel prize just six years later. The scanning tunneling microscope (STM) consists of a fine metallic tip, made of a material such as tungsten, which is brought close to the surface. Since the tunneling current from the metal tip to the semiconductor surface (or in the reverse direction) depends exponentially on the distance between the tip and the surface, the tunneling current is dominated by small protrusions on the tip surface. This leads to resolutions that are on the atomic scale. Therefore, this provides an effective tool for actually studying the electronic states within a few eV of the Fermi level, since the magnitude of the tunneling current is a measure of the local density of states.

In fact, the surfaces of typical cleaved semiconductors (or even more so for semiconductors prepared by other techniques) do not demonstrate the smooth surfaces that are normally conjectured, since any heating will cause the surface to change the local geometry to one in which the energy is in a lower state. In general, one can refer to the restructuring as *relaxation* if the surface primitive cell is not changed, and *reconstruction* if the surface primitive cell is changed. By primitive cell of the surface, the reference is to the projection of the normal three-dimensional unit cell of the diamond or zinc-blende lattice. The surface cell is just the face appropriate to the particular surface. Thus for a (100) surface, the primitive cell is a square with an additional atom (corresponding to the face-centered atom) at its center.

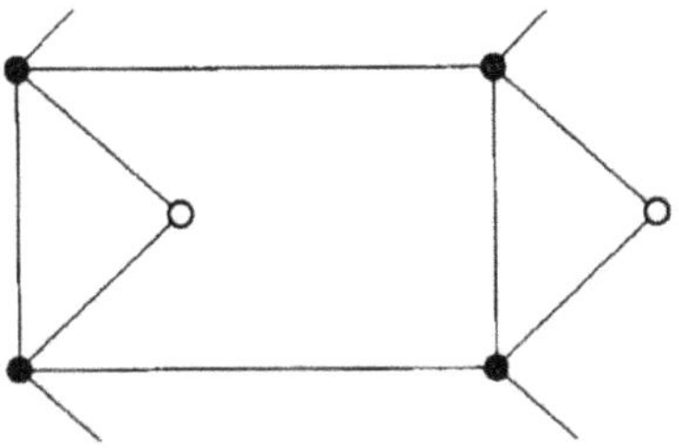

Figure 2.13 The surface primitive cell for a (110) surface, appropriate to the cleavage plane of GaAs. The long axis (horizontal here) is for the edge of the fcc cell, while the short axis is from the corner to the face center. The white atoms are the second atom in the basis at each lattice site.

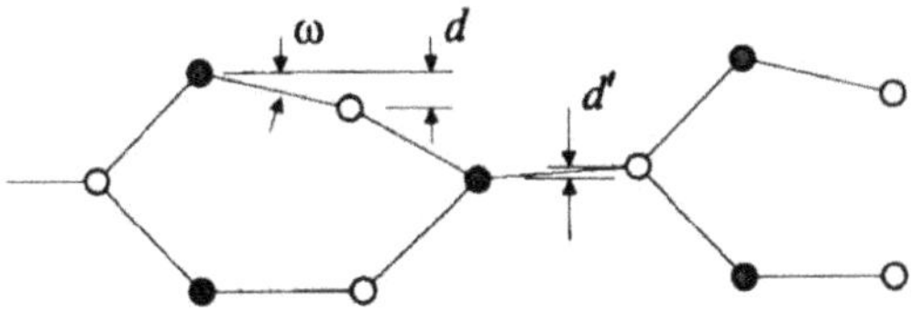

Figure 2.14 The relaxed atoms in a plane normal to that shown in Figure 2.12. That is, the vertical direction here is normal to the surface. The upper white atoms have moved inward, while the upper black atoms have moved slightly outward. The second row black atoms also relax slightly inward.

For the (110) direction, however, the primitive cell is a rectangle with a basis at each lattice site (corresponding to the basis at each lattice site in the face-centered cubic structure). This is shown in Figure 2.13. The next layer of atoms below this is shifted one-half of the rectangle edge in both the lateral and vertical directions, so that a "black" atom can sit on the (111) bond coming up from the "white" atom. The latter points to the right. Surface relaxation does not change this basic unit cell of the surface atoms. Instead, there can be a relaxation of the atoms. In Figure 2.14, this is shown for a plane taken through the white atoms and normal to the surface. The structure of Figure 2.14 may be relatively easily understood. The "short" bonds that appear horizontal are the diagonal bonds in Figure 2.13, and lie in the surface plane. The longer bonds, which are diagonal in Figure 2.14, are the orbitals which point out of the surface. The relaxation occurs in most zinc-blende structures with the cation (the "black" atoms) moving *out of the surface* and the anions (the "white" atoms) moving *into the semiconductor below the surface* by about $d = 0.5$ Å in the case of GaAs (Duke, 1983). In addition, there is a slight distortion of the

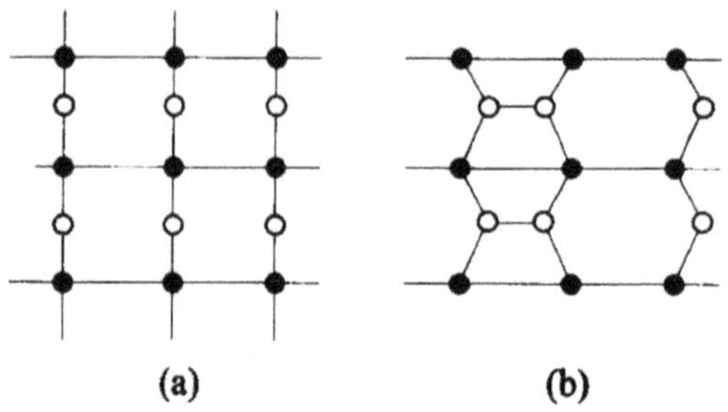

Figure 2.15 (a) The (100) surface appropriate to Si, used in the integrated circuit industry. The white atoms are in the surface layer, while the black atoms are in the second layer, below the surface. (b) Reconstruction resulting from hybridization of the surface atoms, when they bond covalently to one another.

second layer by about $d' = 0.1$ Å. The distortion of the top layer leads to an angle in the "surface" bond of about 28 ± 3° with respect to the true surface plane. These variations are quite systematic across the zinc-blende structures in complete analogy with the bulk chemical trends of the semiconductors. These relaxations also correlate well with the surface electronic structure.

Reconstruction, on the other hand, is a more complex reordering which changes the surface primitive cell of the crystal lattice. The classic example is the (111) cleavage plane of Si. When heated, a great variety of reconstructions can occur in the surface, which create a "superlattice" cell that extends over many projections of the bulk primitive cell. The dominant reconstruction is thought to be the 7 × 7 structure (which refers to the number of unit cells in each of the two surface directions that make up the "new" primitive cell of the surface). This was the first surface of Si observed by STM, and this reconstruction was confirmed by this method.

In the case of the ideal Si (100) surface, the atomic arrangement also results in a dominant reconstruction, which is the 2 × 1 dimerization reconstruction. In this surface plane, each Si atom has two bonds to the next lower layer of atoms, and two free bonds. These two free bonds create a plane which is rotated 90° from the plane defined by the two back bonds. The reconstruction arises when two neighboring Si atoms form hybrids between them. These hybrids are not at the normal bond angles, and this pulls these two atoms closer to each other, as shown in Figure 2.15. Here, it may be seen that this creates a surface super-cell which is twice the normal cell size in the dimerization direction. In Figure 2.16, an STM micrograph of a 2 × 1 surface is shown. One can clearly see the dimer rows running from lower right to upper left in the grey area, and toward the upper right in the lighter area (which is presumed to be an additional layer with a step separating the two regions).

Figure 2.16 An STM image, taken in an ultra-high vacuum environment, of the (100) surface of Si showing the reconstruction due to the dimerization. (Courtesy of M. Hersam and J. Lyding of the University of Illinois at Urbana-Champaign.)

PROBLEMS

1. The Kronig–Penney model utilizes a rectangular potential well, for which

$$V(x) = \begin{cases} V_0, & 0 < x < d, \\ 0, & d < x < a, \end{cases}$$

with

$$V(x \pm a) = V(x).$$

Solve the Schrödinger equation for this system in each of the regions and match the wave functions and derivatives at each boundary. Using the assumptions that V_0 is allowed to grow large and d small in a manner that $V_0 d = Q$ is held constant, show that a set of allowed and forbidden energy bands results.

2. If **k** is a vector in the first Brillouin zone and **K** is an arbitrary lattice vector in **k**-space, then for $\mathbf{k}' = \mathbf{k} + \mathbf{K}$, show that if $\mathbf{k}'$ is in the first Brillouin zone, $\mathbf{K} = 0$ is required, whereas if $\mathbf{k}'$ is outside the first Brillouin zone, $\mathbf{K} \neq 0$.
3. Calculate the ratio of the kinetic energy at the corner of the Brillouin zone (111) to that at the face center (100) for the simple cubic and for the body-centered cubic lattices.
4. A particular semiconductor has $m^* = 0.015m$ and $E_{gap} = 0.22$ eV. Calculate the interaction potential U_G and the free-electron energy E_G for these states in the simple nearly-free electron model.
5. A family of surfaces of constant energy has the form

$$E = \frac{\hbar^2}{2m^*}\left(Ak^2 - B\frac{k_x^2 k_y^2 + k_y^2 k_z^2 + k_z^2 k_x^2}{k^2}\right),$$

where A, $B > 0$. Find the elements of the effective mass tensor. Verify that there are points in the reciprocal space for which components of the mass tensor can have opposite signs.
6. A prototypical semiconductor has $E_{ss} = 2$ eV and $E_{pp} = 0.43$ eV, with $E_p = -5.2$ eV and $E_s = -9.2$ eV. Compute the fundamental optical energy gap at $\mathbf{k} = 0$ and the average energy gap between bonding and antibonding states. Also determine the homopolar energy gap and the metallicity factor. Assume that $V_2 = 6.06$ eV.
7. Construct a simple computer program to compute the band structure throughout the Brillouin zone for Si and Ge using the LCAO approach with only first neighbors. Use the data contained in the text for the parameters. Where are the minima of the conduction band and the maximum of the valence band located, and what is the value of the fundamental optical gap for these materials? Is the conduction-band minimum located at the proper place? Why?
8. Construct the equivalent Bloch sums for the second-neighbor interactions in the LCAO theory. Using only the two dominant additional constants discussed by Slater and Koster, compute the band structure of Si and Ge with the second-neighbor interactions included. Can the conduction-band minima be placed in the proper place by adjusting these constants? What if you couple the sp^3s^* and the second-neighbor interactions? Will this give better band structure?
9. Construct a simple computer program to compute the band structure throughout the Brillouin zone for GaAs and InP using the LCAO approach with only first nearest neighbors. Where are the minima of the conduction band and the maximum of the valence band located, and what is the value of the fundamental optical gap at X and L for

these materials? If you adjust the latter valleys by the inclusion of the s^* interaction, what are these gaps now, and how do they compare with experiment?

10. Compute the optical gaps at Γ, X, and L for the alloy InAlAs using the LCAO theory including the s^* interaction. First, adjust the band structure for the two end compounds, and then that of the alloy. How does the band gap compare with the virtual crystal approximation, and what is the bowing parameter found from this calculation?

REFERENCES

Bennig, G., and Rohrer, H., 1982, *Hevetica Physica Acta*, **55**, 726.

Braun, F., 1874, *Ann. Phys. Pogg.*, **153**, 556.

Brillouin, L., 1953, *Wave Propagation in Periodic Structures* (New York: Dover).

Chelikowsky, J. R., and Cohen, M. L., 1976, *Physical Review B*, **14**, 556.

Dresselhaus, G., Kip, A. F., and Kittel, C., 1955, *Physical Review*, **98**, 368.

Duke, C. B., 1983, *Journal of Vacuum Science and Technology B*, **17**, 989.

Faraday, M., 1833, *Experimental Researches in Electricity*, Ser. IV, pp. 433–9.

Faraday, M., 1834, *Beibl. Ann. Phys.*, **31**, 75.

Ferry, D. K., 1996, *Quantum Mechanics* (Bristol: Inst. Phys. Publ.).

Frensley, W. R., and Kroemer, H., 1977, *Physical Review B*, **16**, 2642.

Harrison, W. A., 1980, *Electronic Structure and the Properties of Solids* (San Francisco: W. H. Freeman).

Jen, H. R., Cherng, M. J., and Stringfellow, G. B., 1986, *Applied Physics Letters*, **48**, 782.

Kane, E. O., 1957, *Journal Physics and Chemistry of Solids*, **1**, 249.

Kane, E. O., 1966, *Semiconductors and Semimetals*, Vol. 1, Ed. by Willardson, R. K. and Beer, A. C. (New York: Academic Press) pp. 75–100.

Kuan, T. S., Kuech, T. F., Wang, W. I., and Wilkie, E. L., 1985, *Physical Review Letters*, **54**, 201.

Langer, J. M., and Heinrich, H., 1985, *Phys. Rev. Lett.*, **55**, 1414.

Martin, T., Ed., 1932, *Faraday's Diary*, Vol. 2 (London: G. Bell and Sons, Ltd.) pp. 55–6.

Martins, J. L., and Zunger, A., 1986, *Physical Review Letters*, **56**, 1400.

Matthews, J. W., and Blakeslee, A. E., 1974, *Journal of Crystal Growth*, **27**, 118.

Mbaye, A. A., Zunger, A., and Wood, D. M., 1986, *Applied Physics Letters*, **49**, 782.

Mbaye, A. A., Ferreira, L. G., and Zunger, A., 1987, *Physical Review Letters*, **58**, 49.

McCaldin, J. O., McGill, T. C., and Mead, C. A., 1976, *Physical Review Letters*, **36**, 56.

Mikkelson, J. C., and Boyce, J. B., 1983, *Physical Review Letters*, **49**, 1412.

Nakayama, H., and Fujita, H., 1986, *Institute of Physics Conference Series*, **79**, 289.

Nye, J. F., 1957, *Physical Properties of Crystals* (Oxford: Clarendon Press).

Pandey, K. C., and Phillips, J. C., 1976, *Physical Review B*, **13**, 750.

Phillips, J. C., *Bonds and Bands in Semiconductors* (New York: Academic Press, 1973).
Slater, J. C., and Koster, G. F., 1954, *Physical Review*, **94**, 1498.
Smith, W., 1873, *Journal of the Society of Telegraph Engineering*, **2**, 31.
Srivastava, G. P., Martins, J. L., and Zunger, A., 1985, *Physical Review B*, **31**, 2561.
Tersoff, J., 1986, *Physical Review Letters*, **56**, 2755.
Vogl, P., Halmerson, H. P., and Dow, J. D., 1983, *Journal of the Physics and Chemistry of Solids*, **44**, 365.
Zunger, A., and Jaffe, E., 1984, *Physical Review Letters*, **51**, 662.

Chapter 3

The Boltzmann equation and the relaxation time approximation

Essentially all theoretical treatments of electron and hole transport in semiconductors are based upon a one-electron transport equation, which usually is the Boltzmann transport equation. As with most transport equations, this equation determines the distribution function under the balanced application of the driving and dissipative forces. How do we arrive at a one-electron (or one-hole) transport equation when there are some 10^{15}–10^{20} carriers per cubic centimeter in the device? Even in so doing, the distribution function is not the end product, as transport coefficients arrive from integrals over this distribution. What are these integrals, and how are they determined? We begin to study the answers to these, and other, questions in this chapter.

In the case of low electric fields, the transport is linear; that is, the current is a linear function of the electric field, with a constant conductivity independent of the field. The approach used is primarily that of the relaxation time approximation, and the distribution function deviates little from that in equilibrium – primarily the Fermi–Dirac distribution or one of its simplifications such as the Maxwellian distribution. In this situation, it must be assumed that the energy gained from the field by the carriers is negligible compared with the mean energy of the carriers.

In this chapter, we achieve essentially three things, and the chapter is so divided. First, we will begin by considering the dynamics of individual carriers and creating ensemble averages of this dynamics. In this sense, we introduce correlation functions and show how they relate to the conductance and diffusion processes. We then show how this may be taken over to the averaging over a one-electron distribution, and how the Boltzmann equation arrives from those assumptions. The relaxation time approximation is defined, and then used to find approximate solutions for transport in electric and magnetic fields. The second part of the chapter is a discussion of the physics of the (almost) elastic scattering processes, which are introduced in this chapter to connect with the relaxation time approximation. We will also discuss the important mesoscopic effects that arise in small structures, first in connection with the linear response of the first part, and then in the third part of the chapter, where we treat transport in a high magnetic field.

3.1 Carrier dynamics and correlation

In general, an electron (or hole) can be thought to respond to an applied electric field F (F is used here to distinguish it from the carrier energy E) through Newton's Law for the change of momentum. In a sense, this can be written as

$$m^*\frac{dv}{dt} = -eF - \frac{m^*v}{\tau}. \tag{3.1}$$

However, there are a great many assumptions inherent in the expression of (3.1). First, it is assumed that the effective mass m^* is a constant, so that it can be brought through the time derivative. In general, in non-parabolic bands, this is incorrect, and one must develop the approach through the momentum $\hbar k$, rather than m^*v. This is a significant approximation, but we will continue to work within this approximation and use (3.1) in this chapter. That is, we will ignore the non-parabolic corrections with the understanding that they can be incorporated as needed. Secondly, a relaxation time has been introduced into (3.1) without any discussion about what it means; an important point since we discussed already in Chapter 1 the entire range of possible time scales and relaxation processes. To address this, we consider the probability that an electron has gone a time t without a scattering event (where t is measured from the last scattering event). Let us write this as

$$P(t) = \frac{e^{-t/\tau}}{\tau}. \tag{3.2}$$

This expression has been normalized so that the entire integral over time is equal to unity, as required for $P(t)$ to be a probability. Then, the average time between collisions is simply

$$\langle t \rangle = \int_0^\infty tP(t)dt = \tau. \tag{3.3}$$

We recognize this τ as the inverse of the scattering rate. In (3.1), however, we are interested in the rate of decay of the momentum. A scattering rate may not always decrease the momentum totally. If we scatter from $\mathbf{v}$ to $\mathbf{v}'$ (where we now use vector notation), then the loss of velocity is $|\mathbf{v} - \mathbf{v}'| = v(1 - \cos\theta)$, where θ is the angle of the scattering process. Hence, only a fraction of the velocity is lost. This is not significant if the scattering is isotropic, as an average over θ yields a result only from the unity term. However, if the scattering rate has a preferred angle, as in impurity scattering, then the relaxation of the momentum can take many scattering processes. Hence, the momentum relaxation time is different from the scattering time. In (3.1), we are interested in the momentum relaxation time, and so we shall use the symbol τ_m to differentiate this from the simple scattering time τ.

3.1.1 Correlation functions and the Kubo formula

An equation like (3.1) needs to be written for each electron in the system. (We can also say that this must be done for each hole, but we will only discuss electron transport in this chapter – the hole transport follows immediately from the results that will be obtained.) We also note that by the introduction of the momentum relaxation time, we have ignored the specific variation of the velocity of a given particle for which the scattering angle can vary for each of the individual scattering events that occur during the momentum relaxation time. To account for this *fluctuation* in velocity, we add a *random* force which accounts for this microscopic variation introduced by the actual scattering processes, rather than the simplified decay time associated with the momentum. The fluctuation inherent in the random force is not a *thermal* fluctuation, but arises from the differences between the detailed scattering of the carrier and the *average* decay expressed by τ_m. Hence, we can now rewrite (3.1) for a "typical" electron as

$$m^*\frac{dv_i}{dt} = -eF - \frac{m^*v_i}{\tau_m} + R_m(t). \tag{3.4}$$

This expression is termed the Langevin equation, and has been used in statistical mechanics to describe random motion, often called Brownian motion, and is most often seen in that context without the electric field F. There are certain *ensemble* properties, in which a summation over the group of electrons of interest yields various properties, such as

$$\frac{1}{N}\sum_i v_i = \langle v_i \rangle = v_d, \quad \langle R_m(t)\rangle = 0, \quad \langle v_i(0)\rangle = 0. \tag{3.5}$$

The first expression defines the *drift* velocity, while the last two express the fact that the random force produces no average force and the average velocity in the absence of a field is zero. As a result, the simple temporal behavior of the drift velocity can be found from (3.4) by taking the ensemble average and solving the resultant differential equation to yield

$$v_d = -\frac{e\tau_m F}{m^*}(1 - e^{-t/\tau_m}). \tag{3.6}$$

From this, we recognize the *mobility* as $\mu = e\tau_m/m^*$.

Let us now Laplace transform the Langevin equation (3.4) to obtain an expression that is more useful for time variations of the more complicated behaviors that can be expected in some cases. This leads to

$$sV_i(s) - v(0) = -\frac{eF}{sm^*} - \frac{V_i(s)}{\tau_m} + R_m(s), \tag{3.7}$$

where it is assumed that the field is initiated at $t = 0$. Here, we can now solve for the transform of the velocity to give

$$V_i(s) = \frac{v_i(0)}{\left(s + \frac{1}{\tau_m}\right)} - \frac{eF/m^*}{s\left(s + \frac{1}{\tau_m}\right)} + \frac{R_m(s)}{\left(s + \frac{1}{\tau_m}\right)}. \tag{3.8}$$

If we now ensemble average, only the second term on the right-hand side remains, and the inverse transform yields simply (3.6), as it should. On the other hand, we can multiply (3.8) by $v_i(0)$ and then ensemble average, in which case the second term vanishes, as does the third term if we add the assumption that

$$\langle v_i(0)R_m(t)\rangle = 0, \tag{3.9}$$

that is, that the random force is uncorrelated to the initial velocity of each particle, at least within the ensemble average. Then we find that

$$\langle V_i(s)v_i(0)\rangle = \frac{\langle v_i^2(0)\rangle}{s + \frac{1}{\tau_m}}, \tag{3.10}$$

or

$$\langle v_i(t)v_i(0)\rangle = \langle v_i^2(0)\rangle e^{-t/\tau_m}. \tag{3.11}$$

In this last expression, we have introduced a *correlation function*, specifically the velocity auto-correlation function (the velocity correlated with itself). If we write the normalized velocity auto-correlation function as

$$\varphi_v(t) = \frac{\langle v_i(t)v_i(0)\rangle}{\langle v_i^2(0)\rangle} = e^{-t/\tau_m}, \tag{3.12}$$

then we can rewrite the expression (3.6) and the inverse transform of (3.8) as

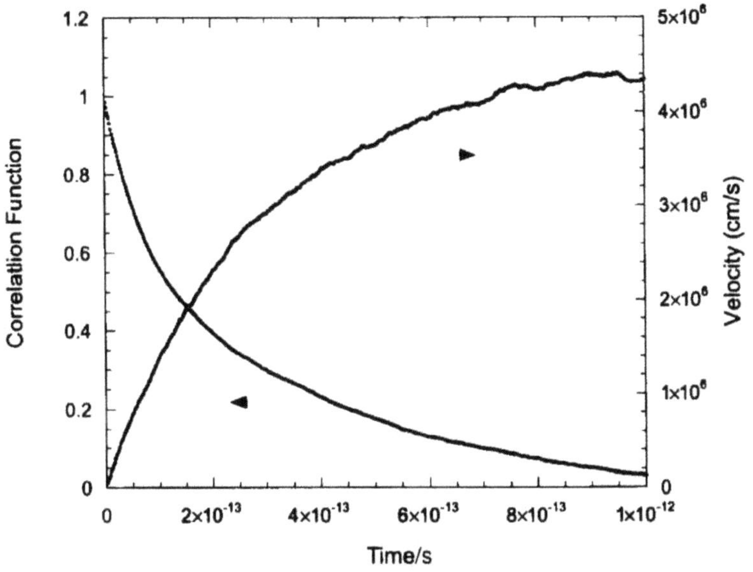

Figure 3.1 The velocity auto-correlation function and the drift velocity for electrons in Si with an applied field of 3 kV/cm applied at $t = 0$. This field is in the ohmic range (300 K), and the plot demonstrates the initial rise of the velocity and decay of the auto-correlation function.

$$v_d(t) = -\frac{eF}{m^*}\int_0^t \varphi_v(t')dt'. \tag{3.13}$$

This last result is quite important, as it is an expression of the *Kubo formula* (Kubo, 1957). The velocity auto-correlation function describes the decay of the fluctuations in velocity, while the drift velocity is the average velocity. This relationship between the average velocity and the correlation function of the fluctuations in the velocity is quite important. The net average that leads to the drift velocity is a representation of the dissipation through τ_m, while the correlation function describes the fluctuations. Hence, the Kubo formula is a *fluctuation–dissipation theorem*, at least in equilibrium. Clearly, the decay of the correlation function, and therefore of the correlated fluctuations, is determined by the momentum relaxation time, which also describes the mobility of the carriers. A comparison of (3.11) and (3.13) is shown in Figure 3.1.

3.1.2 Random forces and diffusion

The above description assumed that we initiated the correlation function at $t = 0$, where there was no average velocity. Consider now a more general case, evaluated at a time well beyond the initial transient, $t \gg \tau_m$. Then, (3.4) may be solved to give

$$v_i(t) = v_d(t) + \frac{1}{m^*}\int_0^t R(t')e^{(t'-t)/\tau_m}dt', \tag{3.14}$$

where we have introduced the drift velocity from the long-time limit of (3.6). We can now formulate the ensemble average for the velocity fluctuation by taking (3.14) at a different time, say t_2, and multiplying the two equations together. Using the fact that the ensemble average of the random force is zero, the result is

$$\langle v_i(t)v_i(t_1)\rangle = v_d(t)v_d(t_1) + \frac{1}{(m^*)^2}\int_0^{t_1}\int_0^{t}\langle R(t')R(t'')\rangle e^{(t'+t''-t-t_1)/\tau_m}dt'dt''. \tag{3.15}$$

The drift velocity is constant, but the time variables have been retained to better illustrate how this last equation is derived. The quantity within the integral is the *random force correlation function*, and in equilibrium is taken to be a delta function in the difference of the two time variables. That is,

$$\langle R(t')R(t'')\rangle = I_R\delta(t'-t''). \tag{3.16}$$

There is some ambiguity in the integral of (3.15). It can only be assured that the delta function lies within the range of both integrals, when the first integration is taken over the earliest time. Here, we shall assume that $t_1 > t$. For the time interval $t < t' < t_1$, there is no contribution from the delta function, and the first integral will vanish. Hence, we may evaluate the resulting integrals as

$$Integral = \frac{I_R}{(m^*)^2}\int_0^{t}\int_0^{t}\delta(t'-t'')e^{(t'+t''-t-t_1)/\tau_m}dt'dt''$$

$$= \frac{I_R}{(m^*)^2}\int_0^{t}e^{(2t'-t-t_1)/\tau_m}dt' = \frac{\tau_m I_R}{(m^*)^2}\left(e^{(t-t_1)/\tau_m} - e^{-(t+t_1)/\tau_m}\right). \tag{3.17}$$

Since we have already taken the long-time limit, the second term in the parentheses can be ignored, and the velocity auto-correlation function becomes

$$\Phi_v(t_d) = \langle v_i(t)v_i(t+t_d)\rangle - v_d^2 = \frac{\tau_m I_R}{(m^*)^2}e^{-t_d/\tau_m}. \tag{3.18}$$

(The $\phi_v(t)$ introduced above is the normalized value of $\Phi_v(t)$ given here, where the normalization is obtained by dividing by the mean-squared velocity.) If we now compare this result with (3.12), and assert that the autocorrelation function is not dependent upon when the initial time is evaluated, in equilibrium, then we see that

$$\frac{\tau_m I_R}{(m^*)^2} = \langle v_i^2(t) \rangle = \frac{k_B T}{m^*}, \tag{3.19}$$

where the last term has been evaluated assuming that the semiconductor is non-degenerate. Hence, the strength of the random force correlation function may be quickly seen to be, for a non-degenerate semiconductor,

$$I_R = \frac{m^* k_B T}{\tau_m}. \tag{3.20}$$

In a degenerate semiconductor, the average energy is not given by the thermal energy, but by the Fermi energy. In this case, the factor of $k_B T$ in (3.20) is replaced by $m^* v_F^2/d$, where v_F is the Fermi velocity and d is the dimensionality of the device. The latter factor arises in the degenerate case because we have been talking about just one degree of freedom in the velocity. In the non-degenerate case, we have $k_B T/2$ per degree of freedom, but in the degenerate case, we have to connect our $\langle v^2 \rangle = \langle (v_T)^2 \rangle / d$, and it is the total velocity which converts to the Fermi energy, and thus to the Fermi velocity.

Let us now turn to the case of random motion, in the absence of an electric field, that arises from the random forces in the system. In this, we are still interested in the initial buildup of this process, so we return the initial velocity to (3.14), and then integrate it to give

$$\begin{aligned} x_i(t) - x_i(0) &= \tau_m v_i(0)(1 - e^{-t/\tau_m}) + \frac{1}{m^*} \int_0^t \int_0^{t'} R(t'') e^{(t''-t')/\tau_m} dt' dt'' \\ &= \tau_m v_i(0)(1 - e^{-t/\tau_m}) + \frac{\tau_m}{m^*} \int_0^t R(t'')[1 - e^{-(t-t'')/\tau_m}] dt''. \end{aligned} \tag{3.21}$$

In the last integral, the order of the two integrations has been reversed, with a corresponding change in the limits, in order to evaluate the integral over t'. If we now further assume that the initial velocity is uncorrelated with the random force (that is, we recognize $\langle v_i(0) R(t) \rangle = 0$), we can then solve for the ensemble averaged value of the displacement as (Kubo, 1974)

$$\langle [x_i(t) - x_i(0)]^2 \rangle = \tau_m^2 \langle v_i^2(0) \rangle (1 - e^{-t/\tau_m})^2$$

$$+ \frac{2\tau_m^2 I_R}{(m^*)^2} \left[t - 2\tau_m (1 - e^{-t/\tau_m}) + \frac{\tau_m}{2}(1 - e^{-2t/\tau_m}) \right]$$

$$= 2\tau_m \langle v_i^2(0) \rangle [t - \tau_m (1 - e^{-t/\tau_m})]. \tag{3.22}$$

The diffusion coefficient may be defined from the above positional auto-correlation function through

$$D = \frac{1}{2}\frac{d}{dt} \langle [x_i(t) - x_i(0)]^2 \rangle = \tau_m \langle v_i^2(0) \rangle (1 - e^{-t/\tau_m}). \tag{3.23}$$

Clearly, the diffusion coefficient only becomes a constant in the long-time limit, and itself is related to an auto-correlation function. By using (3.19), we can relate the diffusion coefficient to the mobility, for the non-degenerate case, as

$$D = \frac{\tau_m}{m^*} k_B T = \mu \frac{k_B T}{e}. \tag{3.24}$$

In a degenerate case, we have

$$D = \frac{\tau_m v_F^2}{d}, \tag{3.25}$$

where d is once again the dimensionality of the device. The mean-square fluctuations, and the diffusion coefficient, are shown as a function of time in Figure 3.2.

3.1.3 Conductivity

The conductivity is clearly related to the average velocity, and thus to the force acting upon the electrons. We obtain this relationship through the connection with the current density, as

$$J = -n_d e v_d = \sigma_d F. \tag{3.26}$$

Here, n_d is the appropriate density for a d-dimensional device. In three dimensions, n is the bulk carrier density, and σ is the normal conductivity. In two dimensions, n_s becomes the sheet carrier density, and σ_s is the sheet conductance. Similarly, in one dimension, n_l is the line density and σ_l is the line conductivity. Using (3.13), we can relate this to the velocity auto-correlation function as

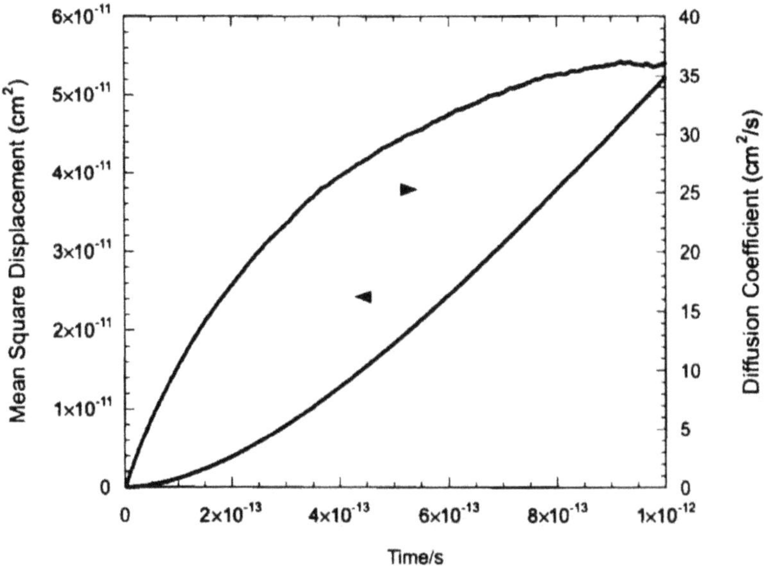

Figure 3.2 The rise of the diffusion coefficient to its steady-state value and the rise of the mean-square fluctuations in position, according to (3.22) and (3.23). For long times, the mean-square fluctuations approach a linear increase, whose derivative is the steady-state value of the diffusion coefficient.

$$\sigma_d(t) = \frac{n_d e^2}{m^*}\int_0^t \varphi_v(t')dt' \rightarrow \frac{n_d e^2 \tau_m}{m^*}. \tag{3.27}$$

Naturally, one takes the long-time limit of this for large semiconductor structures.

There are particularly interesting limits of (3.27) in low-dimensional systems at low temperature. For $d < 3$ (e.g., $d = 1$ or 2), the density can be written as $k_F^d/d\pi$. In the long-time limit, the conductivity then becomes

$$\sigma_d = \frac{e^2}{\pi\hbar} v_F \tau_m \left(\frac{k_F^{d-1}}{d}\right), \tag{3.28}$$

and, in three dimensions,

$$\sigma_3 = \frac{e^2}{\pi\hbar} v_F \tau_m \left(\frac{2k_F^2}{3}\right), \tag{3.29}$$

which is a factor of 2 larger than would be expected from (3.28). The conductance is found from the conductivity as $G = \sigma_d A_d/L$, where A_d is the "area" appropriate to the dimensionality and L is the length of the sample. If we set $A_d = L^{d-1}$, for equivalence of the scaling relationship, then we can rewrite these equations as

$$G = g_0 \times \begin{cases} \dfrac{v_F \tau_m}{L} \dfrac{(k_F L)^{d-1}}{d}, & d = 1, 2, \\ \dfrac{v_F \tau_m}{L} \dfrac{2(k_F L)^2}{3}, & d = 3. \end{cases} \tag{3.30}$$

Here, we have introduced the fundamental unit of conductance

$$g_0 = \frac{e^2}{\pi \hbar} = \frac{2e^2}{h}. \tag{3.31}$$

The remaining factors correspond to a "transmission" probability in the sense introduced by Landauer (1970); that is, they represent a probability that a carrier can make it through the system. This probability is reduced with scattering and enhanced with higher Fermi velocities. We note that, in each dimension, this "transmission" probability is dimensionless. We will return to this quantity shortly, in our discussion of "small" devices, where the transport length is small compared with the transverse dimensions and with $v_F \tau_m$. One notes that the conductance in (3.30) scales with L^{d-2}, and many people working in the mesoscopic area have put a great deal of importance on this exponent. This is probably only significant in the case of a one-dimensional conductor where any potential fluctuation will cut down the conductance through the need for the carriers to tunnel through the barrier. Such effects are not as important in higher dimensions, even though much discussion has appeared in this regard (see e.g., Abrahams *et al.*, 1979; Anderson *et al.*, 1980).

In general, each of the factors we have found (drift velocity, diffusion coefficient, conductivity, etc.) can be obtained with the Boltzmann transport equation. In the present treatments, it is required that we know the velocity and position of each and every particle in order to carry out the ensemble averages. This is quite difficult to achieve. Instead, one wants to describe these velocities and positions through a one-particle distribution function, the Boltzmann distribution function, and to then describe an equation which will provide the evolution of this distribution under the influence of random forces (the scattering processes, which also provide the momentum relaxation process) and fields. To do so automatically introduces additional approximations, as it means ignoring two-particle correlations which must be described by a two-particle distribution function. In general, this is not a major concern so long as we are not dealing directly with inter-particle forces such as the Coulomb interaction. We will begin to treat transport with the Boltzmann distribution function, and the Boltzmann transport equation, in the next section. First, however, we wish to examine what happens to the above transport coefficients when the length is smaller than the momentum relaxation length, which is defined by an effective average thermal (or Fermi) velocity and the momentum relaxation time, as described in Chapter 1.

3.1.4 *Quasi-ballistic transport*

In a short channel, such as that depicted in Figure 1.1(b), the carriers transit the active region in a time short compared to the relaxation time τ_m. This is the so-called quasi-ballistic regime (we term this quasi-ballistic, rather than ballistic by the presence of a few scattering events). We may estimate the conductivity behavior for this regime by returning to (3.6), and writing the velocity as

$$v_d(t) = \frac{d\langle x\rangle}{dt} = \frac{eF\tau_m}{m^*}(1 - e^{-t/\tau_m}). \tag{3.32}$$

There are many problems with solving this, however. The first is that the field is a function of the position and, secondly, this will lead to a non-uniform carrier density through the structure, as will be required for continuity of the current. The actual solutions for density, velocity, and field must be found self-consistently if absolute accuracy is required. Here, we will only estimate the results with a "constant field" model. That is, we will assume that the field throughout the active part of the device is constant, which will be a good approximation to that actually found in the structure. If we now integrate the structure over the length of the device, we find

$$\frac{L}{\mu F} = t_{tr} - \tau_m(1 - e^{-t_{tr}/\tau_m}) \approx \frac{t_{tr}^2}{2\tau_m} + \ldots. \tag{3.33}$$

The leading terms in the exponential cancel with preceding terms, so that the quadratic term is the first non-vanishing term on the right-hand side, and the expansion is terminated here on the assumption that $t_{tr} < \tau_m$. The transit time through the device, t_{tr}, is then found to be

$$t_{tr} = \sqrt{\frac{2L\tau_m}{\mu F}} = \sqrt{\frac{2m^*L}{eF}} = \frac{2L}{v_f}, \tag{3.34}$$

where

$$v_f = \sqrt{\frac{2eV}{m^*}} \tag{3.35}$$

is the final velocity of the carriers attained after traveling through a voltage of V. It is clear from (3.34) that the average velocity is one-half of this final velocity, and this defines the transit time through the structure. The drift velocity at the end of the device is then found from (3.32) to be

$$v_d \sim \mu F(1 - e^{-t_{tr}/\tau_m}) \sim \mu F\left(\frac{2L}{v_f \tau_m}\right). \tag{3.36}$$

Since this is the average velocity at the end of the device, we can use the d-dimensional density at this point to define the current, as the excess charge density will have fairly well damped out by this point. Hence, the conductivity is just

$$\sigma_d = \frac{n_d e^2 \tau_m}{m^*}\left(\frac{2L}{v_f \tau_m}\right) = \frac{2n_d e^2 L}{m^* v_f}. \tag{3.37}$$

This conductivity can be much higher than that found for the longer structures where scattering plays a major role.

For a degenerate, quasi-two-dimensional system, such as commonly found in the conducting inversion layer at an appropriate interface (such as at the Si–SiO_2 interface in a MOSFET), (3.36) is modified because the incoming electrons have a velocity roughly defined by the Fermi velocity. Then, the first-order term in the expansion is not canceled, and

$$t_F \sim L/v_F.$$

We now can use the results of (3.28) to write the conductivity, for the degenerate case, as

$$\sigma_2 = \frac{e^2}{\pi\hbar} v_F \tau_m \left(\frac{k_F}{2}\right)\left(\frac{L}{v_F \tau_m}\right) = \frac{e^2}{\pi\hbar}\frac{k_F L}{2}, \tag{3.38}$$

and

$$G_2 = g_0 \frac{k_F W}{2}. \tag{3.39}$$

Here, W is the width of the structure and $k_F W/2$ is related to the number of transverse modes that exist in the structure. Hence, in the quasi-ballistic case, carriers move through the device at the Fermi velocity (assuming the applied potential is smaller than that corresponding to the Fermi energy), and scattering does not play a significant role in this situation. The result (3.39) is a form of the *Landauer formula* (Landauer, 1957), in which the conductance is given by g_0 times the number of transverse modes, which is the factor $k_FW/2$. This particular form is a two-terminal version of the formula, which is thought to be applicable only in mesoscopic systems. In fact, while it has been obtained here for ballistic devices, it is quite generally applicable once one can relate the transmission coefficient T to the transport parameters. Here, $T = k_FW/2$ is the *total* transmission, or the number of transverse modes, each of which has unity transmission in this quasi-ballistic system. We will encounter a multi-terminal version of this formula in Section 3.4 below.

3.2 The Boltzmann transport equation

In the above discussion, an equation was actually obtained for the conductivity and the conductance assuming that we merely knew the position and velocity of each and every particle. The concept was introduced in which the average motion is obtained by introducing an average over the particle ensemble. Here, we would like to carry out the same averaging process by integrating over a one-electron distribution function that describes the ensemble. Just what is meant by this distribution function? There are various manners in which this quantity can be defined. For example, it is possible to say that the distribution function $f(\mathbf{v},\mathbf{x},t)$ is the probability of finding a particle in the box of volume $\Delta\mathbf{x}$, centered at $\mathbf{x}$, and $\Delta\mathbf{v}$, centered at $\mathbf{v}$, at time t. Here, $\mathbf{v}$ is the particle momentum and $\mathbf{x}$ is the position, now taken to be vector quantities. In this sense, the distribution function is described in a six-dimensional phase space, and the quantities $\mathbf{x}$ and $\mathbf{v}$ do not refer to any single carrier but to the position in this phase space. This is to be compared with (3.1) in which the N particles are defined in a $6N$-dimensional phase space, where we have $3N$ velocity variables and $3N$ position variables even though only the former are shown. With the above definition, then,

$$\iint d^3\mathbf{x}d^3\mathbf{v}f(\mathbf{x},\mathbf{v},t) = 1. \tag{3.40}$$

As with all probability functions, the integral over the measure space must sum to unity. However, this is not the only definition that can be made.

An alternative definition is to define the distribution function as the "average" (the concept of this average will be defined below) number of particles in a phase space box of size $\Delta\mathbf{x}\Delta\mathbf{v}$ located at the phase space point $(\mathbf{x},\mathbf{v})$. In this regard, the distribution function then satisfies

$$\iint d^3\mathbf{x}d^3\mathbf{v}f(\mathbf{x},\mathbf{v},t) = N(t). \tag{3.41}$$

In a sense, this definition is more in keeping with that given in (3.5). Here, $N(t)$ is the *total number of particles* in the entire system at time t. At first view it might be supposed that the Fermi–Dirac distribution satisfies (3.41) and in fact defines the Fermi energy level as a function of time. However, this is not correct for two reasons. First, the normalization is wrong; recall that the Fermi–Dirac distribution has a maximum value of unity for energies well below the Fermi energy. Hence the integral in (3.41) must be modified to account for the density of states in the incremental volume. Moreover, one must convert the velocity integration into an energy integration, and this adds additional numerical and variable factors. An additional objection is more serious. The Fermi–Dirac distribution is a point function, and its application to inhomogeneous systems must be handled quite carefully. The Fermi energy is related to the electrochemical potential, which may vary (relative to one of the band edges) with position. Then the Fermi energy is position dependent in this view. Yet it is well known from simple theory that the Fermi energy must be position independent if the system is to be in equilibrium (no currents flowing). For this to be the case, the band edges must themselves become position dependent in the inhomogeneous system. Thus, while we can equate (3.41) with the use of the Fermi–Dirac distribution function, this must be done with quite some care in inhomogeneous systems. However, it is mainly this definition of the distribution function that will be utilized throughout this book.

In either of these definitions of the distribution function, quantum mechanics further complicates the situation in at least two ways. First, the uncertainty relation requires that $\Delta x \Delta p > \hbar^3/8$, or $\Delta x \Delta v > \hbar^3/8(m^*)^3$, and the *quantum* distribution function can in fact have negative values for regions of smaller extent than this limit. So, we are constrained over just how finely we can examine the position and momentum coordinates. In addition, the distribution function to be dealt with here is an equivalent one-electron distribution function, so that the many electron aspects are averaged out. In both of these cases, the distribution function is said to be *coarse grained* in phase space, in the first case averaging over small regions in which significant local quantization is significant and in the second case averaging out the many-electron properties that modify the one-electron distribution function. This coarse graining in the latter case is the process of the *Stosszahl ansatz*, or molecular chaos, introduced by Boltzmann to justify the use of the one-particle functions or, more exactly, the process by which correlation with early times is forgotten on the scale of the one-particle scattering time τ. The exact manner, by which the multi-electron ensemble of (3.1) is projected onto the one-electron distribution function of (3.41), is best described through the BBGKY Heirarchy. (The letters are taken from the authors Bogoliubov, 1946; Born and Green, 1946; Kirkwood, 1946; Yvon, 1937. The projection approach is described in Ferry, 1991.)

The variation of the distribution function is governed by an equation of motion, and it is this equation that is of interest here. In equilibrium, no transport takes place since the distribution function is symmetric in v-space (more properly, **k**-space). Since the probability of a carrier having the wave vector **k** is the same as for the wave vector −**k** (recall that the Fermi–Dirac distribution depends only on the energy of the carrier, not specifically on its momentum), these balance one another. Since there are equal numbers of carriers with these oppositely directed momenta, the net current is zero. Hence, for electrical or thermal transport, the distribution function must be modified by the applied fields (and made asymmetric in phase space). It is this modification that must now be calculated. In fact, the forcing functions, such as the applied field, are themselves reversible quantities and the evolution of the distribution function in phase space is unchanged by these fields. It is only the presence of the scattering processes that can change this evolution, and the classical statement of this fact is (the right-hand side represents changes due to scattering processes)

$$\frac{df(\mathbf{x},\mathbf{v},t)}{dt} = \left.\frac{\partial f(\mathbf{x},\mathbf{v},t)}{\partial t}\right|_{collisions} . \tag{3.42}$$

By expanding the left-hand side with the chain rule of differentiation, the Boltzmann transport equation is obtained as

$$\frac{\partial f}{\partial t} + \mathbf{v}\cdot\nabla f + \frac{d\mathbf{k}}{dt}\cdot\frac{\partial f}{\partial \mathbf{k}} = \left.\frac{\partial f(\mathbf{x},\mathbf{v},t)}{\partial t}\right|_{collisions} . \tag{3.43}$$

The first term is the explicit time variation of the distribution function, while the second term accounts for transport induced by spatial variation of the density and distribution function. The third term describes the field induced transport. These three terms on the left-hand side are collectively known as *streaming* terms. Here, the third term has been written with respect to the momentum wave vector rather than the velocity to account for the role of the former in the crystal momentum. Still, in keeping with the discussion above, the change of the distribution function with position must be sufficiently slow that the variation of the wave function is very small in one unit cell. This ensures that the band model developed in the previous chapter is valid, and a true statistical distribution can be considered. In addition, the force term must be sufficiently small that it does not introduce any mixing of wave functions from different bands, so that the response can be considered semi-classically within the effective mass approximation. Finally, the time variation must be sufficiently slow that the distribution evolves slowly on the scale of either the mean free time between collisions or on the scale of any hydrodynamic relaxation times (still to be developed, although we have already discussed the momentum relaxation time). The force term, the third term on the left-hand side, is just the Lorentz force in the presence of electric and magnetic fields.

The scattering processes are all folded into the term on the right-hand side of (3.43). Any scattering process induces carriers to make a transition from some initial state **k** into a final state **k**′ with a probability $P(\mathbf{k},\mathbf{k}')$. Then, the number of electrons scattered depends on the latter probability as well as on the probabilities of the state **k** being full [given by $f(\mathbf{k})$] and the state **k**′ being empty [given by $1 - f(\mathbf{k}')$]. (If the volume in **k**-space contains only a single pair of spin degenerate states, the number of carriers in this volume is given by the Fermi–Dirac distribution. If the scattering process can flip the spin then a factor of 2 is also included for this spin degeneracy. It is clear that this can now be seen as mixing the two definitions for the distribution function given above, but it is in fact the latter of the two that is being used.) The rate of scattering *out* of state **k** is then found to be given by putting these three factors together, as

$$P(\mathbf{k},\mathbf{k}')f(\mathbf{k})[1 - f(\mathbf{k}')]. \tag{3.44}$$

But there are also electrons being scattered *into* the state **k** from the state **k**′ with a rate given by

$$P(\mathbf{k}',\mathbf{k})f(\mathbf{k}')[1 - f(\mathbf{k})]. \tag{3.45}$$

The latter two equations are the basis for the scattering term on the right-hand side of (3.43), which is finally obtained by summing over all states **k**′, as

$$\left.\frac{\partial f}{\partial t}\right|_{collisions} = \sum_{\mathbf{k}'}\{P(\mathbf{k}',\mathbf{k})f(\mathbf{k}')[1 - f(\mathbf{k})] - P(\mathbf{k},\mathbf{k}')f(\mathbf{k})[1 - f(\mathbf{k}')]\}. \tag{3.46}$$

In fact, $P(\mathbf{k},\mathbf{k}')$ also contains a summation over all possible scattering mechanisms by which electrons (or holes) can move from **k** to **k**′.

In detailed balance, the two scattering probabilities differ by, for example, differences in the density of final states (the second argument) in energy space and by the phonon factor difference between emission and absorption processes. In fact, (3.46) encompasses four processes when phonons are involved. Carriers can leave by either emitting a phonon (and going to a state of lower energy) or by absorbing a phonon (and going to a state of higher energy). By the same token, they can scatter into the state of interest either by phonon emission from a state of higher energy, or by phonon absorption from a state of lower energy. In equilibrium, the processes connecting our primary state with each of the two sets of levels (of higher and lower energy) must balance. This balancing in equilibrium is referred to as *detailed balance*. Under this condition, the right-hand side of (3.43) vanishes in equilibrium.

When the distribution function is driven out of equilibrium by the *streaming* forces on the left-hand side of (3.43), the collision terms work to restore the system to equilibrium. Interactions between the carriers within the distribution work to randomize the energy and momentum of the distribution by redistributing these quantities within the distribution. This is known to lead to a Maxwellian distribution in the non-degenerate case through a process that can be shown to maximize the entropy of the distribution. However, it is the phonon interactions that cause the overall distribution function to come into equilibrium with the lattice. If the lattice itself is in equilibrium, it may be considered as the *bath* and the phonons serve to couple the electron distribution to this thermal bath. Under high electric fields, it is also possible for the phonon distribution to be driven out of equilibrium and this makes for a very complicated set of equations to be solved. In the following paragraphs, solutions of the Boltzmann transport equation (3.43) will be obtained for a number of simplified cases. More complicated solutions will be dealt with in later chapters.

3.2.1 The relaxation time approximation

If no external fields are present, the collisions tend to randomize the energy and the momentum of the carriers and return them to their equilibrium state. In linear response, it is often useful to assume that the rate of relaxation is proportional to the deviation from equilibrium and that the distribution function decays to its normal equilibrium value in an exponential manner. For this approximation, a relaxation time τ may be introduced by means of the equation

$$\left.\frac{\partial f}{\partial t}\right|_{collision} = -\frac{f - f_0}{\tau}. \tag{3.47}$$

Here f_0 is the equilibrium distribution function, either a Fermi–Dirac distribution or the Maxwellian approximation to this distribution. This is a fairly common approximation that is easily made if the scattering processes are either elastic, or are isotropic and inelastic. If there are no applied fields and the system is homogeneous, (3.43) and (3.47) lead to

$$f(t) = f_0 + (f - f_0)e^{-t/\tau}. \tag{3.48}$$

Even when the relaxation time approximation, as (3.47) is called, holds, it is necessary to be able to calculate τ from the scattering rates (the elastic scattering rates will be dealt with in the second part of this chapter). As will be seen later in this chapter, this entails an average over the distribution function. Here, it is desired to consider the case of the elastic scattering process in further detail. If the scattering process is elastic, the states **k** and **k′** lie on the same energy shell [that is, $E(\mathbf{k}) = E(\mathbf{k}')$], and it is feasible to assume that $P(\mathbf{k},\mathbf{k}') = P(\mathbf{k}',\mathbf{k})$, so that the relaxation term in (3.46) becomes

$$[f(\mathbf{k}) - f(\mathbf{k}')]P(\mathbf{k},\mathbf{k}'). \tag{3.49}$$

This then leads to the loss due to collisions out of the state **k** as

$$\left.\frac{\partial f}{\partial t}\right|_{collision} = -\int d^3\mathbf{k}' P(\mathbf{k},\mathbf{k}')[f(\mathbf{k}) - f(\mathbf{k}')]. \tag{3.50}$$

Now, consider just the acceleration term in (3.43), so that the combination of this term and (3.47) leads to the simple form for the distribution function, for a homogeneous semiconductor sample,

$$f(\mathbf{k}) = f_0(\mathbf{k}) - \frac{\tau}{\hbar}\mathbf{F}\cdot\frac{\partial f(\mathbf{k})}{\partial \mathbf{k}} = f_0(\mathbf{k}) - \tau\mathbf{F}\cdot\mathbf{v}\frac{\partial f(\mathbf{k})}{\partial E}$$
$$\approx f_0(\mathbf{k}) - \tau\mathbf{F}\cdot\mathbf{v}\frac{\partial f_0(\mathbf{k})}{\partial E}. \tag{3.51}$$

It has been assumed in the last form that the deviation from equilibrium is sufficiently small that the entire right-hand side can be represented as a functional of f_0. If this form is introduced into (3.50), it is found that

$$\left.\frac{\partial f}{\partial t}\right|_{collision} = -\int d^2S_{k'}P(\mathbf{k},\mathbf{k}')\tau\mathbf{F}\cdot(\mathbf{v}-\mathbf{v}')\frac{\partial f_0}{\partial E}$$
$$= \tau\mathbf{F}\cdot\mathbf{v}\frac{\partial f_0}{\partial E}\int_{S_{k'}} P(\mathbf{k},\mathbf{k}')\left(1 - \frac{\mathbf{v}\cdot\mathbf{v}'}{\mathbf{v}^2}\right)d^2S_{k'}$$
$$= -(f - f_0)\int_{S_{k'}} P(\mathbf{k},\mathbf{k}')(1-\cos\theta)\,d^2S_{k'}. \tag{3.52}$$

In the first line, the fact that the scattering is elastic has been used to assure that the integration is over a single energy shell, hence the integration is only over the surface corresponding to this energy shell. In the second and third lines of (3.52), the angular variation has been utilized to write the integral in terms of the angle between the two velocity vectors, and the angular weighting discussed already in the early paragraphs of this chapter is recovered. In truly elastic, isotropic scattering such as acoustic phonons in spherically symmetric bands, the cos(θ) term integrates to zero with the shell integration. However, (3.52) now allows us to compute the *momentum* relaxation time for elastic scattering processes as

$$\frac{1}{\tau_m} = \int_0^{\pi} d\theta \sin\theta \int_0^{2\pi} d\phi P(\theta,\phi)(1 - \cos\theta), \tag{3.53}$$

where $P(\theta,\phi) = k^2 P(\mathbf{k},\mathbf{k}')$, and the latter quantity is calculated by the Fermi golden rule, as will be described later in this chapter.

It is obvious that it will be difficult to compute a simple relaxation time in the case of inelastic scattering, as the two distribution functions in (3.46) come from different energy shells. Thus, they cannot readily be separated from the scattering integrals in order to identify just the relaxation rate. It will be shown later that it is in fact possible to calculate the average momentum relaxation rate if a specific form for the non-equilibrium distribution function is assumed. The latter rate can be extrapolated to the equilibrium situation to give an *effective* relaxation rate $1/\tau$, and thus utilize the relaxation time approximation in subsequent calculations. In general, however, this can be done only in very special cases, and more complicated approaches to computing the distribution function in the presence of forces must be utilized.

3.2.2 Conductivity

When the external force is just an electric field, the distribution function is given by just the field streaming term and the scattering term as

$$f(E) = f_0(E) + e\tau \mathbf{F} \cdot \mathbf{v}\frac{\partial f_0(E)}{\partial E}. \tag{3.54}$$

By knowing this distribution function, the electric current carried by these carriers (in this case, electrons because of the sign used in the force) can be found by summing over the electron states as

$$\mathbf{J} = -e\int d^3\mathbf{k}\rho(\mathbf{k})\mathbf{v}f(E) = -e\int dE\rho(E)\mathbf{v}f(E), \tag{3.55}$$

where $\rho(\mathbf{k})$, or $\rho(E)$, is the appropriate density of states. Introducing the distribution function from (3.51) gives

$$\mathbf{J} = -e\int dE\rho(E)\mathbf{v}f_0(E) - e^2\int dE\tau_m\rho(E)\mathbf{v}(\mathbf{F}\cdot\mathbf{v})\frac{\partial f_0(E)}{\partial E}$$

$$= -e^2\mathbf{F}\int dE\tau_m\rho(E)v_F^2\frac{\partial f_0(E)}{\partial E}. \tag{3.56}$$

In the last form, explicit account has been taken of the isotropic nature of linear conduction in semiconductors and the vector product was rearranged to involve just the velocity in the direction of the electric field. The first term, involving just f_0, vanishes due to the symmetry of the equilibrium distribution function in **k**-space (the average momentum of the equilibrium distribution must be zero). Now, it is known that the number of carriers in the band is just determined by the distribution function as

$$n = \int dE \rho(E) f_0(E), \tag{3.57}$$

so that

$$\mathrm{J} = -ne^2 \mathrm{F} \frac{\int dE \tau_m \rho(E) v_F^2 \frac{\partial f_0(E)}{\partial E}}{\int dE \rho(E) f_0(E)}. \tag{3.58}$$

In general, for linear transport, it may be assumed that the energy involved in the drift velocity is negligible in comparison with the thermal energy, which implies that the drift velocity is small in comparison with the thermal velocity. This means that the drift velocity can be ignored in the integral and it may be assumed that $v^2 = v_x^2 + v_y^2 + v_z^2$, or that $v_F^2 = v^2/3 = 2E/3m^*$, where the mass is the appropriate effective mass. This finally leads to the simple form

$$\mathrm{J} = \frac{ne^2 \langle \tau_m \rangle}{m^*} \mathrm{F}. \tag{3.59}$$

From this expression, we can recognize the *mobility*, as $\mu = e\langle \tau_m \rangle / m^*$, and the conductivity $\sigma = ne\mu$. We can integrate the denominator of (3.58) by parts as

$$\int_0^\infty E^{\frac{1}{2}} f_0(E) dE = -\frac{2}{3} \int_0^\infty E^{\frac{3}{2}} \frac{\partial f_0(E)}{\partial E} dE, \tag{3.60}$$

and the average momentum relaxation time can be defined by

$$\langle \tau_m \rangle = \frac{\int_0^\infty E^{\frac{3}{2}} \tau_m(E) \frac{\partial f_0(E)}{\partial E} dE}{\int_0^\infty E^{\frac{3}{2}} \frac{\partial f_0(E)}{\partial E} dE}. \tag{3.61}$$

The same approach can be used in other than three dimensions. In general, the density of states varies as $E^{(d/2)-1}$. Then, the general steps in (3.60) can be followed for an arbitrary dimension. However, note that the prefactor ($\frac{2}{3}$) also arose from the dimensionality, and the argument leading to it gives $2/d$ as the prefactor. Then it may readily be shown that a quite general result is that

$$\langle \tau_m \rangle = \frac{\int_0^\infty E^{d/2} \tau_m(E) \frac{\partial f_0(E)}{\partial E} dE}{\int_0^\infty E^{d/2} \frac{\partial f_0(E)}{\partial E} dE}. \tag{3.62}$$

In writing the limits on the above integrals as infinity, it has been assumed that the conduction-band upper edge is sufficiently far removed from the energy range of interest so that the distribution function is zero at this point. In this case, the upper limit does not affect the final result if the limit is taken as infinity rather than just the upper edge of the band (lower edge if we are dealing with holes). In extremely degenerate cases, the upper limit of the integral may be taken as the Fermi energy, and the relaxation time evaluated at the Fermi energy, as the derivatives of the distribution function are sharply peaked at this energy.

If the constant-energy surfaces are not spherical, some complication of the problem arises. To begin with, the energy is no longer a function of the single effective mass and becomes expressed as

$$E = \frac{\hbar^2}{2}\left(\frac{k_x^2}{m_x} + \frac{k_y^2}{m_y} + \frac{k_z^2}{m_z}\right) \tag{3.63}$$

for each ellipsoid. To simplify this approach, the following transformation is introduced:

$$k_i' = \sqrt{\frac{m^*}{m_i}} k_i \tag{3.64}$$

for each direction within a single ellipsoid. This then re-scales the energy to be

$$E = \frac{\hbar^2}{2m^*}(k_x'^2 + k_y'^2 + k_z'^2). \tag{3.65}$$

By introducing the same transformations on the velocity $\mathbf{v}$ ($= \hbar\mathbf{k}/m$) in each of the ellipsoids, the simple result above is still achieved for the current, but this must be un-transformed to achieve the current in the real coordinates. Carrying out this process for a single ellipsoid yields

$$J_x = \frac{ne^2\langle\tau_m\rangle}{m_x} F_x, \tag{3.67}$$

and so on for each of the other two directions. In most cases of non-spherical energy surfaces, multiple minima are involved, and a summation over these equivalent minima must still be carried through. For example, in silicon with six equivalent ellipsoids in the conduction band, the total conductivity is a sum over the six valleys. However, two of the valleys are oriented in each of the three principal directions and contribute the appropriate amount to each current direction. The total current is then (with 1/6 of the total carrier density in each valley)

$$\sigma = \frac{J}{F} = \frac{n}{6} e^2 \langle\tau_m\rangle \left(\frac{2}{m_1} + \frac{2}{m_2} + \frac{2}{m_3} \right). \tag{3.68}$$

The subscripts 1, 2, 3 cyclically permute through the values of x, y, z for the six ellipsoids. In general, this sum can be reduced to one that arises from replacing the above subscripts with those appropriate for the principal values for one ellipsoid. Then, we may recognize that these are ellipsoids of revolution, so that we can assign $m_1 = m_L$, and $m_2 = m_3 = m_T$, which introduces the longitudinal and transverse masses. With these definitions, (3.68) becomes

$$\sigma = \frac{n}{3} e^2 \langle\tau_m\rangle \left(\frac{1}{m_L} + \frac{2}{m_T} \right) = \frac{ne^2\langle\tau_m\rangle}{m_L} \frac{2K+1}{3}, \quad K = \frac{m_L}{m_T}. \tag{3.69}$$

For silicon, $m_L = 0.91m_0$ and $m_T = 0.19m_0$, so that the conductivity mass $m_c = 3m_L/(2K+1)$ is about $0.26m_0$. This value is different from either of the two curvature masses and is different from the density-of-states mass. This mass, called the ***conductivity mass***, arises from a proper conduction sum over the various ellipsoids. This sum is relatively independent of the actual shape and position of the ellipsoids (the same sum arises in germanium with its four ellipsoids), but arises solely for the sums used in computing the conduction current that is parallel to the electric field. Different sums will arise if a magnetic field is present.

3.2.3 *The diffusion coefficient*

There are many cases where it is the distribution function that varies with position, either through a change in the normalization as the doping concentration changes, or through the presence of a temperature gradient. Here, we consider the former, leaving the latter to a later discussion. Let us consider only the second term on the left-hand side of (3.43), which leads to the expression, in the relaxation time approximation,

$$\mathbf{v}\cdot\nabla f = -\frac{f-f_0}{\tau_m}, \tag{3.70}$$

for which we can now write

$$f(E) = f_0 - \tau_m \mathbf{v}\cdot\nabla f \approx f_0 - \tau_m \mathbf{v}\cdot\nabla f_0(\mathbf{x}), \tag{3.71}$$

following the approximations introduced in the last section. The current is given by (3.55), just as previously, and we have

$$\begin{aligned} \mathrm{J} &= e\int dE\rho(E)\tau_m \mathrm{v}(\mathrm{v}\cdot\nabla f_0(x)) = e\nabla\int dE\rho(E)\tau_m \mathrm{v}_j^2 f_0 \\ &= e\nabla(Dn), \end{aligned} \tag{3.72}$$

where

$$D = \left\langle\frac{\mathrm{v}^2\tau_m}{3}\right\rangle = \frac{\int dE\rho(E)\frac{\mathrm{v}^2\tau_m}{3}f_0}{\int dE\rho(E)f_0}. \tag{3.73}$$

Here, we have used the connection between the component of the velocity along the current to the total velocity that was introduced just before (3.59), and the factor of "3" is the dimensionality d of the system. In general, this definition of the diffusion constant is relatively independent of position due to the normalization with respect to the velocity, so that it can be brought through the gradient operation to produce the more usual (with the sign for electrons)

$$\mathrm{J} = eD\nabla n. \tag{3.74}$$

Clearly, the diffusion coefficient arises from an ensemble average just as the mobility does. However, it should be noted that *the two averages are, in fact, different*! In (3.62), the denominator was integrated by parts in order to get the energy derivative of the distribution function into both the numerator and the denominator. Here, this is not done. While the denominator is technically the same, in fact they differ by the factor $d/2$. One cannot carry out the integration of the numerator by parts to overcome this difference, because we don't know the energy dependence of the momentum relaxation time. Hence, the two averages are quite likely to differ by a numerical factor, except for the case of a Maxwellian distribution where they yield the same amount. In fact, if we assume that $f_0 = A\exp(-E/k_BT)$, then it is simple to show that

$$D = \mu \frac{k_B T}{e}, \tag{3.75}$$

which we had previously found. This result, of course, holds only for the non-degenerate case for which a Maxwellian is valid. In the degenerate case, (3.75) is multiplied by the ratio of two Fermi–Dirac integrals, which provides the correction between the average energy and the fluctuation represented by k_BT.

3.2.4 *Magnetoconductivity*

It is now time to introduce the magnetic field into the discussion of the conductivity. This produces what is called a magnetoconductivity. To simplify the notation somewhat, the incremental distribution function is defined through the results found above as $f_1 = f - f_0$, so that the relaxation-time approximation operates only on this incremental quantity. We return to consideration of a homogeneous semiconductor, under steady-state conditions, so that the Boltzmann transport equation becomes

$$-\frac{e}{\hbar}(\mathbf{F} + \mathbf{v} \times \mathbf{B}) \cdot \frac{\partial f}{\partial \mathbf{k}} = -\frac{f_1}{\tau_m}, \tag{3.76}$$

and τ_m is the momentum relaxation time we have used in the previous sections. It will still be assumed that the incremental distribution function is small compared to the equilibrium one, so that the latter can be used in the gradient (with respect to momentum) term. However, it is known that for the equilibrium distribution function, the derivative produces a velocity, which yields zero under the dot product with the $\mathbf{v} \times \mathbf{B}$ term $[\mathbf{v} \cdot (\mathbf{v} \times \mathbf{B}) = \mathbf{B} \cdot (\mathbf{v} \times \mathbf{v}) = 0]$. Hence, we must keep the first-order contribution to the distribution function in this term. Then, the Boltzmann equation becomes

$$-e\mathbf{v} \cdot \mathbf{F}\frac{\partial f_0}{\partial E} = -\frac{f_1}{\tau_m} + \frac{e}{\hbar}(\mathbf{v} \times \mathbf{B}) \cdot \frac{\partial f_1}{\partial \mathbf{k}}. \tag{3.77}$$

In analogy to (3.54), the incremental distribution function is written as

$$f_1 = e\tau_m(\mathbf{v} \cdot \mathbf{A})\frac{\partial f_0}{\partial E}, \tag{3.78}$$

where **A** plays the role of an equivalent electric field vector that must still be determined. If higher-order terms in the distribution function are neglected, as indicated in (3.78), the force functions can be written as

$$-\mathbf{v} \cdot \mathrm{F} = -\mathbf{v} \cdot \mathrm{A} + \frac{e\tau_m}{m^*}(\mathbf{v} \times \mathrm{B}) \cdot \mathrm{A}, \tag{3.79}$$

for which, for an arbitrary value of the velocity, one finds that the generalized field vector is related to the actual electric field through

$$\mathrm{F} = \mathrm{A} - \frac{e\tau_m}{m^*}\mathrm{B} \times \mathrm{A}. \tag{3.80}$$

By elementary geometry, it can be shown that the general solution for the vector **A** must be (one can back substitute this into the previous equation to show that it is the proper solution)

$$\mathrm{A} = \frac{\mathrm{F} + \dfrac{e\tau_m}{m^*}\mathrm{B} \times \mathrm{F} + \left(\dfrac{e\tau_m}{m^*}\right)^2 \mathrm{B}(\mathrm{B} \cdot \mathrm{F})}{1 + \left(\dfrac{e\tau_m}{m^*}\right)^2 B^2}. \tag{3.81}$$

This equation can now be used in the incremental distribution function, and the forms slightly rearranged to give the result

$$f_1 = e\tau_m\mathbf{F} \cdot \frac{\mathbf{v} + \dfrac{e\tau_m}{m^*}\mathbf{v} \times \mathbf{B} + \left(\dfrac{e\tau_m}{m^*}\right)^2 \mathbf{B}(\mathbf{B} \cdot \mathbf{v})}{1 + \left(\dfrac{e\tau_m}{m^*}\right)^2 B^2}\frac{\partial f_0}{\partial E}. \tag{3.82}$$

We now consider the case where the magnetic field is perpendicular to the electric field, and to the plane in which the transport is to take place. We take $\mathrm{B} = \mathrm{Ba}_z$, and consider the x- and y-directed transport, with the current in the x-direction.

The averaging of the distribution function with the current is a straightforward procedure once we have f_1, as given in (3.82). This averaging leads to the equations

$$J_x = \frac{ne^2}{m^*}\left[\left\langle\frac{\tau_m}{1+\omega_c^2\tau_m^2}\right\rangle F_x - \left\langle\frac{\omega_c\tau_m^2}{1+\omega_c^2\tau_m^2}\right\rangle F_y\right]$$

$$J_y = \frac{ne^2}{m^*}\left[\left\langle\frac{\omega_c\tau_m^2}{1+\omega_c^2\tau_m^2}\right\rangle F_x + \left\langle\frac{\tau_m}{1+\omega_c^2\tau_m^2}\right\rangle F_y\right]. \tag{3.83}$$

Here, we have introduced the *cyclotron frequency* $\omega_c = eB/m^*$. Instead of a simple average over the relaxation time, there is now a complicated average that must be carried out.

The result (3.83) can be handled in a simpler manner if the magnetic field is sufficiently small (i.e., $\omega_c\tau_m \ll 1$). In this case, (3.83) reduces to the more tractable form

$$J_x = \frac{ne^2}{m^*}[\langle\tau_m\rangle F_x - \omega_c\langle\tau_m^2\rangle F_y]$$

$$J_y = \frac{ne^2}{m^*}[\omega_c\langle\tau_m^2\rangle F_x + \langle\tau_m\rangle F_y]. \tag{3.84}$$

Of interest here is a long, filamentary semiconductor – one whose length is much larger than its width so that contact effects at the ends are not important. As discussed above, the current is assumed to be flowing in the x-direction. If we set the y-component of current to zero, a transverse field, the *Hall field*, will develop in the y-direction. This field is related to the longitudinal field as

$$\frac{F_y}{F_x} = -\omega_c\frac{\langle\tau_m^2\rangle}{\langle\tau_m\rangle} = -\omega_c\langle\tau_m\rangle\frac{\langle\tau_m^2\rangle}{\langle\tau_m\rangle^2} = -\mu B r_H, \quad r_H = \frac{\langle\tau_m^2\rangle}{\langle\tau_m\rangle^2}. \tag{3.85}$$

In the last relation, we have introduced the *Hall scattering factor* r_H. Now, however, if we introduce this transverse field into the x-component of current, the transverse field has little effect on the conductivity, and

$$J_x = \frac{ne^2\langle\tau_m\rangle}{m^*}F_x, \tag{3.86}$$

as in the absence of the magnetic field. This is merely a consequence of our assumption that the field is small. On the other hand, if the momentum relaxation time is independent of the energy, then the result that the conductivity is unaffected by the magnetic field becomes exact.

The Hall factor is a scattering factor and takes account of the energy spread of the carriers. It is, of course, unity in the case of an energy-independent scattering mechanism or for degenerate semiconductors, where the transport is at the Fermi energy. We will evaluate this for some scattering processes later in this chapter.

The *Hall coefficient* R_H is now defined through the relation $F_y = R_H J_x B$, which may be determined by combining the last two equations above, as

$$R_H = \frac{F_y}{J_x B} = -\frac{r_H}{ne}. \tag{3.87}$$

This is, of course, just the case for electrons (due to the sign assumed on the force earlier). If the Hall factor is not known, it provides a source of error in determining the carrier density from the Hall effect. Worse, in the case of multiple scattering mechanisms, the scattering factor is usually complex and varies in magnitude with both temperature and carrier concentration. However, its value is typically in the range 1 to 1.5 so that the absolute measurement of the density is not critically upset by lack of knowledge of this factor.

If both holes and electrons are present, one cannot set the transverse currents to zero separately but must combine the individual particle currents prior to invoking the boundary conditions. The second of equations (3.84) can be rewritten, with both carriers present, in the form

$$J_y = e^2\left\{\left[\frac{n\omega_{ce}\langle\tau_{me}^2\rangle}{m_e} - \frac{p\omega_{ch}\langle\tau_{mh}^2\rangle}{m_h}\right]F_x + \left[\frac{n\langle\tau_{me}\rangle}{m_e} + \frac{p\langle\tau_{mh}\rangle}{m_h}\right]F_y\right\}. \tag{3.88}$$

The Hall angle is given now by

$$\frac{F_y}{F_x} = -\frac{n\mu_e^2 B r_{He} - p\mu_h^2 B r_{Hh}}{n\mu_e + p\mu_h} = -\mu_h B\frac{nb^2 r_{He} - p r_{Hh}}{bn + p}, \tag{3.89}$$

and the mobility factor $b = \mu_e/\mu_h$ has been introduced. Since $J_x = (ne\mu_e + pe\mu_h)F_x$, the Hall coefficient now becomes

$$R_H = -\frac{1}{e}\frac{b^2 n r_{He} - p r_{Hh}}{(bn + p)^2}. \tag{3.90}$$

It may be ascertained from this equation that the sign of the Hall coefficient identifies the carrier type, but it is important to note that the presence of equal numbers of electrons and holes does not equate to a zero Hall effect. Rather, the difference in the ability of the two types of carriers to move directly affects their lateral motion, and it is a cancellation of the transverse currents, rather than an equality in the carrier concentrations, that is important for a vanishing Hall effect. In fact, it is quite usual to observe a change of sign of the Hall coefficient in p-type semiconductors as they become intrinsic at high temperature due to the fact that the electron mobility is usually greater than the hole mobility. In higher magnetic fields, other effects begin to appear, even in non-quantizing magnetic fields. But, the most common is that for quantizing magnetic fields, where $\omega_c\tau_m > 1$ and $\hbar\omega_c > k_BT$. It is not usually recognized that both of these conditions are required to see quantization of the magnetic orbits. The first is required so that complete orbits are formed, and the latter is required to have the quantized levels separated by more than the thermal smearing. We shall return to this in a later section, where we encounter edge states and the quantum Hall effect.

3.2.5 *Thermoelectric effects*

In the paragraphs above, where the Boltzmann equation was introduced, the term involving the spatial variation of the distribution function has so far been ignored, other than for the diffusion coefficient. This term can lead to an additional effect. This phenomenon can occur if the temperature and/or the Fermi energy are spatially varying. To be sure, spatial variation of the carrier density will lead to a variation of the Fermi energy, but here the primary interest is in the effects raised by a variation in the local temperature with position in the crystal. For this case, the spatially varying term can be written

$$\nabla f(E) = \frac{\partial f}{\partial T}\nabla T \rightarrow \frac{E}{k_BT^2}f(E)\nabla T, \tag{3.91}$$

for a case in which the distribution function is Maxwellian in shape. If we had, instead, a Fermi–Dirac distribution, the factor of f is replaced by $f(1-f)$. The first term on the right-hand side normally is used in the Boltzmann equation, for a steady-state situation, and the result is

$$\mathbf{v}\cdot\left(\nabla T\frac{\partial f_0}{\partial T} + \mathbf{F}\frac{\partial f_0}{\partial E}\right) = -\frac{f-f_0}{\tau_m}. \tag{3.92}$$

As indicated above, the thermal forcing term can be simplified somewhat, but a form that can also accommodate the Fermi–Dirac distribution will be utilized here. Noting that reduced variables have been utilized in the past, it is possible to write the partial derivative with respect to the temperature in the form

$$\frac{\partial f_0}{\partial T} = \frac{\partial f_0}{\partial u}\frac{\partial u}{\partial T} = k_BT\frac{\partial f_0}{\partial E}\frac{\partial}{\partial T}\left(\frac{E-E_F}{k_BT}\right), \quad u = \frac{E-E_F}{k_BT}, \tag{3.93}$$

and the dependence upon u allows the inclusion of variations in the Fermi energy that arise from variations in the temperature. The Boltzmann equation now becomes

$$f(E,T) = f_0(E,T) + \tau_m \mathbf{v} \cdot \left\{ \mathbf{F} + \nabla T \left[\frac{E}{T} + T \frac{\partial}{\partial T} \left(\frac{E_F}{k_B T} \right) \right] \right\} \frac{\partial f_0(E,t)}{\partial E}. \quad (3.94)$$

The primary effect that is of interest here is the case for which an electric field has been introduced, but for which no current flows. It is apparent that the thermal forces create an additional "effective field." This thermal force can be directed either along the electric field, which modifies the resulting current, or it can be directed transverse to the electric field, in which case it creates additional currents that also are transverse to the applied field. The normal case considered is the former. The electric field and the thermal gradient are taken to be directed along the x-axis, so that

$$J_x = \frac{ne^2}{m^*} \left\{ \left[F_x + \frac{T}{e} \frac{\partial T}{\partial x} \frac{d}{dT} \left(\frac{E_F}{k_B T} \right) \right] \langle \tau_m \rangle + \frac{1}{eT} \frac{\partial T}{\partial x} \langle \tau_m E \rangle \right\}. \quad (3.95)$$

In the above equation, the averaging procedure has been used to obtain the final result. In addition to the current, however, it is necessary to consider an equation that describes the flow of heat in the semiconductor. This heat flow is composed of a thermal energy flow W_x (Smith, 1959). The latter flow arises from the transport of energy, in response to the thermal gradient rather than the transport of charge. Therefore, instead of multiplying the distribution function by $-e\mathrm{v}$ as was done to calculate the current, we multiply by a factor $E\mathrm{v}$, which is the local energy flow. The heat flux is then

$$W_x = -\frac{ne}{m^*} \left\{ \left[F_x + \frac{T}{e} \frac{\partial T}{\partial x} \frac{d}{dT} \left(\frac{E_F}{k_B T} \right) \right] \langle E\tau_m \rangle + \frac{1}{eT} \frac{\partial T}{\partial x} \langle \tau_m E^2 \rangle \right\}. \quad (3.96)$$

In the previous two equations, it has been assumed that the energy bands are spherically symmetric, so that the effective carrier mass is a simple scalar quantity.

Many times, a thermal gradient can exist without the flow of electric current, and this is a standard simple tool in determining whether a slice of semiconductor material is n-type or p-type. The effect that arises in this case is known as the thermoelectric effect, and is sometimes called the Seebeck effect. The converse effect, in which a current flow (presumably in the presence of an additional electric field) produces a temperature gradient, is known as the Peltier effect. The combination of these two effects – namely, the absorption of thermal power due to a current flow through a material in which a temperature gradient exists – is termed the Thompson effect. The resultant electric field can be expressed in terms of the thermoelectric power and the thermoelectric force, which are defined below. It should be remarked that, in many solid-state books, each of these effects is described as occurring at a *junction* between dissimilar materials. They can occur in the latter situation, but the effects of interest here are *bulk* effects occurring within semiconductors. In metals, because of the extremely high number of free carriers, the effects are orders of magnitude weaker than in semiconductors, so that metals physicists often are aware only of the effects at junctions rather than in the bulk. In semiconductors, however, it is the bulk effects that are of most interest.

Consider the Seebeck effect, for which no electric current flows. For this to occur, the current equation (3.95) is set to zero, so that the term in the curly braces must vanish. This leads to an electric field induced by the thermal gradient, and

$$F_x = -\frac{1}{e}\left[T\frac{\partial}{\partial T}\left(\frac{E_F}{T}\right) + \frac{\langle\tau_m E\rangle}{T\langle\tau_m\rangle}\right]\frac{dT}{dx}$$

$$= \frac{T}{e}\frac{\partial}{\partial T}\left[\frac{\langle\tau_m E\rangle}{T\langle\tau_m\rangle} - \frac{E_F}{T}\right]\frac{dT}{dx} = -S\frac{dT}{dx}. \tag{3.97}$$

The last line serves to define the thermoelectric power S.

In extrinsic semiconductors, for very small electric fields, the magnitude of the Fermi energy is generally much larger than the free carrier kinetic energy, so that the second term dominates the bracketed terms of (3.97). In these extrinsic semiconductors, the slope of the Fermi energy as a function of temperature is predominantly negative for n-type material and positive for p-type material. Thus the thermoelectric power has opposite signs for the two types of extrinsic material, and the electric field produced will have an opposite sign. This provides a simple determination of the type of semiconductor by measuring the direction of the field produced by a thermal gradient induced in the semiconductor.

Normally, a force derives from an energy (or a power). The same is true here, and the thermoelectric power is related to the thermoelectric force through the definition

$$S = T\frac{dA}{dt}, \tag{3.98}$$

so that

$$A = \frac{1}{eT}\left[E_F - \frac{\langle\tau_m E\rangle}{\langle\tau_m\rangle}\right] \tag{3.99}$$

in an n-type semiconductor. Because there is no derivative with respect to temperature involved in the thermoelectric power, it is sensitive to the definition of the reference level for the energy. In p-type material, the energies are referenced to the valence-band edge so that

$$A_h = \frac{1}{eT}\left[E_{gap} - E_F - \frac{\langle\tau_m E\rangle}{\langle\tau_m\rangle}\right]. \tag{3.100}$$

Because of the averages of the relaxation time that appear in the equations for thermoelectric power, measurement of these quantities can give information on the nature of the scattering mechanisms in the semiconductor.

So far, only the current equation has been discussed. It is possible that, although there is no current flow, there can be a thermal power flow in the semiconductor if a temperature gradient is present, and in fact this is how the excess heat is carried to the reservoir. This effect leads to the carrier contribution to the thermal conductivity. The thermal gradient produces a thermal heat flow proportional to the temperature gradient. The constant of proportionality is termed the thermal conductivity. Since $J_x = 0$, the electric field found in (3.97) can be used in (3.96) for the heat flow. After some simplifications, the resulting flow is described by

$$W_x = -\frac{n}{m^*T}\left[\langle\tau_m E^2\rangle - \frac{\langle\tau_m E\rangle^2}{\langle\tau_m\rangle}\right]\frac{dT}{dx}. \tag{3.101}$$

The heat flow in (3.101) is just the electronic contribution to the thermal conductivity in the material and must be added to the lattice contribution. The latter will dominate the total thermal conductivity except for very high temperatures and/or very high carrier concentrations. The electronic contribution to the thermal conductivity is defined by the coefficient of the temperature gradient in (3.101), and is

$$\kappa_e = \frac{n}{m^*T}\left[\langle\tau_m E^2\rangle - \frac{\langle\tau_m E\rangle^2}{\langle\tau_m\rangle}\right] = L\sigma T = \Lambda\left(\frac{k_B}{e}\right)^2\sigma T. \tag{3.102}$$

The last two terms define the Lorenz ratio,

$$L = \left(\frac{1}{eT}\right)^2\left[\frac{\langle\tau_m E^2\rangle}{\langle\tau_m\rangle} - \frac{\langle\tau_m E\rangle^2}{\langle\tau_m\rangle^2}\right], \tag{3.103}$$

and the Lorenz number

$$\Lambda = \frac{\langle\tau_m\rangle\langle\tau_m E^2\rangle - \langle\tau_m E\rangle^2}{k_B^2 T^2\langle\tau_m\rangle^2}. \tag{3.104}$$

The Lorenz ratio describes the ratio of the electronic contribution to the thermal conductivity to the product of the electrical conductivity and the temperature. It is a constant that depends only on the details of the averaging of the relaxation time and the energy, hence on the subtle details of the distribution function. For acoustic phonon scattering, Λ is about 2. In the strongly degenerate metal case, a properly weighted average of the various variables gives a value of $\pi^2/3 \sim 3.33$, which is not too far different from the case of acoustic phonon scattering in a non-degenerate semiconductor.

3.2.6 *The energy dependence of the relaxation time*

In the last few sections, various averages of the relaxation time τ_m have appeared in which it is necessary to average over the distribution function. These averages give simple relationships that are necessary for computing various transport coefficients. The energy dependence of the scattering rates, for most processes, is quite complicated. In this section, it is desired to examine a general form for the dependence of the momentum relaxation time on the energy, which is taken to be $\tau_m = AE^{-s}$, where A and s are constants that are different for the different scattering mechanisms. The average relaxation time is determined by carrying out the integrations inherent in (3.62) and (3.73) for a specific distribution function. For the latter, we take a Maxwellian so that

$$\frac{\partial f_0}{\partial E} \sim \frac{1}{k_B T} \exp\left(-\frac{E}{k_B T}\right). \tag{3.105}$$

In addition, reduced units will be defined as $x = E/k_B T$. Then (3.62) becomes

$$\langle \tau_m \rangle = \frac{A}{(k_B T)^s} \frac{\int_0^\infty x^{d/2-s} e^{-x} dx}{\int_0^\infty x^{d/2} e^{-x} dx} = \frac{A}{(k_B T)^s} \frac{\Gamma(1 - s + d/2)}{\Gamma(1 + d/2)}, \tag{3.106}$$

and the gamma function has been introduced in the last form. Usually, the semiconductor of interest is a bulk material, and therefore a three-dimensional solid. As a consequence, the value $d = 3$ will be used in the remainder of this section.

A second important average of the relaxation time is that arising in the Hall scattering factor, where the average of the square of the relaxation time is required. This can be obtained merely by extending (3.106) to

$$\langle\tau_m^2\rangle = \frac{A^2}{(k_BT)^{2s}}\frac{\Gamma(5/2+2s)}{\Gamma(5/2)}, \tag{3.107}$$

and the scattering factor can readily be determined to be

$$r_H = \frac{\langle\tau_m^2\rangle}{\langle\tau_m\rangle^2} = \frac{\Gamma(5/2-2s)\,\Gamma(5/2)}{\Gamma^2(5/2-s)}. \tag{3.108}$$

As an example, consider the case of acoustic phonon scattering for which $s = \frac{1}{2}$. Then the scattering factor is easily shown to be $r_H = 3\pi/8 \sim 1.18$. Although this value is often used, it arises *only* for the particular case of $s = \frac{1}{2}$, which is limited primarily to acoustic phonon scattering.

Another average that is important is that of the diffusion coefficient in (3.73). Using the above parameterizations in (3.73) leads to

$$\begin{aligned} D = \left\langle \frac{v^2\tau_m}{3} \right\rangle &= \frac{2A}{3m^*(k_BT)^{s-1}} \frac{\int_0^\infty x^{d/2-s}e^{-x}dx}{\int_0^\infty x^{d/2-1}e^{-x}dx} \\ &= \frac{2A}{3m^*(k_BT)^{s-1}}\frac{\Gamma(d/2+1-s)}{\Gamma(d/2)} = \frac{2A}{3m^*(k_BT)^{s-1}}\frac{\Gamma(5/2-s)}{\Gamma(3/2)} \\ &= \frac{2\langle\tau_m\rangle}{3m^*}k_BT\frac{\Gamma(5/2)}{\Gamma(3/2)} = \frac{e\langle\tau_m\rangle}{m^*}\frac{k_BT}{e}. \end{aligned} \tag{3.109}$$

In the last line, we have used the results of (3.106) and the properties that $\Gamma(n + 1) = n\Gamma(n)$. Hence, while our average is different than that used for the momentum relaxation time, the result yields the common version of the Einstein relationship between the diffusion coefficient and the mobility.

If more than one scattering mechanism is present, their effects must be combined prior to computing the average, which leads to a very complicated average. The most common manner of adding the effect of various scattering mechanisms is introduced by adding the effective resistances of each, which leads to

$$\frac{1}{\tau_{mT}} = \sum_i \frac{1}{\tau_{mi}}, \tag{3.110}$$

for which the sum is carried out over the different scattering mechanisms. In a typical semiconductor, impurity scattering, acoustic phonon scattering, and a variety of optical phonon scattering processes will all be involved. The average relaxation time, however, introduces a temperature dependence to the mobility through the temperature term arising in the above equations and from any temperature variation that is in the constant A. In fact, (3.110) is an expression of Mathiesen's rule, in which each scattering process is considered to be independent of all others. This is valid only when there is no correlation between the scattering events, such as may occur between carrier–carrier scattering (screening) and impurity scattering. While one must examine this in each case, it is usually true except in very high carrier density situations. In the next section, we begin to calculate the scattering rates for various elastic processes, in which the energy is conserved (or almost conserved) during the scattering event.

3.3 Elastic scattering processes

Scattering of the electrons, or the holes, from one state to another, whether this scattering occurs due to the lattice vibrations or by the Coulomb field of an ionized impurity, is one of the most important processes in the transport of the carriers through the semiconductor. In one sense it is the scattering that limits the velocity of the charge carriers in the applied fields, as discussed above. On the other hand, the carriers that are not scattered will be subject to a uniform increase of the wave vector $\mathbf{k}$ (in an applied dc field) and will cycle continuously through the Brillouin zone and yield a time-average velocity that is zero. In the latter case, it is the scattering that breaks up the correlated, accelerated state and introduces the actual transport process. Transport is again seen as a balance between accelerative forces and dissipative forces (the scattering). In general, the electronic motion is separated from the lattice motion. It is the adiabatic principle, where the atomic motion is supposed to be slow relative to the electronic motion, which allows separation of the electronic motion from the lattice motion. The former was solved for the static energy bands in the last chapter, while the latter yields the lattice dynamics and phonon spectrum. There remains one term in the total system Hamiltonian that couples the electronic motion to the lattice motion. This term gives rise to the electron–phonon interaction. However, there is not a single interaction term. Rather, the electron–phonon interaction can be expanded in a power series in the scattered wave vector $\mathbf{q} = \mathbf{k} - \mathbf{k}'$, and this process gives rise to a number of terms, which correspond to the number of phonon branches and the various types of interaction terms. There can be acoustic phonon interactions with the electrons, and the optical interactions can be either through the polar interaction (in compound semiconductors) or through the non-polar interaction. These are just the terms up to the harmonic expansion of the lattice; higher-order terms give rise to higher-order interactions.

In this section, the basic *elastic* electron–phonon (which may also be hole–phonon) interaction is treated in a general sense and we develop the acoustic phonon scattering rate. The various interactions that provide an inelastic process, where significant energy is exchanged during the scattering, will be left to the next chapter. We will, however, also introduce other elastic scattering processes here, including piezoelectric scattering, impurity scattering, and a few others as well. In general, all of the processes discussed here, except the piezoelectric interaction and interface scattering are found in all semiconductors, and piezoelectric scattering is found in all compound semiconductors.

3.3.1 Acoustic phonon scattering

The treatment followed here is based on the simple assumption that vibrations of the lattice cause small shifts in the energy bands. Deviations of the bands due to these small shifts from the frozen lattice positions lead to an additional potential that causes the scattering process. The scattering potential is then used in time-dependent, first-order perturbation theory to find a rate at which electrons are scattered out of one state **k** and into another state **k**′, while either absorbing or emitting a phonon of wave vector **q**. Each of the different processes, or interactions, leads to a different "matrix element" in terms of its dependence on these three wave vectors and their corresponding energy. These are discussed in the following sections, but here the treatment will retain just the existence of the scattering potential δE which leads to a matrix element

$$M(\mathbf{k},\mathbf{k}') = \langle \Psi_{\mathbf{k}',\mathbf{q}} | \delta E | \Psi_{\mathbf{k},\mathbf{q}} \rangle, \tag{3.111}$$

and the subscripts indicate that the wave function involves both the electronic and the lattice coordinates. Normally, the electronic wave functions are taken to be Bloch functions that exhibit the periodicity of the lattice. In addition, the matrix element usually contains the momentum conservation condition. Here this conservation condition leads to

$$\mathbf{k} - \mathbf{k}' \pm \mathbf{q} = \mathbf{G}, \tag{3.112}$$

where **G** is a vector of the reciprocal lattice. In essence, the presence of **G** is a result of the Fourier transform from the real space lattice to the momentum space lattice, and the result that we can only define the crystal momentum within a single Brillouin zone. (While the reciprocal lattice vector **G** will not be of much importance here, it will become quite important for inelastic scattering in Si, which is discussed in the next chapter.) For the upper sign, the final state lies at a higher momentum than the initial state, and therefore also at a higher energy. This upper sign must correspond to the absorption of a phonon by the electron. The lower sign leads to the final state being at a lower energy and momentum, hence corresponds to the emission of a phonon by the electrons.

Straightforward time-dependent, first-order perturbation theory then leads to the equation for the scattering rate, in terms of the *Fermi golden rule* (Ferry, 1995):

$$P(\mathbf{k},\mathbf{k}') = \frac{2\pi}{\hbar}|M(\mathbf{k},\mathbf{k}')|^2\delta(E_{\mathbf{k}} - E_{\mathbf{k}'} \pm \hbar\omega_{\mathbf{q}}), \tag{3.113}$$

and the signs have the same meaning as in the preceding paragraph: for example, the upper sign corresponds to the absorption of a phonon and the lower sign corresponds to the emission of a phonon. A derivation of (3.113) is found in most introductory quantum mechanics texts. Principally, the δ-function limit requires that the collision be fully completed through the invocation of a $t \to \infty$ limit. Moreover, each collision is localized in real space so that use of the well-defined Fourier coefficients **k**, **k**′, and **q** is meaningful. The perturbing potential must be small, so that it can be treated as a perturbation of the well-defined energy bands and so that two collisions do not "overlap" in space or time.

The scattering rate out of the state defined by the wave vector **k** and the energy $E_{\mathbf{k}}$ is obtained by integrating (3.113) over all final states. Because of the momentum conservation condition (3.112), the integration can be carried out over either **k**′ or **q** with the same result (omitting the processes for which the reciprocal lattice vector $\mathbf{G} \neq 0$). For the moment, the integration will be carried out over the final state wave vector **k**′, and ($\Gamma = 1/\tau$ is the scattering rate, whose inverse is the scattering time τ used in previous paragraphs)

$$\Gamma(\mathbf{k}) = \frac{2\pi}{\hbar}\sum_{\mathbf{k}'}|M(\mathbf{k},\mathbf{k}')|^2\delta(E_{\mathbf{k}} - E_{\mathbf{k}'} \pm \hbar\omega_{\mathbf{q}}). \tag{3.114}$$

In those cases in which the matrix element M is independent of the phonon wave vector, the matrix element can be removed from the summation, which leads to just the density of final states

$$\Gamma(\mathbf{k}) = \frac{2\pi}{\hbar}|M(\mathbf{k})|^2\rho(E_{\mathbf{k}} \pm \hbar\omega_{\mathbf{q}}). \tag{3.115}$$

This has a very satisfying interpretation: the total scattering rate is just the product of the square of the matrix element connecting the initial state to the final state and the total number of final states (we note, however, that care must be exercised on the evaluation of the density of states: those scattering processes which conserve spin must not include the "factor of 2 for spin" in the density of states). For these cases the scattering angle is a random variable that is uniformly distributed across the energy surface of the final state. Thus any state lying on the final energy surface is equally likely, and the scattering is said to be isotropic.

When there is a dependence of the matrix element on the wave vector of the phonon, the treatment is somewhat more complicated and this dependence must stay inside the summation and be properly treated. For this case, it is slightly easier to carry out the summation over the phonon wave vectors. At the same time, the summation over the wave vectors is changed to an integration and

$$\Gamma(\mathbf{k}) = \frac{2\pi}{\hbar}\frac{V}{(2\pi)^3}\int_0^{2\pi} d\varphi \int_0^{\pi} d\vartheta \sin\vartheta \int_0^{\infty} q^2 dq |M(\mathbf{k},\mathbf{q})|^2 \delta(E_{\mathbf{k}} - E_{\mathbf{k}\pm\mathbf{q}} \pm \hbar\omega_{\mathbf{q}}). \quad (3.116)$$

Here, it is assumed the semiconductor is a three-dimensional crystal. There is almost no case where the individual angle of q appears in the matrix element. Rather, it is only the relative angle between q and k that is important, so that it is permissible to consider that the latter vector is aligned in the z-direction, or the polar axis of the spherical coordinates used in (3.116). Since there is no reason not to have azimuthal symmetry in this configuration, there is no reason to have any φ variation and this integral can be done immediately, yielding 2π. We return to a treatment of this equation later. Here, however, we will primarily use (3.115) in our study. We now turn to the deformation potential interaction for the acoustic modes.

One of the most common phonon scattering processes is the interaction of the electrons (or holes) with the acoustic modes of the lattice through a deformation potential. Here, a long-wavelength acoustic wave moving through the lattice can cause a local strain in the crystal that perturbs the energy bands due to the lattice distortion. This change in the bands produces a weak scattering potential, which leads to a perturbing energy (Shockley and Bardeen, 1950)

$$\delta E = \Xi_1 \Delta = \Xi_1 \nabla \cdot \mathbf{u}_{\mathbf{q}}. \quad (3.117)$$

Here, Ξ_1 is the *deformation potential* for a particular band and Δ is the *dilation* of the lattice produced by a wave, whose Fourier coefficient is $\mathbf{u}_{\mathbf{q}}$. We note here that any static displacement of the lattice is a displacement of the crystal as a whole and does not contribute, so that it is the wave-like variation of the amplitude within the crystal that produces the local strain in the bands. This variation is represented by the dilation, which is just the desired divergence of the wave. The amplitude $\mathbf{u}_{\mathbf{q}}$ is a relatively uniform Fourier coefficient for the overall lattice wave, and may be expressed as (Ferry, 1991)

$$\mathbf{u}_{\mathbf{q}} = \left(\frac{\hbar}{2\rho_m V \omega_{\mathbf{q}}}\right)^{\frac{1}{2}} [a_{\mathbf{q}} e^{i\mathbf{q}\cdot\mathbf{r}} + a_{\mathbf{q}}^{+} e^{-i\mathbf{q}\cdot\mathbf{r}}]\mathbf{e}_{\mathbf{q}} e^{-i\omega_{\mathbf{q}} t}, \quad (3.118)$$

where ρ_m is the mass density, V is the volume, $a_{\mathbf{q}}$ and $a_{\mathbf{q}}^{+}$ are annihilation and creation operators for phonons, $\mathbf{e}_{\mathbf{q}}$ is the polarization vector, and the plane wave factors have been incorporated along with the normalization factor for completeness. Because the divergence produces a factor proportional to the component of q in the polarization direction (along the direction of propagation), only the longitudinal acoustic modes couple to the carriers in a spherically symmetry band (the case of ellipsoidal bands will be treated later). The fact that the resulting interaction potential is now proportional to q (i.e., to first order in the phonon wave vector) leads to this term being called a *first-order* interaction.

The matrix element may now be calculated by considering the proper sum over both the lattice and the electronic wave functions. The second term in (3.118), the term for the emission of a phonon by the carrier, leads to the matrix element squared, as

$$|M(\mathbf{k},\mathbf{q})|^2 = \frac{\hbar \Xi_1^2 q^2}{2\rho_m V \omega_q}(N_q + 1) I_{\mathbf{k},\mathbf{q}}^2, \qquad (3.119)$$

where $N_{\mathbf{q}}$ is the Bose–Einstein distribution function for the phonons, and

$$I_{\mathbf{k},\mathbf{q}} = \int_{\Omega} u_{\mathbf{k}-\mathbf{q}}^{+} u_{\mathbf{k}} \, d^3\mathbf{r} \qquad (3.120)$$

is the *overlap* integral between the cell portions of the Bloch waves (2.10) (unfortunately, similar symbols are used, but the $u_{\mathbf{k}}$ in this equation is the cell periodic part of the Bloch wave and not the phonon amplitude given above) for the initial and final states, and the integral is carried out over the cell volume Ω. For elastic processes, and for both states lying within the same "valley of the band," this integral is unity. Essentially, exactly the same result (3.119) is obtained for the case of the absorption of phonons by the electrons, with the single exception that $(N_{\mathbf{q}} + 1)$ is replaced by $N_{\mathbf{q}}$.

One thing that should be recalled is that the acoustic modes have very low energy. If the velocity of sound is 5×10^5 cm/s, a wave vector corresponding to 25 percent of the zone edge yields an energy only of the order of 10 meV. This is a very large wave vector, so for most practical cases the acoustic mode energy will be less than a millivolt. This will be important later when this matrix element is introduced into the scattering formulas above. Scattering processes in which the phonon energy may be ignored are termed *elastic* scattering events. Of more interest here is the fact that these energies are much lower than the thermal energy except at the lowest temperatures, and the Bose–Einstein distribution can be expanded under the equipartition approximation as

$$N_q = \frac{1}{\exp\left(\dfrac{\hbar\omega_q}{k_B T}\right) - 1} \sim \frac{k_B T}{\hbar\omega_q} \gg 1. \tag{3.121}$$

Since this distribution is so large and the energy exchange so small, it is quite easy to add the two terms for emission and absorption together, and use the fact that $\omega_q = q\mathrm{v}_s$, where v_s is the velocity of sound, to achieve

$$|M(\mathbf{k})|^2 \approx \frac{\Xi_1^2 k_B T}{\rho_m V \mathrm{v}_s^2}. \tag{3.122}$$

The final form (3.122) is independent of the wave vector of the phonons, so that the simple form of (3.115) can be used. For electrons in a simple, spherical energy surface and parabolic bands, this leads to

$$\begin{aligned}\Gamma(k) &= \frac{2\pi}{\hbar}\frac{\Xi_1^2 k_B T}{\rho_m V \mathrm{v}_s^2}\frac{V}{4\pi^2}\left(\frac{2m^*}{\hbar^2}\right)^{\frac{3}{2}} E^{\frac{1}{2}} \\ &= \frac{\Xi_1^2 k_B T (2m^*)^{\frac{3}{2}}}{2\pi\hbar^4 \rho_m \mathrm{v}_s^2} E^{\frac{1}{2}}.\end{aligned} \tag{3.123}$$

It has been assumed that the interaction does not mix spin states, and this factor is accounted for in the density of states. Although most of the parameters may easily be obtained for a particular semiconductor, it is found that the deformation potential itself is almost universally of the order of 7 to 10 eV for nearly all semiconductors.

In the treatment of spherical energy surfaces above, it was found that the matrix element was independent of the direction in momentum space and was independent of the wave vector (in the equipartition limit). In a many-valley semiconductor, such as the conduction band of silicon or germanium, this is no longer the case. Because the constant energy surfaces are ellipsoidal, shear strains as well as *dilational* strains can produce deformation potentials. The shear strain still leads to a term that depends on the vector direction of **q**, and it should be expected that band edge shifts will depend on all six components of the shear tensor. Thus there might be as many as six deformation potentials. However, in the semiconductors of interest, the valleys are ellipsoidal and centered on the high symmetry ⟨100⟩ and ⟨111⟩ axes, so that the symmetry properties allow a reduction to just two independent potentials. These are the *dilational potential* Ξ_d and the *uniaxial shear potential* Ξ_u. In terms of these potentials, the deformation energy is just (Herring, 1955)

$$\delta E = \Xi_d(e_{xx} + e_{yy} + e_{zz}) + \Xi_u e_{zz} \tag{3.124}$$

for an ellipsoid whose major axis is aligned with the z-axis. For longitudinal waves in an arbitrary direction q, the factor $\Xi_1^2(\mathbf{e}_q \cdot \mathbf{q})^2$ goes over into

$$\Xi_{LA}^2 q^2 = (\Xi_d^2 + \Xi_u^2 \cos^2\vartheta)q^2, \tag{3.125}$$

and ϑ is the angle between the z-axis (major ellipsoid axis) and the vector q. For transverse waves, only the e_{zz} term couples, and the proper form is just

$$\Xi_{TA}^2 q^2 = \Xi_u^2 \sin^2\vartheta \cos^2\vartheta q^2. \tag{3.126}$$

It should be remarked that both transverse modes are incorporated here in the general treatment. The differences above lead to different scattering rates for each principal axis within a single ellipsoidal valley. The summation over the multiple valleys (for the current) returns the overall system to cubic symmetry (unless the valleys are taken out of equilibration with each other, a point discussed in a later chapter). To achieve the latter result, each valley must be treated separately in the summation over q, and the separate results summed. When numerical evaluations of the angular averages are carried out for Si and Ge, it is found that it is a fairly good approximation to use a single energy-dependent scattering rate for the combined longitudinal and transverse acoustic modes. For the case of Ge, for example, it is found that

$$\Xi_1^2 \approx \tfrac{3}{4}(1.31\Xi_d^2 + 1.61\Xi_u\Xi_d + 1.01\Xi_u^2) \approx 0.99\Xi_d^2. \tag{3.127}$$

Thus the use of a single deformation potential is not a bad approximation in most cases, especially if the set of ellipsoids remains equivalent under application of the fields. Values of Ξ_d and Ξ_u that are accepted for Si are −6 and 9 eV, respectively (Ridley, 1982).

3.3.2 *Piezoelectric scattering*

The piezoelectric effect arises from the polar nature of compound materials, such as GaAs and other III–V compounds. These materials lack a center of inversion (sitting between the Ga and As atoms, one can understand why there is no inversion symmetry – look one way and you see a Ga atom, look the other and you see an As atom). Strain applied in certain directions in the lattice will produce a built-in electric field, which arises from the distortion of the basic unit cell. This creation of an electric field by the strain is called the piezoelectric effect. In materials with large piezoelectric coefficients, such as quartz, one can use the effect to provide oscillators at precise frequencies.

In most semiconductors, the effect is small, but can lead to scattering of the carriers, particularly at low temperatures where other scattering mechanisms are weak. For the purpose here, it is the presence of the acoustic mode that induces a local electric field. The carriers are deflected by this field and are therefore scattered by it. The crystals of interest have a single piezoelectric constant d (in the tensor notation by which stress and strain are discussed in a general cubic material, this is the element d_{14}, and this notation will be used below). By expanding the displacement waves, the polarization components can be found as follows (in Fourier transform form)

$$P_x = i\frac{d_{14}}{\varepsilon_\infty}(e_{q_y}q_z + e_{q_z}q_y)u_{\mathbf{q}}$$

$$P_y = i\frac{d_{14}}{\varepsilon_\infty}(e_{q_z}q_x + e_{q_x}q_z)u_{\mathbf{q}}$$

$$P_z = i\frac{d_{14}}{\varepsilon_\infty}(e_{q_y}q_x + e_{q_x}q_y)u_{\mathbf{q}}. \tag{3.128}$$

Here, ε_∞ is the high frequency dielectric permittivity. The interaction energy shift can be found by

$$\delta E = -\varepsilon_\infty \mathbf{F} \cdot \mathbf{P}, \tag{3.129}$$

where the electric field arises from the induced potential. The polarization leads to this potential, which couples to form the perturbing energy. For the potential, we shall use a standard screened Coulomb form (the derivation of this form appears in Chapter 7)

$$\Phi(\mathbf{r}) = \frac{e}{4\pi\varepsilon_\infty r}e^{-q_d r}, \tag{3.130}$$

where q_d is the reciprocal of the Debye screening length. The exponential factor provides a cut-off in the Coulomb interaction. This potential may be Fourier transformed (in three dimensions) to give

$$\Phi_{\mathbf{q}} = \frac{e}{\varepsilon_\infty}\frac{1}{q^2 + q_d^2}. \tag{3.131}$$

This, in turn, yields the electric field

$$\mathbf{F} = -i\frac{e}{\varepsilon_\infty}\frac{\mathbf{q}}{q^2 + q_d^2}. \tag{3.132}$$

The perturbing potential can now be written, with the definitions above, as

$$\delta E = -i\frac{2ed_{14}}{\varepsilon_\infty}\frac{q^2}{q^2+q_d^2}(\beta\gamma e_{q_x} + \gamma\alpha e_{q_y} + \alpha\beta e_{q_z})u_{\mathbf{q}}. \tag{3.133}$$

Here α, β, and γ are the directional cosines between the wave vector q and the three axes x, y, and z, respectively.

The role of the screening is interesting. In examining (3.133), it is clear that for small q (large distances), the interaction potential vanishes with q^2. On the other hand, for large q (small distances), the central q-dependent factor becomes unity. There is a natural cutoff value for q, which is determined by q_d, the reciprocal of the Debye screening length. From this it appears that piezoelectric scattering is a short range effect, much like other Coulomb scatterers to be discussed later. There is a complication in this screening, however. The actual potential, which arises from the electron–phonon interaction, is not a true Coulomb potential because of the harmonic variation at frequency ω_q. In Chapter 7, the full dynamic screening effects will be discussed, but if the frequency is sufficiently high, the screening is significantly reduced. This effect strengthens the piezoelectric interaction at longer wave vectors. However, the de-screening is fully effective only when the phonon energy is comparable to the electron energy. Since we are dealing with elastic scattering, this event seldom occurs, and the formulas above may be used freely.

The results of the preceding section can be used to evaluate the matrix element. Equation (3.133) can be compared directly to (3.122) to yield

$$|M(\mathbf{k},\mathbf{q})|^2 = \frac{4e^2d_{14}^2k_BT}{\varepsilon_\infty^2\rho_m V\omega_q^2}\left(\frac{q^2}{q^2+q_d^2}\right)^2(\beta\gamma e_{q_x} + \gamma\alpha e_{q_y} + \alpha\beta e_{q_z})^2, \tag{3.134}$$

in the equipartition limit. The last term can be averaged over the various directions to produce a spherically symmetric average, which gives 12/35 for longitudinal waves and 16/35 for transverse waves (Hutson, 1961; Zook, 1964). With this in mind, we can introduce average lattice strain constants through

$$c_L = \tfrac{2}{5}(c_{12} + 2c_{44}) + \tfrac{3}{5}c_{11}, \quad c_T = \tfrac{1}{5}(c_{11} - c_{12}) + \tfrac{3}{5}c_{44}. \tag{3.135}$$

This allows us to define an effective coupling constant as

$$K^2 = \frac{d_{14}^2}{\varepsilon_\infty}\left(\frac{12}{35c_L} + \frac{16}{35c_T}\right), \tag{3.136}$$

This result may now be used to calculate the scattering rate. However, this scattering is essentially elastic, as it involves the acoustic modes, which have very low energy in comparison with the carriers. Thus the limits can be simplified by ignoring the phonon energy, but the anisotropic approximation (3.116) must be used. With the above considerations, the scattering rate can be written as

$$\Gamma(\mathbf{k}) = \frac{2m^*e^2K^2k_BT}{\pi\varepsilon_\infty\hbar^3k}\int_0^{2k}\frac{q^3dq}{(q^2+q_d^2)^2}$$

$$= \frac{m^*e^2K^2k_BT}{\pi\varepsilon_\infty\hbar^3k}\left[\ln\left(1+4\frac{k^2}{q_d^2}\right)-\frac{4k^2}{4k^2+q_d^2}\right]. \tag{3.137}$$

This result assumes that the scattering can flip the spin (e.g., it is thought that in piezoelectric scattering, the electron may scatter into either of the final two possible spin states at $\mathbf{k}'$), although this is not well understood and may not be the case. Piezoelectric scattering predominantly occurs at relatively low temperatures (where the equipartition approximation may not be valid).

3.3.3 *Impurity scattering*

In any treatment of electron scattering from the Coulomb potential of an ionized impurity atom, it is necessary to consider the long-range nature of the potential. If the interaction is summed over all space, the integral diverges and a cut-off mechanism must be invoked to limit the integral. One approach is just to cut off the integration at the mean impurity spacing, the so-called Conwell–Weisskopf (1950) approach. A second approach is to invoke screening of the Coulomb potential by the free carriers, which was done for piezoelectric scattering. In this case, the potential is induced to fall off much more rapidly than a bare Coulomb interaction, due to the Coulomb forces from the neighboring carriers. The screening is provided by the other carriers, which provide a background of charge. This is effective over a distance on the order of the Debye screening length (we return to this in Chapter 7) in non-degenerate materials. This screening of the repulsive Coulomb potential results in an integral (for the scattering cross section) which converges without further approximations (Brooks, 1955).

For spherical symmetry about the scattering center, or ion location, the potential is screened in a manner that gives rise to a screened Coulomb potential, as

$$\Phi(\mathbf{r}) = \frac{e^2}{4\pi\varepsilon_\infty r}e^{-q_d r}, \tag{3.138}$$

where the Debye wave vector q_d is the inverse of the screening length, and is given by

$$q_d^2 = \frac{ne^2}{\varepsilon_\infty k_B T}. \tag{3.139}$$

Here ε_∞ is the high-frequency permittivity. Generally, if both electrons and holes are present, n is replaced by the summation $n + p$. The above results are for a non-degenerate semiconductor. A similar result can be found in degenerate systems, for which the Fermi-Thomas screening wave vector is found to be

$$q_{FT}^2 = \frac{3ne^2}{2\varepsilon_\infty E_F}. \tag{3.140}$$

In treating the scattering from the screened Coulomb potential, we will take a slightly different approach. We use a wave scattering approach and compute the *scattering cross section* $\sigma(\theta)$, which gives the angular dependence of the scattering. It is assumed the incident wave is a plane wave, and the scattered wave is also a plane wave. Thus the total wave function can be written as

$$\Psi(\mathbf{r}) = e^{ikz} + v(\mathbf{r})e^{i\mathbf{k}'\cdot\mathbf{r}}. \tag{3.141}$$

In this, $\mathbf{k} = k\mathbf{a}_z$ orients the incident wave along the polar axis, and the second term represents the scattered wave. This may then be inserted into the Schrödinger equation, neglecting terms of second or higher order in the scattered wave, and

$$\nabla^2 v(\mathbf{r}) + k'^2 v(\mathbf{r}) = \frac{2m^*}{\hbar^2}\frac{Ze^2}{4\pi\varepsilon_\infty r} e^{-q_d r} e^{ikz}. \tag{3.142}$$

The factor Z has been inserted to account for the charge state of the impurity (normally $Z = 1$). If the terms on the right-hand side are treated as a charge distribution, the normal results from electromagnetic field theory can be used to write the solution as

$$v(\mathbf{r}) = -\frac{m^* Ze^2}{8\pi^2\varepsilon_\infty}\int \frac{d^3\mathbf{r}'}{r'|\mathbf{r} - \mathbf{r}'|} e^{ikz' - q_d r'} e^{ik'|\mathbf{r}-\mathbf{r}'|}. \tag{3.143}$$

To proceed, it is assumed that $r \gg r'$, and the polar axis in real space is taken to be aligned with r. Further, the scattering wave vector is taken to be $\mathbf{q} = \mathbf{k} - \mathbf{k}'$, so that

$$\int_0^\pi \sin\theta d\theta e^{i\mathbf{q}\cdot\mathbf{r}'} = \frac{\sin(qr')}{qr'}, \tag{3.144}$$

the ϕ integration is immediate, and the remaining integration becomes

$$v(\mathbf{r}) \cong -\frac{m^* Ze^2}{2\pi\varepsilon_\infty \hbar^2 qr}\int_0^\infty \sin(qr')e^{-q_d r'}dr' = \frac{m^* Ze^2}{2\pi\varepsilon_\infty \hbar^2 qr(q^2+q_d^2)}. \tag{3.145}$$

Now $\mathbf{q} = \mathbf{k} - \mathbf{k}'$, but for elastic scattering $k = k'$, and $q = 2k\sin(\theta/2)$, where θ is the angle between $\mathbf{k}$ and $\mathbf{k}'$. If we write the scattered wave function $v(\mathbf{r})$ as $f(\theta)/r$, then we recognize that the factor $f(\theta)$ is the matrix element, and the cross-section is defined as

$$\sigma(\theta) = |f(\theta)|^2 = \left(\frac{m^* Ze^2}{8\pi\hbar^2 k^2\varepsilon_\infty}\right)^2 \frac{1}{[\sin^2(\theta/2) + q_d^2/4k^2]^2}. \tag{3.146}$$

The total scattering cross-section (*for the relaxation time*) is found by integrating over θ, weighting each angle by an amount $(1 - \cos\theta)$. The last factor accounts for the momentum relaxation effect discussed in the earlier part of this chapter. The dominance of small-angle scattering prevents each scattering event from relaxing the momentum so this factor is inserted by hand (this factor is not necessary for inelastic processes, and averages to zero for isotropic elastic processes). This is one of the few scattering processes where this factor is included, primarily because each scattering event lasts for a quite long time, and it is necessary to calculate the average momentum loss rate. Thus

$$\sigma_c = 2\pi\int_0^\pi \sigma(\theta)(1-\cos\theta)d\theta = 16\pi\int_0^{\pi/2}\sigma\left(\frac{\theta}{2}\right)\sin^3\left(\frac{\theta}{2}\right)d\left(\sin\frac{\theta}{2}\right)$$

$$= \frac{\pi}{2}\left(\frac{m^* Ze^2}{2\pi\hbar^2 k^2\varepsilon_\infty}\right)^2\left[\ln\left(\frac{1+\beta^2}{\beta^2}\right) - \frac{1}{1+\beta^2}\right], \tag{3.147}$$

where $\beta = q_d^2/2k$. The scattering rate is now the product of the cross-section, the number of scatterers, and the velocity of the carrier, or

$$\Gamma(k) = N\sigma_c v = \frac{Z^2e^4m^*}{8\pi\varepsilon_\infty^2\hbar^3k^3}\left[\ln\left(\frac{4k^2+q_d^2}{q_d^2}\right) - \frac{4k^2}{4k^2+q_d^2}\right]. \tag{3.148}$$

In some situations, the actual scattering rate, and not the relaxation rate for momentum, is required. This is true in Monte Carlo simulation programs, as discussed later. In this situation, (3.147) must be modified by removal of the $(1 - \cos\theta)$ term, which results in the terms in square brackets in (3.148) being replaced by the factor

$$\left[\frac{4k^2}{q_d^2(4k^2 + q_d^2)}\right], \tag{3.149}$$

which dramatically changes the energy dependence for small k ($\ll q_d$). The form (3.148) is the one normally found when discussions of mobility and diffusion constants are being evaluated for simple transport in semiconductors. However, when weighting various random processes for Monte Carlo approaches, it is the total scattering rate that is important, and this is given by the use of (3.149).

3.3.4 *Coulomb scattering in two dimensions*

If the Coulomb scatterer is near an interface, the problem becomes more complicated. This is particularly the case for charged scattering centers near the Si–SiO_2 (or any semiconductor–insulator) interface as well as in mesoscopic structures. In general, there are always a large number of Coulomb centers near the interface, due to disorder and defects in the crystalline structure in the neighborhood of the interface. In many cases, these defects are associated with dangling bonds and can lead to charge trapping centers which scatter the free carriers through the Coulomb interaction. The Coulomb scattering of carriers lying in an inversion layer (or a quantized accumulation layer) at the interface differs from the case of bulk impurity scattering, due to the reduced dimensionality of the carriers.

Coulomb scattering of surface quantized carriers was described first by Stern and Howard (1967) for electrons in the Si–SiO_2 system. Since then, many treatments have appeared in the literature, which differ little from the original approach. In general, the interface is treated as being abrupt and as having an infinite potential discontinuity in the conduction (or valence) band, so that problems with interfacial non-stoichiometry and roughness are neglected (the latter is treated below as an additional scattering center). In treating this scattering, it is most convenient to use the electrostatic Green's function for charges in the presence of a dielectric interface, so that the image potential is properly included in the calculation. The scattering matrix element involves integration over plane-wave states for the motion parallel to the interface, and thus one is led to consider only the two-dimensional transform of the Coulomb potential

$$G(\mathbf{q}, z - z') = \begin{cases} \dfrac{1}{2q\varepsilon_s}\left(e^{-q|z-z'|} + \dfrac{\varepsilon_s - \varepsilon_{ox}}{\varepsilon_s + \varepsilon_{ox}} e^{-q|z-z'|}\right), & z' > 0, \\[2ex] \dfrac{1}{q(\varepsilon_s + \varepsilon_{ox})} e^{-q|z-z'|}, & z' < 0. \end{cases} \tag{3.150}$$

Here, q is the two-dimensional scattering vector and ε_s and ε_{ox} are the high-frequency total permittivities for the semiconductor and the oxide, respectively. Equation (3.150) assumes that the scattering center is located a distance z' from the interface (the semiconductor is located in the space $z > 0$) and has an image at $-z'$. If the interface were non-abrupt, the ratio of dielectric constants that appears in the second term on the right of the first line of (3.150) would be a function of q.

The Coulomb potential in (3.150) is still unscreened, and it is necessary to divide this equation by the equivalent factor that appears in (3.144): for example,

$$q \to \sqrt{q^2 + q_0^2}\,, \tag{3.151}$$

where q_0 is the appropriate two-dimensional screening vector in the presence of the interface, for example

$$q_0 = \frac{n_s e^2}{2\pi(\varepsilon_s + \varepsilon_{ox})k_B T} \tag{3.152}$$

for a non-degenerate semiconductor (Ferry and Goodnick, 1997). Here, n_s is the sheet carrier concentration in the inversion (or accumulation) layer. More complicated screening approaches are possible, and were considered by Stern and Howard (1967), but these are beyond the introductory scope of the present approach. Some will be discussed further in Chapter 7.

For two-dimensional scattering, the scattering cross-section is determined by the matrix element of the screened Coulomb interaction for a charge located at z', with $\mathbf{q} = \mathbf{k} - \mathbf{k}'$ being the difference in the incident and the scattered wave vector as previously. Now, however, the motion in the direction normal to the interface must also be accounted for, as it is not in the two-dimensional Fourier transform. This leads to

$$\langle \mathbf{k}|V(z')|\mathbf{k}'\rangle = e^2\int_0^{\infty} |\zeta(z)|^2 G(q, z - z')dz\,, \tag{3.153}$$

where $\zeta(z)$ is the z portion of the wave function; that is, we write this wave function as

$$\psi(x, y, z) = \zeta(z)e^{i(k_x x + k_y y)}\,. \tag{3.154}$$

It is not a bad assumption to consider only scattering from charges located at the interface itself (i.e., $z' = 0$), since this is the region at which the density of scattering charges is usually large. In this idealization, charges are assumed to be uniformly distributed in the plane $z' = 0$. Then, we need only the second line of (3.150), and the scattering rate is

$$\Gamma(\mathbf{k}) = \frac{N_{sc} e^4 m^*}{4\pi\hbar^3(\varepsilon_s + \varepsilon_{ox})^2} \int_0^{2\pi} A^2(\theta) \frac{(1-\cos\theta)d\theta}{q^2 + q_0^2}, \tag{3.155}$$

where

$$A(\theta) = \int_0^\infty |\zeta(z)|^2 e^{-qz} dz. \tag{3.156}$$

Here again, the scattering is elastic and $q = 2k\sin(\theta/2)$. At this point it is necessary to say something about the envelope function $\zeta(z)$ in order to proceed. In the lowest subband, it is usually acceptable to take the wave function as

$$\zeta(z) = 2b^{\frac{3}{2}} z e^{-bz}, \tag{3.157}$$

which leads to an average thickness of the inversion layer of $3b/2$. With this form (3.155) becomes

$$\Gamma(\mathbf{k}) = \frac{8b^3 N_{sc} e^4 m^*}{\pi\hbar^3(\varepsilon_s + \varepsilon_{ox})^2} \int_0^\pi \frac{\sin^2(\vartheta) d(\vartheta)}{[4k^2\sin^2(\vartheta) + q_0^2][2k\sin\vartheta + 2b]^3}, \tag{3.158}$$

where the substitution $\vartheta = \theta/2$ has been made. [Although (3.157) is usually applied to the triangular potential (infinite wall at the interface, and linear rising potential inside the semiconductor), it can be applied to any potential shape.] In general, the peak of the wave function lies only a few nanometers from the interface, and then dies off exponentially, so that it represents electrons localized in a plane parallel to the interface. The factor b is a significant fraction of the Brillouin zone boundary distance. For this reason it can generally be assumed that $b \gg k$, so that $A(q)$ is near unity.

Two limiting cases may be found from (3.158), with the approximation of $A(q)$ near unity. For $q_0 \ll q$, the behavior is essentially unscreened, and the integral yields $\pi/4k^2$. In this case, the scattering rate is inversely proportional to the square of the wave vector, which may be assumed to be near the Fermi wave vector for a degenerate inversion layer. Thus the mobility actually increases as the inversion density increases, since the average energy (and hence the average wave vector) increases with the density. At the other extreme, $q_0 \gg q$, the scattering is heavily screened by the charge in the inversion layer. In the latter case, the wave-vector dependence disappears, and the integral yields only $\pi/2q_0^2$, so that the density dependence also disappears from the equation and thus the scattering rate becomes constant.

What if it is not desired to omit the dependence on b above? How does one determine the value for b? It was remarked above that b is a variational parameter. This means that the assumed form of the wave function (3.157) is inserted into the Schrödinger equation and the resulting energy is minimized by varying the parameter b (Ferry, 1995). One problem is the form of the potential (band bending) in the semiconductor. Stern and Howard (1967) used a Hartree potential, in which the band bending was determined self-consistently by including the potential of the charge itself through Poisson's equation. This is beyond the level of the approach desired here. Instead, the potential will be taken as a linear potential described by a constant field, which is the effective field $F = e(N_{dep} + n_s/2)/\varepsilon_s$. The procedure is a standard one in mathematical physics, and proceeds by (1) inserting the assumed wave function into the Schrödinger equation, (2) multiplying by its complex conjugate and integrating over all space, and (3) varying b to minimize the energy. This procedure yields (Ferry, 1995)

$$b = \left(\frac{3eFm^*}{2\hbar^2}\right)^{\frac{1}{3}}. \tag{3.159}$$

This relationship then gets the dependence on the inversion density into (3.158) and the resulting scattering rate. This gives a density dependence over and above that from the average value of the wave vector k. Some numbers may give further insight. If we assume an inversion layer in Si, with $n_s = 10^{12}$ cm^{-2}, then the effective field is about 0.15 MV/cm and $b \sim 4 \times 10^6$/cm. On the other hand, k_F is about 2.5×10^6/cm. The approximation of assuming a very large value for b may not be appropriate for such situations. However, the situation improves at lower densities.

3.3.5 *Surface roughness scattering*

In addition to Coulomb scattering, short-range scattering associated with the interfacial disorder also limits the mobility of quasi-two-dimensional electrons at the interface. In Figure 3.3, a high-resolution transmission electron micrograph of the interface between Si and SiO_2 is shown (Goodnick *et al.*, 1985). The crystalline Si atomic rows are visible in the bottom half of the figure, while the random amorphous structure of the oxide is apparent in the top half of the figure. The plane of the image is a (111) plane, while the interface is along a (100) plane. It is clear that the interface is not abrupt on the

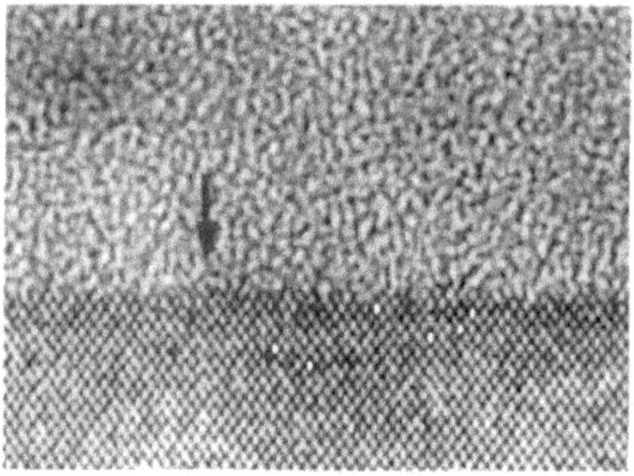

Figure 3.3 High resolution transmission electron micrograph of the interface between Si and SiO_2. The oxide is the top half of the picture. The rows of Si atom pairs can be observed in the bottom half. The image is a "lattice-plane image" lying in the (111) plane, while the interface is a (100) plane. (After Goodnick *et al.* (1985), with permission.)

atomic level, but that variation in the actual position of the interfacial plane can extend over one or two atomic layers along the surface. The local atomic interface actually has a random variation which, coupled with the surface potential, gives rise to fluctuations of the energy levels in the quantum well formed by the potential barrier to the oxide and the band bending in the semiconductor. The randomness induced by the interfacial roughness has some similarity to alloy scattering, treated in the next section, and can lead to limitations on the mobility of the carriers in the inversion layer. At present, calculation of the scattering rate based on the microscopic details of the roughness does not exist. Instead, the usual models rely on a semi-classical approach in which a phenomenological surface roughness is parameterized in terms of its height and the correlation length.

In current surface roughness models, displacement of the interface from a perfect plane is assumed to be described by a random function $\Delta(\mathbf{r})$, where $\mathbf{r}$ is a two-dimensional position vector parallel to the interface (the average interface). This model assumes that $\Delta(\mathbf{r})$ varies slowly over atomic dimensions so that the boundary conditions on the wave functions can be treated as abrupt and continuous. This assumption is obviously in error when surface fluctuations occur on the atomic level. However, the model has proven to provide quite good agreement with measured mobility variations in a variety of materials and interfaces.

The scattering potential may be obtained by expanding the surface potential in terms of $\Delta(\mathbf{r})$ as

$$\delta V(\mathbf{r}) = V[z + \Delta(\mathbf{r})] - V(z) \approx eF(z)\Delta(\mathbf{r}), \tag{3.160}$$

where $F(z)$ is the electric field in the inversion layer itself. The scattering rate for the perturbing potential must include the role of the correlation between the scattering centers along the interface. This correlation is described by a Fourier transform $\Delta(\mathbf{q})$. This leads to the scattering matrix element

$$M(\mathbf{k},\mathbf{q}) = e\Delta(\mathbf{q})\int_0^\infty F(z)|\zeta(z)|^2 dz$$
$$= e^2\Delta(\mathbf{q})\frac{N_{depl} + n_s/2}{\varepsilon_s}, \tag{3.161}$$

where the orthonormality of the wave function has been used and the *average* electric field has been introduced. The inversion density appears with the factor of 2 since it is the field at the interface that is of interest. The inversion charge appears almost entirely at the interface, so that it creates a field on each side, of which only one-half of the total field discontinuity appears at the interface. The factor $\Delta(\mathbf{q})$ is the Fourier transform of $\Delta(\mathbf{r})$.

In the matrix element, only the statistical properties of $\Delta(\mathbf{q})$ need be considered. Thus the descriptors discussed earlier may be introduced. There is some debate, and the experimental results are not clear, about the form of the positional auto-correlation function for the interface roughness. In most of the early work, it was assumed to be describable by a Gaussian, given by

$$\langle \Delta(\mathbf{r})\Delta(\mathbf{r}-\mathbf{r}')\rangle = \Delta^2 e^{-r'^2/L^2}, \tag{3.162}$$

for which

$$|\Delta(\mathbf{q})|^2 = \pi\Delta^2 L^2 \exp\left(-\frac{q^2L^2}{4}\right). \tag{3.163}$$

The quantity Δ is the rms height of the fluctuation in the interface and L is the correlation length for the fluctuations. In a sense, L is the average distance between "bumps" in the interface. It must be remembered that the actual interface used for the TEM picture has a finite thickness, and some averaging of the roughness will occur in the image. Nevertheless, typical values obtained from this approach are in the range 0.2 to 0.4 nm for Δ and 1.0 to 3.0 nm for L at the Si–SiO_2 interface (Goodnick *et al.*, 1985). It was pointed out by Goodnick *et al.* (1985), however, that there was significant evidence that the correlation function was not Gaussian, but had a more exponential character to it. Subsequent measurements by Yoshinobu *et al.* (1993) with atomic-force microscopy, and by Feenstra (1994) with cross-sectional scanning–tunneling microscopy, have confirmed that the correlation function is quite likely to be an exponential, given by

$$\langle \Delta(\mathbf{r})\Delta(\mathbf{r}-\mathbf{r}')\rangle = \Delta^2 e^{-\sqrt{2}r/L}, \tag{3.164}$$

for which

$$|\Delta(\mathbf{q})|^2 = \frac{\pi\Delta^2 L^2}{[1+(q^2L^2/2)]^{\frac{3}{2}}}. \tag{3.165}$$

The matrix element is now found by combining the preceding equations to incorporate the correlation function as

$$|M(\mathbf{k},\mathbf{q})|^2 = \pi\left(\frac{\Delta L e^2}{\varepsilon_s}\right)^2\left(N_{depl}+\frac{n_s}{2}\right)^2\frac{1}{[1+(q^2L^2/2)]^{\frac{3}{2}}}. \tag{3.166}$$

The actual scattering rate is calculated for two-dimensional scattering as in the preceding section. The scattering is elastic, so that $|\mathbf{k}| = |\mathbf{k}'|$, and $q = 2k\sin(\theta/2)$ arises from the delta function, which produces energy conservation. Thus one finds that

$$\Gamma(\mathbf{k}) = \frac{1}{2\pi\hbar}\int_0^{2\pi} d\theta \int_0^{\infty} q\,dq\,|M(\mathbf{k},\mathbf{q})|^2\delta(E_{\mathbf{k}} - E_{\mathbf{k}+\mathbf{q}})$$

$$= \frac{m^*}{2\pi\hbar^3}\int_0^{2\pi} d\theta \frac{\pi\Delta^2L^2e^4}{\varepsilon_s^2}\left(N_{depl}+\frac{n_s}{2}\right)^2\frac{1}{[1+(k^2L^2\cos^2(\theta/2))]^{\frac{3}{2}}}$$

$$= \frac{m^*\Delta^2L^2e^4}{\hbar^3\varepsilon_s^2}\left(N_{depl}+\frac{n_s}{2}\right)^2\frac{1}{\sqrt{1+k^2L^2}}E\left(\frac{kL}{\sqrt{1+k^2L^2}}\right). \tag{3.167}$$

Here, E is a complete elliptic integral. The explicit dependence of the scattering rate on the square of the effective field at the interface results in a decreasing mobility with increasing surface field (and increasing inversion density), which agrees with the trends observed in the experimental mobility data of most materials. This decrease in the experimental mobility with surface density qualitatively arises from the increased electric field dispersion around interface discontinuities at higher surface fields, which in turn gives rise to a larger scattering potential. In general, the entire mobility behavior in inversion layers at low temperature is explainable in terms of surface-roughness scattering and Coulomb scattering from interfacial charge, as discussed in the preceding section. This is shown in Figure 3.4. The decrease in mobility at low inversion density is due to the Coulomb scattering.

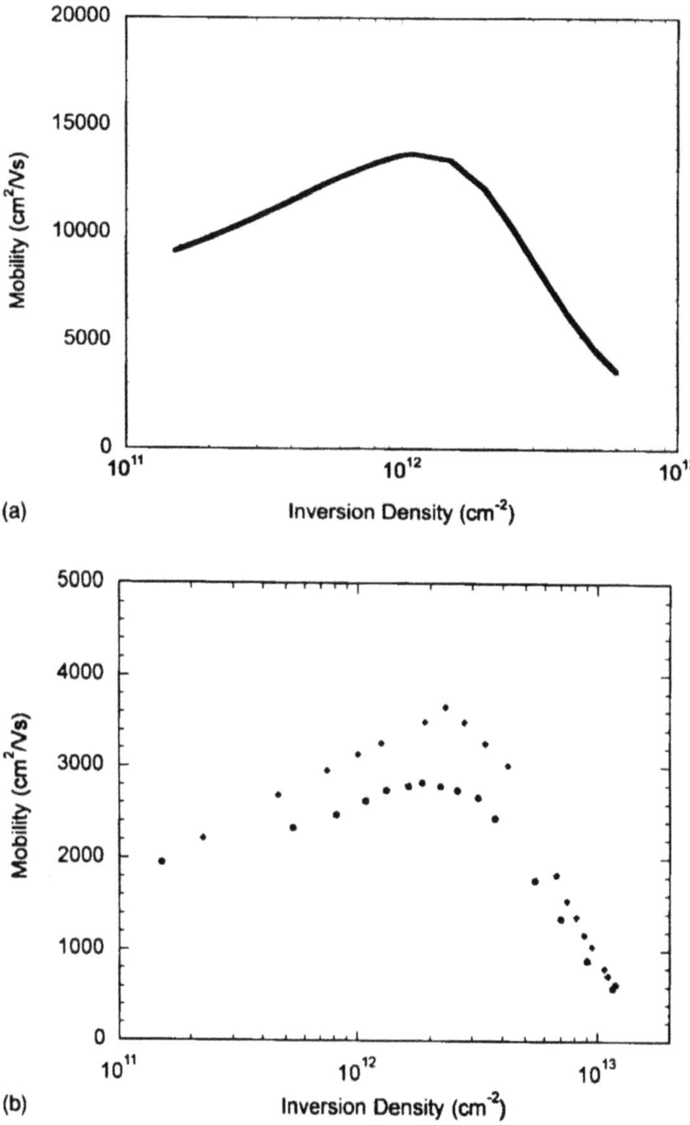

Figure 3.4 The mobility as a function of the inversion density in an Si inversion layer at 4.2 K (a) and at 77 K (b). The solid curves are computed using surface roughness and charged impurity scattering. The data in (a) is taken from Goodnick *et al.* (1985), while that in (b) is taken from Takagi *et al.* (1994). Two different dopings are shown in the latter figure; the upper curve (diamonds) is for $n_i = 2 \times 10^{16}$ cm^{-3} while the lower curve (circles) is for $n_i = 7.2 \times 10^{16}$ cm^{-3}. (Simulation courtesy of Vasileska (1995), with permission.)

3.3.6 Alloy scattering

In a semiconductor alloy, the scattering of free carriers due to the deviations from the virtual crystal model has been termed alloy scattering. The virtual crystal concept was introduced in Chapter 2 in connection with the alloys of various semiconductor materials. The general treatment of alloy scattering has usually followed an unpublished, but well-known, result due to Brooks, and extended by Makowski and Glicksman (1976). Although this scattering mechanism generally supplements the normal phonon and impurity scattering, it has on occasion been conjectured to be sufficiently strong as to be the dominant scattering mechanism in alloys. The work of Makowski and Glicksman, however, showed that the scattering was in general quite weak. They found that it was probably important only in the InAsP system, although even here it was likely to be much weaker than experimental data would suggest (which itself contains additional scattering due to defects in the alloy material, which always seem to be overlooked). These authors utilized a scattering potential given by the difference in the band gaps of the constituent semiconductors. Harrison and Hauser (1970) suggested that the scattering potential is related to the differences in the electron affinities. However, as pointed out by Kroemer (1975), the electron affinity is a true surface property, but not a qualitatively useful quantity in the bulk and is even a very bad indicator of bulk band offsets. Its use in scattering theories for carrier transport in bulk materials should therefore be treated with a degree of skepticism. A subsequent effort has suggested that the proper estimator for the disorder potential can be calculated from the previously specified heteropolar potential C, introduced in equation (2.30), as this quantity tends to affect strongly the bowing parameter and should therefore affect the random potential that leads to the scattering.

The electron scattering rate for alloy scattering is determined directly by the scattering potential δE, which is the topic of the discussion above. The scattering is elastic, and the matrix element can therefore be given simply by

$$M(\mathbf{k},\mathbf{q}) = \delta E e^{i\mathbf{q}\cdot\mathbf{r}}, \tag{3.168}$$

which can be used immediately in (3.115) to give

$$\Gamma(\mathbf{k}) = \frac{(\delta E)^2}{2\pi\hbar}\left(\frac{2m^*}{\hbar^2}\right)^{\frac{3}{2}} E^{\frac{1}{2}}. \tag{3.169}$$

This result is for parabolic bands, of course. A factor describing the degree of ordering has been omitted, which assumes that the alloy is a perfect random alloy. The scattering is reduced if there is any ordering in the alloy.

Table 3.1 Alloy scattering parameters

Alloy	U_C *(eV)*	U_{BG} *(eV)*	*Alloy*	U_C *(eV)*	U_{BG} *(eV)*
GaAlAs	0.12	0.7	InGaP	0.56	1.08
GaInAs	0.5	1.07	InAlP	0.54	1.08
InAsP	0.36	1.0	InGaSb	0.44	0.52
GaAsP	0.43	0.83	InPSb	1.32	1.17
InAsSb	0.82	0.18	GaPSb	1.52	1.57
InAlAs	0.47	1.49	InGaAsP	0.29	0.54
AlAsP	0.64	0.27	InGaPSb	0.54	0.56
InAlAsP	0.28	0.58			

Besides the effect of ordering, the most significant parameter in (3.169) is the scattering potential δE. One could use the difference in the actual values of C, but this would ignore the effect that the change in the lattice constant in the alloy would have on the general value of the average C. The scattering potential that leads to disorder scattering is just the aperiodic contribution to the crystal potential that arises from the disorder introduced into the lattice by the random siting of constituent atoms. In the virtual crystal approximation, the perfect zinc-blende lattice is retained in the solid solution. Thus the bond lengths are equal and the homopolar energy E_h does not make a contribution to the random potential δE. The random potential arises solely from the fluctuations in the heteropolar energy C. For a binary compound, the heteropolar energy is given by (2.30). But, the binary solid does not usually have an alloy scattering. In a ternary solid $A_xB_{1-x}C$, it has been suggested that this form should be replaced by (Ferry, 1978)

$$C_{AB} = sZ_{AB}\left(\frac{1}{r_A} - \frac{1}{r_B}\right)\exp\left(-q_{FT}\frac{r_A + r_B}{2}\right), \tag{3.170}$$

where the r_i are the atomic radii, q_{FT} is the Fermi–Thomas screening vector mentioned earlier (but for the entire set of valence electrons rather than the free carriers), s is a factor ~1.5 to account for the typical over-screening of the Fermi–Thomas approximation, and it has been assumed that the valence of the A and B atoms is the same. This can now be used to calculate the aperiodic potential used for alloy scattering. The results for a number of alloys are presented in Table 3.1. Also shown, for comparison, are the equivalent values estimated by the discontinuity in the band gap. There is a weak dependence on the scattering potential δE from the composition as well, but this is small compared to the $x(1 - x)$ term.

As mentioned above, it is generally found that the role of alloy scattering is very weak, although there are often strong assertions from experimentalists that reduced mobility found in alloys must be due to "alloy scattering."

In fact, only the work of Makowski and Glicksman (1976) was sufficiently careful to exclude other mechanisms. In general, little effort is taken to include the proper strength of optical phonon scattering (discussed in the next chapter), due to the complicated multimode behavior of the dielectric function in the alloy or to include dislocation or cluster scattering that can arise in impure crystals.

3.4 Transport in high magnetic fields

In Section 3.2.4, it was generally assumed that $\omega_c\tau < 1$, so that we did not have to worry about closed orbits around the magnetic field. When the magnetic field is large, however, the electrons can make complete orbits about the magnetic flux lines. In this case the orbits behave as harmonic oscillators and the energy of the orbit is quantized (Ferry, 1995). We will assume that the magnetic field is large (e.g., $\omega_c\tau \gg 1$), so that the relaxation effects in (3.83) can be ignored. Then, in the plane perpendicular to the magnetic field,

$$\frac{dv_x}{dt} = -\frac{eF_x}{m^*} - \omega_c v_y$$

$$\frac{dv_y}{dt} = -\frac{eF_y}{m^*} + \omega_c v_x. \tag{3.171}$$

If the derivative with respect to time of the first of these equations is taken and then the last term is replaced with the second of these equations, we arrive at

$$\frac{d^2v_x}{dt^2} = \frac{\omega_c e}{m^*} F_y - \omega_c^2 v_x. \tag{3.172}$$

For the purposes here, the electric field may be taken as $F = 0$ without any loss of generality. Then the two velocity components are given by

$$v_x = v_0 \cos(\omega_c t + \varphi), \quad v_y = v_0 \sin(\omega_c t + \varphi). \tag{3.173}$$

Here, φ is a reference angle that gives the orientation of the electron at $t = 0$. For the steady-state case of interest here, this quantity is not important and it can be taken as zero without loss of generality. The quantity v_0 is a term that will be equated to the energy, but is the linear velocity of the particle as it describes its orbit around the magnetic flux lines.

The position of the particles is found by integrating (3.172). This gives the result in the simple form

$$x = \frac{v_0}{\omega_c}\sin\omega_c t, \quad y = -\frac{v_0}{\omega_c}\cos\omega_c t. \tag{3.174}$$

These equations describe a circular orbit with the radius

$$r^2 = x^2 + y^2 = \frac{v_0^2}{\omega_c^2}. \tag{3.175}$$

This is the radius of the cyclotron orbit for the electron as it moves around the magnetic field. In principle, this motion is that of a harmonic oscillator in two dimensions, and the motion for this becomes quantized in high magnetic fields when $\hbar\omega_c > k_B T$ and $\omega_c\tau > 1$. We introduce the quantization by writing the energy in terms of the energy levels of a harmonic oscillator as

$$E = \tfrac{1}{2}m^* v_0^2 = (n + \tfrac{1}{2})\hbar\omega_c. \tag{3.176}$$

Thus the size of the orbit is also quantized. In the lowest level, (3.176) gives the radial velocity and this may be inserted into (3.175) to find the *Larmor radius,* more commonly called the ***magnetic length*** l_B,

$$r_L = l_B = \sqrt{\frac{\hbar}{eB}}. \tag{3.177}$$

This is the quantized radius of the harmonic oscillator and is the minimum radius, as the higher-energy states involve a larger energy, which converts to a larger radial velocity and then to a larger radius. As the magnetic field is raised, the radius of the harmonic oscillator orbit is reduced and the radial velocity is increased. In fact, we can define the cyclotron radius at the Fermi surface as

$$r_c = k_F r_L^2 = \frac{\hbar k_F}{eB} = \sqrt{2n_{max} + 1}\, l_B. \tag{3.178}$$

Here, n_{max} is the highest occupied Landau level (that in which the Fermi level resides).

In the following paragraphs, several magneto-transport effects will be discussed. These may be used in many applications to determine various properties of the electron "gas." It should be emphasized, however, that these measurements are *always* made with very small values of the electric field, as the heating from the latter field can destroy much of the quantization used to determine the parameters of interest.

3.4.1 The Shubnikov–de Haas effect

The quantized energy levels described by (3.176) are termed *Landau levels*, after the original work on the quantization carried out by Landau. Transport across the magnetic field (e.g., in the plane of the orbital motion) shows interesting oscillations due to this quantization. Let us consider this motion in the two-dimensional plane to which the magnetic field is normal. In looking at (3.175), one must consider the fact that the electrons will fill up the energy levels to the Fermi energy level. Thus there are in fact several Landau levels occupied, and therefore the electrons exhibit several distinct values of the orbital velocity v_0. In computing the effective radius, it is then necessary to sum over these levels. If there are n_s electrons per square centimeter, this sum can be written as

$$n_s\langle r^2\rangle = \sum_{i=0}^{i_{max}} r_i^2 = \sum_{i=0}^{i_{max}} \frac{(2i+1)\hbar}{eB} = 2\sum_{i=0}^{i_{max}} \frac{(i+1/2)\hbar\omega_c}{m^*\omega_c^2}, \tag{3.179}$$

or

$$\langle r^2\rangle = \sum_{i=0}^{i_{max}} (2i+1)\frac{\hbar}{eBn_s}. \tag{3.180}$$

How are the quantities B, i_{max}, and $\langle r^2\rangle$ related for this equation? The number of levels that are filled depends upon the degeneracy that is formed in each Landau level. This depends upon the magnetic field B, and thus connects directly to the radius r. But all of this depends on the areal density n_s as well.

As the magnetic field is raised, each of the Landau levels rises to a higher energy. However, the Fermi energy remains fixed, so at a critical magnetic field, the highest filled Landau level will cross the Fermi energy. At this point, the electrons in this level must drop into the lower levels, which reduces the number of terms in the sum in (3.180). This can only occur due to the increase of the density of states in each Landau level, so a given Landau level can hold more electrons as the magnetic field is raised. As a consequence, the average radius (obtained from the squared average of the radius) is modulated by the magnetic field, going through a maximum each time a Landau level crosses the Fermi level and is emptied. From (3.180), it appears that the radius is periodic in the inverse of the magnetic field. At least in two dimensions, this periodicity is proportional to the areal density of the free carriers and can be used to measure this density. The effect, commonly called the Shubnikov–de Haas effect, is normally applied by measuring the conductivity in the plane normal to the field. The Landau level must cross the Fermi energy, as mentioned above, so it can be argued that the magnetic field must move the Landau level this far. Thus the amount the reciprocal field must move is just

$$\Delta\left(\frac{1}{B}\right) = \frac{1}{B}\frac{\hbar\omega_c}{E_F} = \frac{e\hbar}{m^* E_F}. \tag{3.181}$$

The Fermi energy must now be related to the carrier density. The value for the density of states in two dimensions is needed, for which the areal density may be found to be

$$n_s = 2\int \frac{d^2\mathbf{k}}{(2\pi)^2} = \int_0^{k_F} k\frac{dk}{\pi} = \frac{k_F^2}{2\pi} = \frac{m^* E_F}{\pi\hbar^2}, \tag{3.182}$$

at low temperatures. The Fermi energy follows from (3.182) and this may be used in (3.181) to find the periodicity for the (spin-degenerate) Shubnikov–de Haas effect as

$$\Delta\left(\frac{1}{B}\right) = \frac{e}{\pi\hbar n_s}. \tag{3.183}$$

We can achieve this same result another way. That is, the separation of the Landau levels is $\hbar\omega_c$, and all the states in the range $E_n \pm \hbar\omega_c/2$ are coalesced into the *nth* Landau level. In cylindrical coordinates, the density of states (with the factor of 2 for spin degeneracy) in two dimensions is obtained from (3.182) as $m^*/\pi\hbar^2$, so the number of carriers in each Landau level is the product of this density of states and $\hbar\omega_c$, or $m^*\omega_c/\pi\hbar = 1/\pi l_B^2$. Thus, the number of filled Landau levels is $\pi l_B^2 n_s$ and this leads to (3.183) for the periodicity in $1/B$. This is for the case for two dimensions, which occurs in quantum wells and inversion layers – semiconductor systems, or metallic systems, in which one dimension is no longer a free dimension. The third dimension can be quantized due to barriers or potential wells or by physical thickness of the material.

At this point the reader should still be somewhat skeptical, as we have really not shown why the conductivity should oscillate, only that the square of the Larmor radius oscillates. In fact, the Fermi energy is not a constant. The reason for this lies in the degeneracy of the Landau level discussed above. Notice that the Landau level spacing remains proportional for all values of the magnetic field. The reason for this is, of course, that the levels are spaced evenly by $\hbar\omega_c$. There are (at this point in the discussion) no states lying in the energy range between the Landau levels. But in the absence of the magnetic field, the density of states is given by $m^*/\pi\hbar^2$, after dropping the differential in energy. The states that lie between the Landau levels are

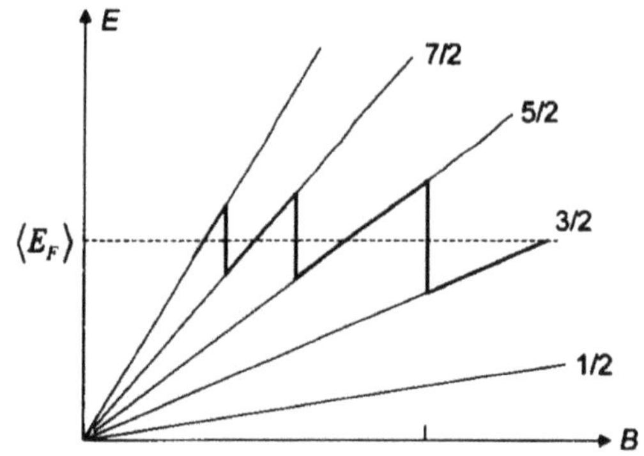

Figure 3.5 The Landau levels "fan out" as the magnetic field is raised. For a fixed areal density, the Fermi level oscillates between these levels (described in the text).

gathered into these levels themselves, and their degeneracy rises accordingly. Thus, as the magnetic field increases, each Landau level increases the number of states that it can hold. If there are n_s electrons per unit area, then at an arbitrary magnetic field, all of these electrons will fill several Landau levels and a few left over will be in the next-highest Landau level. If a small increase in the magnetic field occurs, the number of electrons that can be accommodated in the lower Landau levels increases correspondingly. At a critical value of the magnetic field, all of the electrons can be accommodated in the lower lying levels and the partially filled level above will empty. At this magnetic field, the Fermi energy must *drop from the upper Landau level, which is now empty, to the next-lowest Landau level, which is full.* This is shown in Figure 3.5. At a magnetic field corresponding to the "tick" mark on the axis, all of the electrons are accommodated in the $\frac{3}{2}$ and $\frac{1}{2}$ levels. At a slightly lower value of magnetic field, the transition of the Fermi energy from the $\frac{5}{2}$ level to the $\frac{3}{2}$ level has occurred. The transition is required to be as shown, since the Fermi energy must be in the $\frac{5}{2}$ level if there is a single electron in this level (near absolute zero in temperature where we are working). On the other hand, if we raise the magnetic field sufficiently to add just four more states to the $\frac{1}{2}$ and $\frac{3}{2}$ levels (two to each), then there is a single empty level in the $\frac{3}{2}$ level and the Fermi energy must lie in this level. Thus the Fermi energy oscillates about the average value of the Fermi energy.

From (3.180), n_s may be replaced by the Fermi energy and some fundamental constants. Thus the square of the radius is a sum over terms that vary with the inverse of both the magnetic field and the Fermi energy. When the Fermi energy drops, as in Figure 3.5, the square of the radius increases, but this increase is moderated by the loss of one of the terms in the summation. As the field increases, the Fermi energy also increases and both of these lead to a reduction in the square of the radius. This explains in somewhat greater detail the oscillation in the radius, but why does the conductivity oscillate?

To understand the conductivity oscillations, it is necessary to reintroduce the scattering process. Without this process, the electrons remain in the closed Landau orbits. However, the scattering process can cause the electrons to "hop" slowly from one orbit to the next in real space by randomizing the momentum. This leads to a slow drift of the carriers in the direction of the applied field (we will see below that the edge states are mainly responsible for this). The drift is slower than in the absence of the magnetic field because the tendency is to have the carriers remain in the orbits. Here the scattering induces the motion instead of retarding the motion as in the field-free case. Thus the conductivity is expected to be less in the presence of the magnetic field.

When the Fermi level lies in a Landau level, away from the transition regions, there are many states available for the electron to gain small amounts of energy from the applied field and therefore contribute to the conduction process. On the other hand, when the Fermi level is in the transition phase, the upper Landau levels are empty and the lower Landau levels are full. Thus there are no available states for the electron to be accelerated into, and the conductivity drops to zero in two dimensions. In three dimensions it can be scattered into the direction parallel to the field (the z-direction), and this conductivity provides a positive background on which the oscillations ride.

The problem with the foregoing argument is that the transition region in Figure 3.5 occurs over an infinitesimal range of magnetic field. If the conductivity is zero only over this small range, it would be almost undetectable, and the oscillations would be unobservable. In fact, it is the failure of the crystal to be perfect that creates the regions of low conductivity. In nearly all situations in transport in semiconductors, the role of the impurities and defects is quite small and can be treated by perturbation techniques, as with scattering. However, in situations where the transport is sensitive to the position of various defect levels, this is not the case. The latter is the situation here.

Defects in the crystal, such as impurities, vacancies, interstitial atoms, and so on, lead to the presence of localized levels. These levels lie continuously throughout the energy range available to the electrons. In general, however, these levels are noticed only when they lie in the energy gaps (say, between the conduction and valence band), since the existence of normal itinerant electron states masks these local levels. When the normal continuum of states is broken up with a magnetic field, the localized states become unmasked and can contribute an important effect. In the Shubnikov–de Haas effect, the localized electrons will broaden the transition between Landau levels as the Fermi energy moves through these states. In the argument above, only a few electrons were needed to move the Fermi energy from one Landau level to

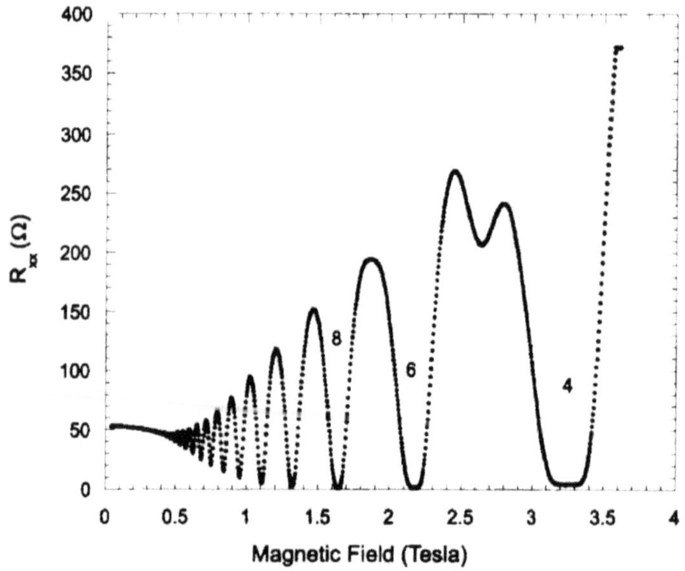

Figure 3.6 The longitudinal resistance for a quasi-two-dimensional electron gas in a high magnetic field. The oscillations are the Shubnikov–de Haas oscillations, and correspond to the sequential emptying of Landau levels. (Data courtesy of D. P. Pivin, Jr., Arizona State University.)

the next-lowest one. Now, however, a sufficiently large increase in magnetic field is also required to empty all the filled, localized levels lying between the two Landau levels.

The end result is that the transition of the Fermi energy between Landau levels is broadened significantly due to the presence of the localized states. Thus, while the Fermi level is passing through the localized levels, the conductivity can drop to zero, since the localized levels also do not contribute to any appreciable conductivity. These levels are essential to the conductivity oscillations but do not contribute to either the periodicity or the conductivity itself. This is an interesting but true enigma of semiconductor physics. Figure 3.6 shows a typical measurement of the longitudinal resistivity, which is related to the conductivity σ_{xx} (σ_{ij} refers to the conductivity measured by a current in the i-th direction in response to an electric field in the j-th direction; symmetry requires that $\sigma_{xx} = \sigma_{yy}$). Now, the "zeros" of the longitudinal resistance (R_{xx} will be described more carefully in the next section) correspond to the magnetic field for which there are full Landau levels. However, the index that is shown (4, 6, 8, etc.) is that for spin-resolved levels, rather than spin-degenerate Landau levels. Hence, for the case of the zero at index 4 ($B = 3.25$ T, termed the $\nu = 4$ level), the $\frac{1}{2}$ and $\frac{3}{2}$ Landau levels (both doubly spin degenerate) are full. The zero corresponds to the transition of the Fermi energy between the $\frac{5}{2}$ and $\frac{3}{2}$ levels in Figure 3.5.

These measurements are for a quasi-two-dimensional electron gas at the interface of a GaAlAs/GaAs heterostructure. From (3.183), we can determine the areal density to be approximately 3.3×10^{11} cm^{-2}.

3.4.2 *The quantum Hall effect*

The zeros of the conductivity that occur when the Fermi energy passes from one Landau level to the next-lowest level are quite enigmatic. They carry some interesting by-products. Consider, for example, the conductivity tensor, which can be written in analogy to (3.83) in the presence of scattering:

$$\bar{\sigma} = \begin{bmatrix} \sigma_{xx} & \sigma_{xy} & 0 \\ -\sigma_{xy} & \sigma_{xx} & 0 \\ 0 & 0 & \sigma_{zz} \end{bmatrix}. \tag{3.184}$$

Here, we have included the possibility of motion along the z-direction, but this term is omitted for the discussion of the two-dimensional system, just as it was omitted in (3.83). Inverting this matrix to find the resistivity matrix, the longitudinal resistivity is given by

$$\rho_{xx} = \frac{\sigma_{xx}}{\sigma_{xx}^2 + \sigma_{xy}^2}. \tag{3.185}$$

In the situation where the longitudinal conductivity σ_{xx} goes to zero, we note that the longitudinal resistivity ρ_{xx} also goes to zero. Thus there is no resistance in the longitudinal direction. Is this a superconductor? No! It must be remembered that the conductivity is also zero, so that there is no allowed motion along that direction. The entire electric field must be perpendicular to the current and there is no dissipation since $\mathbf{E} \cdot \mathbf{J} = 0$, but the material is not a superconductor.

The presence of the localized states, and the transition region for the Fermi energy between one Landau level and the next-lowest level, leads to another remarkable effect. This is the quantum Hall effect, first discovered in silicon metal-oxide-semiconductor (MOS) transistors prepared in a special manner so that the transport properties of the electrons in the inversion layer could be studied (von Klitzing *et al.*, 1980). Klaus von Klitzing was awarded the Nobel Prize for this discovery. The effect leads to quantized resistance, which can be used to provide a much better measurement of the fine structure constant used in quantum field theory (used in quantum relativity studies), and today provides the standard of resistance in the United States, as well as in many other countries.

A full derivation of the quantum Hall effect is well beyond the level at which we are discussing the topic here. However, we can use a consistency argument to illustrate the quantization exactly, as well as to describe the effect we wish to observe. When the Fermi level is in the localized state region, and lies between the Landau levels, the lower Landau levels are completely full. We may then say that

$$E_F \approx \upsilon \hbar \omega_c, \tag{3.186}$$

where υ is an integer giving the number of filled Landau levels. [At first, one might think that (3.186) would place the Fermi energy in the center of a Landau level, but note that no equality is used. The argument desired is to relate the Fermi level to the number of carriers that must be contained in the filled Landau levels. In fact, the magnetic field is usually so high that spin degeneracy is raised and υ measures half-levels rather than full levels. That is, it measures the number of spin-resolved levels, which is why the integers were assigned as they were in Figure 3.6, where the $\upsilon = 5$ level is just beginning to be resolved as a splitting of the $\frac{5}{2}$ Landau level into two spin levels.] Using (3.186) for the Fermi energy, the definition of the cyclotron frequency (n_s here is the electron density per unit area) is given as

$$n_s = \upsilon \frac{eB}{\pi \hbar e}. \tag{3.187}$$

However, the density is constant in the material, so using (3.187) in (3.87) for the Hall resistance gives

$$R_H = \frac{B}{ne} = \frac{h}{\upsilon e^2}. \tag{3.188}$$

The quantity $h/e^2 = 25.81$ kilohms is a ratio of fundamental constants. Thus the conductance (reciprocal of the resistance) increases stepwise as the Fermi level passes from one Landau level to the next-lowest level. Between the Landau levels, when the Fermi energy is in the localized state region, the Hall resistance is constant (to better than 1 part in 10^7) at the quantized value given by (3.188) since the lower Landau levels are completely full. In Figure 3.7, the variation of the Hall resistance as a function of magnetic field for a typical sample is shown. Also shown is the conductance which steps downward as the magnetic field is increased. These measurements were made in the same sample and geometry of Figure 3.6, so that they can be easily compared.

The magnetic field could, of course, be swept to higher values in both Figures 3.6 and 3.7. When this is done, new features appear, and these are not explained by the above theory. In fact, in high quality samples, once the Fermi energy is in the lowest Landau level, one begins to see fractional filling

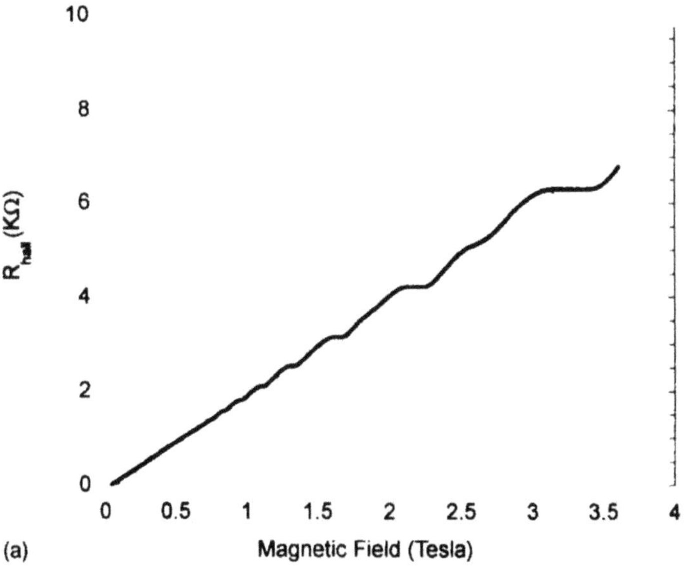

(a)

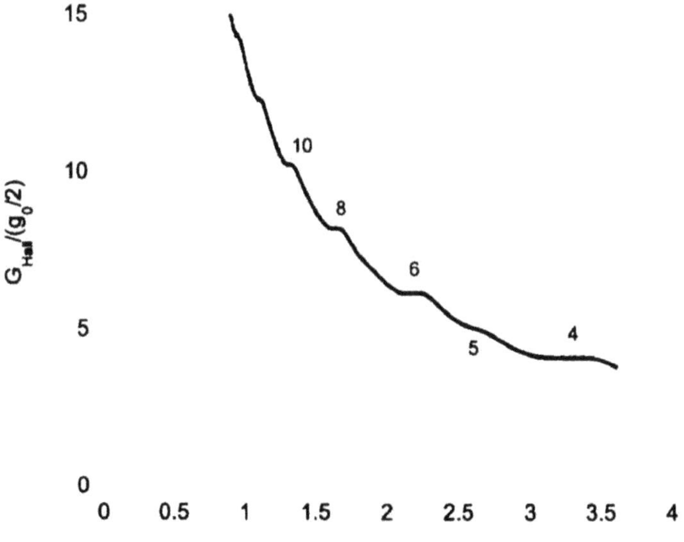

(b)

Figure 3.7 (a) The Hall resistance, and (b) the Hall conductance (normalized to the value of $g_0/2 = e^2/h$). The normalization is chosen so that the index value is of spin-resolved Landau levels. The first spin splitting is just being resolved at $\nu = 5$ (~2.6 T). (Data courtesy of D. P. Pivin, Jr., Arizona State University.)

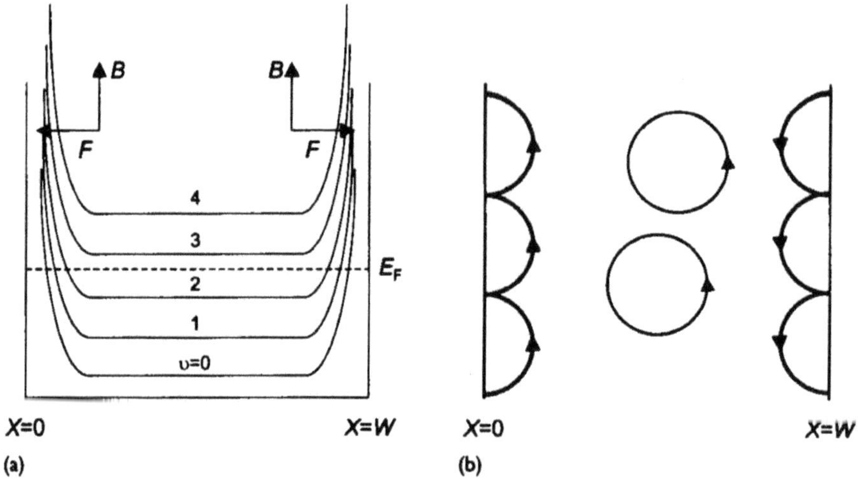

Figure 3.8 (a) The bending of the Landau levels at the edges of a confined sample. The electric field is shown for comparison at the two edges. (b) The confined and bouncing orbits for the situation of (a). The magnetic field is out of the page.

and plateaux, in which the resistance is a fraction of h/e^2 (Tsui *et al.*, 1982). This ***fractional quantum Hall effect*** is theorized to arise from the condensation of the interacting electron system into a new many-body state characteristic of an incompressible fluid (Laughlin, 1983). Tsui, Störmer, and Laughlin shared the Nobel Prize for this discovery. However, the properties of this many-body ground state are clearly beyond the present level, and we leave this topic to discuss more properties of the quantum Hall effect itself.

3.4.3 Edge states and the Landauer–Büttiker formula

The physical picture associated with the states in a high magnetic field becomes more complicated when we consider the lateral confinement of most Hall bar structures. Then, the Landau levels are homogeneous and evenly spaced in the center of the bar, but these energy levels are pushed upwards by the presence of the lateral confinement. This may be seen in Figure 3.8(a). The lowest three (spin resolved) Landau levels are occupied in the center of the structure. Near the edges, however, the *depletion* due to the surface potential causes these levels to be depopulated, which is indicated by the rising property of the levels. We note that the levels actually now cross the Fermi energy, so that they are empty above this energy level. In Figure 3.8(b), we indicate the cyclotron orbits for one of the Landau levels (since each Landau level has a different cyclotron radius, its size will be different). In the center of the structure, these orbits close on themselves, so that these electrons do not contribute to the conductance (the Fermi energy is in the localized state regime between two Landau levels, hence the longitudinal conductance is zero). On the other hand, the orbits near the edge actually strike the potential barrier and are reflected which produces a net motion along the side of the structure. These trajectories are called *edge states.* We can determine just which direction they move from consideration of the Lorentz force (we note that the curves in Figure 3.7(a) are energy curves and not potential curves, which affects the sign of the electric field obtained from the gradient of these curves)

$$\mathbf{F}_{Lorentz} = -e\mathbf{F} \times \mathbf{B} = -e(\nabla V(r)) \times \mathbf{B}. \tag{3.189}$$

Thus, on the left-hand side of the figure ($x \sim 0$), the electric field points in the $-x$ direction, and the net force is into the page (y direction for $\mathbf{B} = \mathrm{B}\mathbf{a}_z$). At the other side, the electric field is in the $+x$ direction, so that the force is out of the page ($-y$ direction). This is in keeping with the rotation directions shown in Figure 3.8(b), where the z direction is out of the page.

We can evaluate the conductance and resistance of a quantum Hall bar through the generalization of the Landauer formula (3.39) to a multi-terminal situation. Here, we assume that the current is fed into (out of) the Hall bar at particular terminals, and the voltage is measured at another set of terminals. Current injecting contacts are ones which are assumed to equilibrate the carriers with the "reservoirs," or contact properties in this case. On the other hand, voltage measuring terminals are assumed to be non-intrusive in that they do not equilibrate the carriers. In this sense, the injected carriers from the contacts (metallic behavior in most cases) quickly form into the edge states in a high magnetic field, while these are destroyed when the carriers re-enter the equilibrating current contacts. The voltage terminals, however, do not affect the population of the edge states (at least, in theory).

We begin first with a two-terminal configuration, for which (3.39) was obtained. It can be assumed, without loss of generality, that carriers are fed into all available modes at the injecting contact. So all modes up to the Fermi energy of the source μ_1 are filled. Similarly, the modes at the drain are filled up to the Fermi energy at the drain μ_2. The applied voltage across the device is $(\mu_1 - \mu_2)/e$. For a particular mode j on the source side, the current injected into the channel i on the drain side is independent of the velocity and density of states, according to (3.39), and is given by

$$I_{ij} = \frac{e}{h} T_{ij}(\mu_1 - \mu_2). \tag{3.190}$$

We note that spin degeneracy has *not* been assumed here, as we are going to discuss the spin-resolved Landau levels below. Hence, the factor of 2 that appears in (3.39), accounting for spin degeneracy has been removed. Since each channel is assumed to be fed equally, the total current due to charge injected into mode i on the drain end is now given by

$$I_i = \frac{e}{h}\left[\sum_{j=1}^{N} T_{ij}\right](\mu_1 - \mu_2) = \frac{e}{h} T_i (\mu_1 - \mu_2), \tag{3.191}$$

where now T_i is the short-hand notation for the sum over the input modes j of the source contact. Since all channels are independent, the total current is given by

$$\begin{aligned} I = \sum_{i=1}^{N} I_i &= \frac{e}{h}(\mu_1 - \mu_2)T = \frac{e}{h}(\mu_1 - \mu_2)\sum_{i=1}^{N} T_i \\ &= \frac{e}{h}(\mu_1 - \mu_2)Tr(tt^+), \end{aligned} \tag{3.193}$$

where, on the right-hand side of this latter equation, the total transmission is written in terms of the complex transmission sub-matrices

$$\sum_{i=1}^{N} T_i = \sum_{i,j=1}^{N} T_{ij} = \sum_{i,j=1}^{N} t_{ij} t_{ji}^{*} = Tr(tt^+). \tag{3.194}$$

We recall that the actual transmission probability in quantum mechanics is the square of the magnitude of the actual wave function transmission (Ferry, 1995), the latter of which is represented in the matrices t. Similarly, the current into the source contact may also be expressed in terms of the reflection coefficients, as

$$I_i = \frac{e}{h}(\mu_1 - \mu_2)\left[1 - \sum_{j=1}^{N} R_{ij}\right] = \frac{e}{h}(\mu_1 - \mu_2)(1 - R_i). \tag{3.195}$$

Just as in (3.193), the total current may now be written as

$$\begin{aligned} I = \sum_{i=1}^{N} I_i &= \frac{e}{h}(\mu_1 - \mu_2)(N - R) = \frac{e}{h}(\mu_1 - \mu_2)\left[\sum_{i=1}^{N}(1 - R_i)\right] \\ &= \frac{e}{h}(\mu_1 - \mu_2)\left[N - \sum_{i=1}^{N} R_i\right]. \end{aligned} \tag{3.196}$$

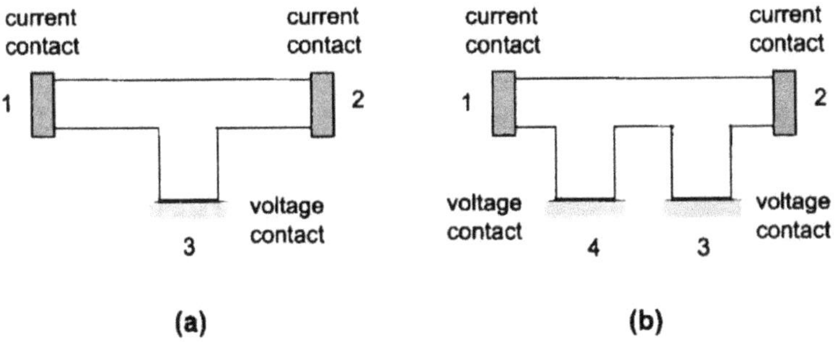

Figure 3.9 Conceptual structure for a (a) 3 terminal and (b) 4 terminal sample, in order to discuss the Landauer–Büttiker formula.

If we compare (3.193) and this equation, we arrive at an important current continuity equation applicable to the system, which is

$$\sum_{i=1}^{N} T_i = N - \sum_{i=1}^{N} R_i. \tag{3.197}$$

It has been assumed that there are equal numbers of modes in each of the two contacts. This is not required, but it is usually assumed.

Consider now the three terminal structure depicted in Figure 3.9(a). *Note that, for the moment, the high magnetic field limit will not be invoked.* Here, we consider that terminal 3 is an ideal voltage probe through which no current flows ($I_3 = 0$). We consider first the resistance $R_{12,13}$, where the first two subscripts describe the current contacts and the second two subscripts denote the contacts between which the voltage will be measured. Here, we will denote T^{ij} as the total transmission from contact j to contact i. This has been summed over the number of modes. A similar notation will be used for the reflection coefficient. The current into a lead is composed of the net transmission from that terminal to all others minus the transmission from the others to that terminal, or

$$I_{i,net} \equiv I_i = \sum_{j=2,3} T^{ji}\mu_i - \sum_{j=2,3} T^{ij}\mu_j. \tag{3.198}$$

Using the continuity of current condition (3.197), this may be rewritten as (Büttiker, 1986)

$$I_i = \frac{e}{h}\left[(N - R^{ii})\mu_i - \sum_{j\neq i} T^{ij}\mu_j\right]. \tag{3.199}$$

We can now use this relationship to write the condition of zero current for terminal 3 as

$$0 = \frac{e}{h}[(N - R^{33})\mu_3 - T^{31}\mu_1 - T^{32}\mu_2]. \tag{3.200}$$

Using the current continuity, we can write $N - R^{33} = T^{31} + T^{32}$, which means that this can be used to determine the Fermi energy at terminal 3 as

$$\mu_3 = \frac{T^{31}\mu_1 + T^{32}\mu_2}{T^{31} + T^{32}}. \tag{3.201}$$

The current flowing into terminal 1 is $I = I_1 = -I_2$. Using (3.199) at terminal 2 leads to

$$\begin{aligned} -I &= \frac{e}{h}[(N - R^{22})\mu_2 - T^{23}\mu_3 - T^{21}\mu_1] \\ &= \frac{e}{h}\left[(T^{21} + T^{23})\left(\frac{T^{31} + T^{32}}{T^{32}}\mu_3 - \frac{T^{31}}{T^{32}}\mu_1\right) - T^{23}\mu_3 - T^{21}\mu_1\right] \\ &= \frac{e}{h}\left[\left(\frac{(T^{21} + T^{23})(T^{31} + T^{32})}{T^{32}} - T^{23}\right)\mu_3\right. \\ &\quad \left. - \left(T^{21} + T^{31}\frac{(T^{21} + T^{23})}{T^{32}}\right)\mu_1\right]. \end{aligned} \tag{3.202}$$

Now, consider the quantity obtained in the numerator of the first parentheses

$$D = T^{21}T^{31} + T^{21}T^{32} + T^{23}T^{31}, \tag{3.203}$$

which is the same as the numerator of the second parentheses. This quantity is symmetric in the magnetic field, since in general the individual transmission must satisfy the asymmetry requirements of (3.184), or $T^{ij}(B) = T^{ji}(-B)$. Hence, D is symmetric in the field, since it is generally the product of two reversal terms. Equation (3.202) may now be rewritten as

$$-I = \frac{e}{h}\frac{D}{T^{32}}(\mu_3 - \mu_1), \tag{3.204}$$

or

$$R_{12,13} = \frac{\mu_1 - \mu_3}{eI} = \frac{h}{e}\left(\frac{T^{32}}{D}\right). \tag{3.205}$$

By an analogous procedure, it may easily be shown that

$$R_{12,32} = \frac{\mu_3 - \mu_2}{eI} = \frac{h}{e}\left(\frac{T^{31}}{D}\right). \tag{3.206}$$

Finally, it may also be shown easily that

$$R_{12,12} = R_{12,13} + R_{12,32} = \frac{\mu_1 - \mu_2}{eI} = \frac{h}{e}\left(\frac{T^{32} + T^{31}}{D}\right). \tag{3.207}$$

Hence, we have the simple addition of resistances.

As the next example, let us consider the four-terminal structure that is shown in Figure 3.9(b). From this example, we will be able to discuss the four-terminal vs. two-terminal versions of the Landauer formula, exhibiting the importance of *contact* resistance. In this example, we will assume that the current passes from terminal 1 to terminal 2, so that $I_1 = -I_2 = I$. The two probes 3 and 4 will be used merely to sample the voltage in a manner that we seek $R_{12,34}$. These two probes do not pass current, so that we can write the current continuity equations for them as

$$\begin{aligned} I_3 = 0 &= \frac{e}{h}[(N - R^{33})\mu_3 - T^{31}\mu_1 - T^{32}\mu_2 - T^{34}\mu_4] \\ I_4 = 0 &= \frac{e}{h}[(N - R^{44})\mu_4 - T^{41}\mu_1 - T^{42}\mu_2 - T^{43}\mu_3]. \end{aligned} \tag{3.208}$$

Rewriting these two equations allows us to solve for the potentials of the other two leads as

$$\begin{aligned} \mu_1 &= \frac{(N - R^{33})T^{42} + T^{43}T^{32}}{T^{31}T^{42} - T^{41}T^{32}}\mu_3 - \frac{(N - R^{44})T^{32} + T^{34}T^{42}}{T^{31}T^{42} - T^{41}T^{32}}\mu_4 \\ \mu_2 &= \frac{(N - R^{33})T^{41} + T^{43}T^{31}}{T^{32}T^{41} - T^{31}T^{42}}\mu_3 - \frac{(N - R^{44})T^{31} + T^{34}T^{41}}{T^{32}T^{41} - T^{31}T^{42}}\mu_4. \end{aligned} \tag{3.209}$$

We note that the two denominators are the same except for a sign change between the two potentials on the left. We now can write the current equation for terminal 2 as

$$I_2 = -I = [(N - R^{22})\mu_2 - T^{21}\mu_1 - T^{23}\mu_3 - T^{24}\mu_4]$$

$$= \frac{e}{h}\left[(N - R^{22})\frac{(N - R^{33})T^{41} + T^{43}T^{31}}{T^{32}T^{41} - T^{42}T^{31}}\right.$$

$$\left. + T^{21}\frac{(N - R^{33})T^{42} + T^{43}T^{32}}{T^{32}T^{41} - T^{42}T^{31}} - T^{23}\right]\mu_3$$

$$-\frac{e}{h}\left[(N - R^{22})\frac{(N - R^{44})T^{31} + T^{34}T^{41}}{T^{32}T^{41} - T^{42}T^{31}}\right.$$

$$\left. + T^{21}\frac{(N - R^{44})T^{32} + T^{34}T^{42}}{T^{32}T^{41} - T^{42}T^{31}} + T^{24}\right]\mu_4. \tag{3.210}$$

While somewhat tedious, it can be shown that the two bracketed terms are the same, so that we may write the conductance as

$$I = \frac{e}{h}\left(\frac{D_4}{T^{32}T^{41} - T^{42}T^{31}}\right)(\mu_4 - \mu_3)$$

$$= \frac{e^2}{h}\left(\frac{D_4}{T^{32}T^{41} - T^{42}T^{31}}\right)V_{43}, \tag{3.211}$$

where V_{43} is the voltage $V_4 - V_3$. The quantity D_4 is the common numerator of the bracketed terms in (3.210), which is different from that found earlier. Büttiker (1986) has shown that

$$R_{ij,kl}(B) = R_{kl,ij}(-B), \tag{3.212}$$

so that the current and voltage leads must be interchanged upon reversal of the magnetic field if the symmetry is to be seen. This is just an application of the *reciprocity theorem*.

The dominant terms in the factor D_4 can be considered to be the transmission from terminal 1 to 2, and the back-scattering from terminals 3 and 4 to the current leads, or

$$D_4 \approx T^{12}(T^{31} + T^{32})(T^{41} + T^{42}) - T^{12}(N - R^{33})(N - R^{44}), \tag{3.213}$$

We may assume that the voltage probes are weakly coupled to the actual device by an attenuation factor ε, and we ignore the magnetic field, so that $T^{12} = T^{21} = T$ to lowest order in this factor ε. On either side of the scattering region (between the probes), it is argued that

$$T^{13} = T^{31} = T^{42} = T^{24} = \varepsilon(1 + R), \tag{3.214}$$

whereas for carriers injected to or from one of the voltage probes through the scattering region, the transmission probabilities are proportional to T, as

$$T^{32} = T^{23} = T^{14} = T^{41} = \varepsilon T. \tag{3.215}$$

Finally, those carriers moving between the two voltage leads are doubly attenuated, and

$$T^{32} = T^{23} = \varepsilon^2 T. \tag{3.216}$$

We may now rewrite (3.211) as

$$I = \frac{2e^2}{h}\left(\frac{T[\varepsilon(1+R)+\varepsilon T]^2}{\varepsilon^2(1+R)^2 - \varepsilon^2(1-R)^2}\right)V_{43} = 2\frac{e^2}{h}\frac{T}{R}V_{43}, \tag{3.217}$$

where use has been made of $T + R = 1$, and the spin degeneracy has been re-inserted (no magnetic field). The conductance in this four-terminal measurement clearly now depends upon the reflection as well as the transmission. In the preceding two-terminal measurement, there was no factor of R in the denominator. We can understand this by writing the resistance as

$$R_4 = \left(\frac{h}{2e^2}\right)\frac{1-T}{T} = R_2 - R_{contact}, \tag{3.218}$$

where the quantum contact resistance is

$$R_{contact} = \frac{h}{2e^2}. \tag{3.219}$$

However, one must be careful about the interpretation of this quantity, as it is not merely a series resistance. If this were the case, then the total resistance would be always larger than this contact resistance of 12.91 kΩ. This would, in turn, limit the conductance available, and this is not what occurs. In fact, this denominator factor is a reflection of the change in the distribution function in the contact region, which in turn modifies the current distribution.

Finally, we invoke the five terminal configuration shown in Figure 3.10. To treat this, we return to the high magnetic field limit, where *the current is carried entirely by the edge states*. With reference to the figure, terminals 1 and 2 are taken to be the current injecting terminals, while terminals 3–5 are voltage measuring terminals. With our understanding of the edge states, only T^{51}, T^{25}, T^{32}, T^{43}, and T^{14} are non-zero; all other T^{ij} are zero! Invoking the current continuity relationship at terminal 5 leads to

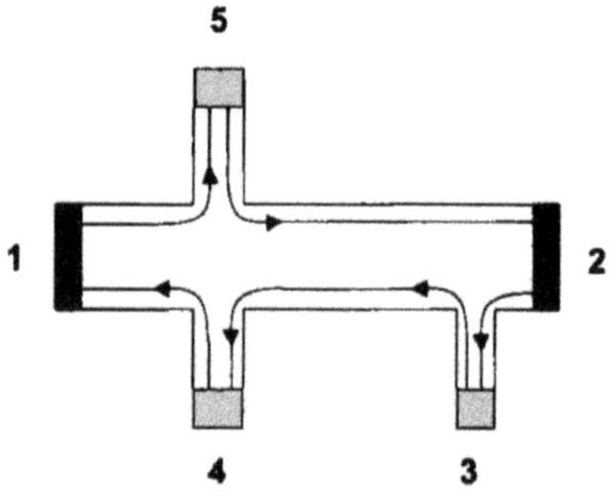

Figure 3.10 A five terminal Hall bar in the edge state regime. The edge states circulate as shown for a magnetic field out of the page.

$$0 = T^{25}\mu_5 - T^{51}\mu_1, \quad T^{51} = T^{25} = N, \tag{3.220}$$

or $\mu_5 = \mu_1$. Similar considerations lead to $\mu_3 = \mu_4 = \mu_2$, and to $T^{32} = T^{43} = T^{14} = N$. Hence, it is immediately clear that

$$R_{12,34} = R_{xx} = 0, \tag{3.221}$$

as required for the Shubnikov–de Haas effect described above. On the other hand, an additional invoking of the current continuity equation at terminal 5 and at terminal 4 leads to

$$T^{51} = T^{32} = N, \tag{3.222}$$

since there is no reflection of the incoming (no incoming) currents. Hence, we may write the Hall resistance as

$$R_{12,45} = \frac{I_1}{-e(\mu_4 - \mu_5)} = -\frac{h}{e^2}\frac{N\mu_1 - T^{14}\mu_4}{(\mu_4 - \mu_5)} = \frac{Nh}{e^2}, \tag{3.223}$$

where the above relationships have been used. The result is the quantized Hall resistance. Similarly, the two-terminal resistance between the two current contacts is

$$R_{12,21} = \frac{I_1}{-e(\mu_2 - \mu_1)} = -\frac{h}{e^2}\frac{N\mu_1 - T^{14}\mu_4}{(\mu_2 - \mu_1)} = \frac{Nh}{e^2}. \tag{3.224}$$

The two-terminal resistance is just the Hall resistance. That is, the entire voltage drop is transverse to the sample everywhere except for a small region adjacent to the current carrying contacts. This means that the entire resistance is just the contact resistance required to inject the carriers into the edge states from the bulk-like "metallic" contacts. One generally assumes that this resistance is equally divided between the two current injecting contacts. From this formulation, it is absolutely clear that the real quantum Hall effect is a contact and edge effect, rather than a bulk effect. When one is off the plateaux in the Hall resistance, the Fermi level is in the Landau level and no edge states play much of a role, as the bulk behavior washes out the quantization in the resistance, longitudinal resistance appears, and the simple behavior is no longer evident.

PROBLEMS

1. The Fermi–Dirac distribution function is defined by

$$f(E) = \frac{1}{1 + \exp\left(\dfrac{E - E_F}{k_B T}\right)}.$$

   Show that the width of the transition is about $3.5k_BT$. (Differentiate this function and then compute the full width at half-maximum of the resulting function. Use the magnitude of the resulting function, as it is negative.)
2. In the text, the Einstein relation is derived for non-degenerate statistics. Repeat the derivation for the case of degenerate statistics using the Fermi integral representations of the carrier density, and show that the result is modified by a ratio of Fermi integrals.
3. Using the temperature-dependent mobility found for silicon, plot both the mobility and the diffusion constant as a function of temperature in the range 50 to 300 K for both electrons and holes.
4. From the mobility data for GaAs, compute the average electron momentum relaxation time as a function of the impurity density.
5. For an electron with $m^* = 0.5\ m_0$, and $\mu = 10^3$ cm^2/Vs, calculate the mean free path at 300 K. For an electric field of 100 V/cm, find the drift length of the electrons between collisions and the mean free time between collisions.
6. The mobility of holes in germanium is about 2000 cm^2/Vs. If the hole masses are 0.3 and 0.04 (in terms of m_0) for the heavy and light holes, respectively, calculate the ratio of the Hall mobility to the drift mobility if (a) the relaxation times are constant, and (b) the mean free paths are constant. Assume that the scattering times are equal for both types of holes and that the densities are related through the density of states functions for the two bands.

7. What is the relative number of electrons and holes in silicon when the Hall voltage disappears in a sample?
8. In many-valley semiconductors, such as germanium, where the valleys are not perpendicular to each other, the simple approach cannot be used, and a tensor approach is necessary, as has been indicated in the text. Consider the case for ellipsoids along the (111) directions; write the effective mass tensor for each valley separately in terms of the direction cosines with the electric field and magnetic field. Find the Hall and magnetoresistance coefficients for various crystal directions of F and B.
9. A semiconductor has the following properties: the energy gap is determined to be 0.16 eV at room temperature. The Hall coefficient is measured as 150 cm^3/C at both 77 K and 120 K. The resistivity at 77 K is 0.03 Ω–cm, and at 120 K it is 0.06 Ω–cm. The thermoelectric power at 120 K is measured to be 400 μV/K. The low-field Hall coefficient vanishes at 250 K. The electron effective mass has been measured to be 0.015 m_0. Determine the hole effective mass, the density of acceptor atoms, the electron and hole mobilities at 300 K, the resistivity at 300 K, and the thermoelectric power at 300 K.
10. Assume that you are given the mobility ratio $b = 10$. Determine a relationship that will yield the value of the acceptor concentration at which the Hall constant is zero at a given temperature.
11. Using only ionized impurity scattering and acoustic deformation potential scattering, so that the average relaxation time can be easily computed using (3.62) and (3.110), analyze the data of W. W. Tyler and H. H. Woodbury, *Phys. Rev.* 102, 647 (1956), for n-type germanium. Treat the impurity concentration and the deformation potential as adjustable parameters for each sample and tabulate the results (you should have a single value for each of these two numbers for each sample and the deformation potential should not vary from sample to sample).
12. A semiconductor sample measures 1 cm by 0.5 cm and is 0.1 cm thick. For an applied electric field of 1 V/cm, 5 mA of current flows. If a 0.5 T magnetic field is applied normal to the broad surface, a Hall voltage of 5 mV is developed. Determine the Hall mobility and the carrier density.
13. Consider a free-electron "gas" with an areal density of 2×10^{12} cm^{-2}. What is the periodicity (in units of $1/B$) expected for the Shubnikov–de Haas oscillation?

REFERENCES

Abrahams, E., Anderson, P. W., Licciardello, D. C., and Ramakrishnan, T. V., 1979, *Phys. Rev. Lett.*, 42, 673.

Anderson, P. W., Thouless, D. J., Abrahams, E., and Fisher, D. S., 1980, *Phys. Rev. B*, 22, 3519.

Bogoliubov, N. N., 1946, *J. Phys. Soviet. U.*, 10, 256.
Born, M., and Green, H. S., 1946, *Proc. Roy. Soc. (London)*, A188, 10.
Brooks, H., 1955, *Adv. Electr. Electron Phys.*, 8, 85.
Büttiker, M., 1986, *Phys. Rev. Lett.*, 57, 1761.
Conwell, E. M., and Weisskopf, V., 1950, *Phys. Rev.*, 77, 388.
Feenstra, R. M., 1994, *Phys. Rev. Lett.*, 72, 2749.
Ferry, D. K., 1978, *Phys. Rev. B*, 17, 912.
Ferry, D. K., 1991, *Semiconductors* (New York: Macmillan) pp. 179–185.
Ferry, D. K., 1995, *Quantum Mechanics* (Bristol: Institute of Physics Publishing).
Ferry, D. K., and Goodnick, S. M., 1997, *Transport in Nanostructures* (Cambridge: Cambridge University Press).
Goodnick, S. M., Ferry, D. K., Wilmsen, C. W., Lilienthal, Z., Fathy, D., and Krivanek, O. L., 1985, *Phys. Rev. B*, 32, 8171.
Harrison, J. W., and Hauser, J. R., 1970, *Phys. Rev. B*, 1, 3351.
Herring, C., 1955, *Bell. Sys. Tech. J.*, 34, 237.
Hutson, A. R., 1961, *J. Appl. Phys.*, 32, 2287.
Kirkwood, J. G., 1946, *J. Chem. Phys.*, 14, 180.
Kroemer, H., 1975, *Crit. Rev. Sol. State Sci.*, 5, 555.
Kubo, R., 1957, *J. Phys. Soc. Jpn.*, 12, 570.
Kubo, R., 1974, Response, Relaxation, and Fluctuation. In *Transport Phenomena*, edited by Ehlers, J., Hepp, K., and Weidenmüller, H. A. (Berlin: Springer-Verlag) pp. 75–125.
Landauer, R., 1957, *IBM J. Res. Develop.*, 1, 233.
Landauer, R., 1970, *Philos. Mag.*, 21, 863.
Laughlin, R. B., 1983, *Phys. Rev. Lett.*, 50, 1395.
Makowski, L., and Glicksman, M., 1976, *J. Chem. Phys. Sol.*, 34, 487.
Matsumoto, Y., and Uemura, Y., 1974, *Jpn. J. Appl. Phys., Suppl.*, 2, 367.
Ridley, B. K., 1982, *Quantum Processes in Semiconductors* (Oxford: The Clarendon Press).
Shockley, W., and Bardeen, J., 1950, *Phys. Rev.*, 77, 407; 80, 72.
Smith, R. A., 1959, *Semiconductors* (Cambridge: Cambridge University Press).
Stern, F., and Howard, W. E., 1967, *Phys. Rev.*, 163, 816.
Takagi, S., Toriumi, A., Iwase, M., and Tango, H., 1994, *IEEE Trans. Electron Dev.*, 41, 2357.
Tsui, D., Störmer, H. L., and Gossard, A. C., 1982, *Phys. Rev. Lett.*, 48, 1559.
Vasileska, D., 1995, dissertation, Arizona State University, unpublished.
von Klitzing, K., Dorda, G., and Pepper, M., 1980, *Phys. Rev. Lett.*, 45, 494.
Yoshinobu, T., Iwamoto, A., and Iwasaki, H., 1993, in Proc. *3rd Intern. Conf. Sol. State Dev. Mater.* (Makuhari, Japan).
Yvon, J., 1937, *Act. Sci. Ind.*, 542, 543 (Paris: Herman).
Zook, J. D., 1964, *Phys. Rev.*, A136, 869.

Chapter 4

Inelastic scattering and non-equilibrium transport

Essentially all theoretical treatments of electron and hole transport in semiconductors are based upon the one-electron transport equation, usually the Boltzmann transport equation. This was illustrated in the previous chapter for near equilibrium transport, although the many-electron, kinetic picture was introduced via the Langevin equation as well. The use of the Boltzmann equation is especially prevalent in the case of high-electric-field transport, where the transport becomes quite non-equilibrium. The overriding theoretical problem in such transport is that of obtaining the solution of this transport equation to ascertain the form of the distribution function in the presence of the high field.

Even in the previous chapter, it was first assumed that the distribution function was quite near its thermal equilibrium form, and a simple modification was made to introduce the relaxation-time approximation. For more complicated scattering processes, in particular the inelastic phonon scattering processes, even this simple approach fails and a more general solution must be found. The general approaches to solving the transport equation in either this inelastic case, or in the high field case, are closely intertwined, and differ only in the level of approximation.

For transport purposes, the distribution function is not an end in itself, since integrals over the distribution function must be performed in order to evaluate the transport coefficients. It turns out in many cases, especially in numerical approaches, the appropriate averages can be computed more easily than the direct computation of the distribution function and subsequent integration for the transport average. This is true in the ensemble Monte Carlo technique, which will be introduced in the latter parts of this chapter, since the transport averages are computed from averages over an ensemble of semi-classical carriers, whose individual trajectories are followed in the numerical simulation.

In the case of low fields, the transport is linear; that is, the current is a linear function of the electric field, with a constant conductivity independent of the magnitude of the applied field. In general, the transport of hot carriers is nonlinear, in that the conductivity is itself a function of the applied electric field. The relationship between the velocity and field is expressed by a mobility, which depends on the average energy of the carriers. For high fields, the latter quantity is a function of the electric field. In normal linear response theory, a linear conductivity is found by a small deviation from the equilibrium distribution function. This small deviation is linear in the electric field, and the equilibrium distribution function dominates the transport properties. Once the carriers begin gaining significant energy from the field, this is no longer the case. The dominant factor in the actual nonlinear transport does not arise directly from higher-order terms in the field, but rather from the implicit field dependence of the non-equilibrium distribution function, such as that of the electron temperature. Thus it is critical to ascertain this non-equilibrium distribution function correctly, because it is the spreading of this function (to higher average energy) in response to the field that dominates nonlinear response in semiconductors.

In this chapter, the more complicated solutions are obtained for these two cases. It must be remembered, though, that the physics is a natural extension of that introduced in the previous chapter. In the next section, we begin by illustrating the two most obvious properties of high electric fields: velocity saturation and velocity overshoot, first discussed in Chapter 1. Here, we will give a more general introduction to these effects, and to the physical processes that lead to their occurrence. Then, a simple descriptive example of solving the Boltzmann equation will be presented to introduce the ideas inherent in the evolving non-equilibrium distribution function. We then turn to the inelastic phonon scattering processes, as their more detailed forms will be necessary for the following section, which begins to discuss real solutions of the Boltzmann equation. A return to the kinetic picture is then introduced to illustrate how the *retarded* Langevin equation can give rise to a number of properties of hot carriers. Finally, the detailed ensemble Monte Carlo technique is introduced.

4.1 Physical observables

Early studies of high-electric-field transport in solids focused mainly on the breakdown studies of dielectrics (Fröhlich and Seitz, 1950). As a result, these studies were carried out at very high fields and generally resulted in the destructive breakdown of the sample. With the studies of Shockley (Shockley, 1951), emphasis shifted to the study of transport in semiconductors, especially as these materials were becoming useful at that time for new electron devices. The earliest studies of semiconductors focused on the dependence of the drift velocity on the applied electric field and the decrease of mobility at high fields, which led to the concept of velocity saturation. This remains one of the ways of evaluating the high-field behavior of semiconductors, since it directly evaluates the effectiveness of the electron–phonon interaction. In the early 1960s, attention was focused on the observation of negative differential conductivity (NDC) in GaAs (Gunn, 1963) (and subsequently in several other semiconductors), and this led to renewed interest in the field for the applicability of this emission to microwave devices. When individual semiconductor devices began to be fabricated with gate lengths in the sub-micron regime, velocity saturation was found to be important in the operation of these devices, as predicted earlier (Grosvalet *et al.*, 1963; Trofimenkoff, 1965). With the possibility of still shorter gate lengths, focus in recent years has shifted to the role of non-stationary transport and velocity overshoot. These effects provide a valuable overview of the general behavior that is associated with carrier dynamics in the far-from-equilibrium situation at high electric fields (Conwell, 1967).

4.1.1 Velocity saturation

As the magnitude of the electric field applied to a semiconductor is raised, the carriers begin to gain energy from the field. To balance this energy gain, there is an increase in the net rate of energy loss to the lattice via the emission of phonons. Due to the increased rate of phonon emission, the actual rate of scattering of the carriers by the phonons increases and the mobility is reduced. As a result, the velocity increases sub-linearly with the electric field. At very high electric fields, the velocity almost saturates, continuing to increase only very slightly with further increases in the electric field. A typical curve for Si was shown previously in Figure 1.2(a).

The saturated (or nearly saturated) velocity is an important parameter for electron device considerations. In many studies concerning the appropriate figures of merit for high-speed, high-frequency, or high-power devices, it is readily apparent that this parameter affects the frequency response (and hence the speed, through the transit time) and the power-handling capabilities (through the peak current). Thus, it is an important parameter for the study of high-field effects, not only for the reasons stated above, but also because it is a direct mirror on the electron–phonon interaction. Indeed, the saturated velocity is a characterization of the far-from-equilibrium carrier system, which is dependent on the lattice interactions in governing the form of the non-equilibrium distribution function. The non-equilibrium distribution deforms to fit the nature of the electron–lattice interactions rather than maintaining the equilibrium Maxwellian form. In fact, the actual non-equilibrium distribution function takes a shape that balances the driving forces and the dissipative (scattering) forces. An additional scattering process is carrier–carrier scattering, which works to return the distribution function to a quasi-Maxwellian form – a Maxwellian form, but with an enhanced carrier temperature, and perhaps with other parameters as well.

It turns out that the saturation velocity is almost scalable from material to material, precisely because it is a property of the carrier–lattice interaction. The rate of energy loss can be estimated to be approximately

$$ev_s F \sim \tfrac{3}{2}\hbar\omega_0(e^x - 1)N_q, \tag{4.1}$$

where v_s is the saturation velocity, ω_0 is the phonon frequency, $x = \hbar\omega_0/k_B T$, and N_q is the phonon distribution function. The factor in parentheses represents the *difference* between emission and absorption, so the right-hand side of (4.1) is proportional to the net energy lost to phonons, while the left-hand side is the energy gained from the field. In a fashion similar to (4.1), the momentum relaxation rate can be characterized by

$$eF \sim 2m^* v_s(e^x + 1)N_q, \tag{4.2}$$

which is now a *sum* of emission and absorption terms, since both cause a relaxation of the drift momentum. The factor of 2 is a numerical factor, which is correct only for non-polar optical phonon scattering. Equations (4.1) and (4.2) can now be solved for the saturated velocity, by dividing the latter into the former (which causes the aforementioned constants to cancel each other), and then solving for the saturated velocity as

$$v_s = \left[\frac{3\hbar\omega_0}{4m^*}\tanh\left(\frac{\hbar\omega_0}{2k_B T}\right)\right]^{\frac{1}{2}}. \tag{4.3}$$

As remarked, (4.3) is correct only for non-polar scattering, but has appeared even for this scattering in slightly modified forms in a variety of other derivations. These various forms differ by only a numerical factor. The result allows one to predict variations in materials, as it depends only on the effective mass, the phonon energy, and the lattice temperature. In Figure 4.1, the observed and/or calculated (by detailed calculations) saturation velocities are plotted against (4.3) for a variety of semiconductors. Only the dominant phonon is used in this comparison.

In materials in which the polar-optical phonon is the dominant scattering mechanism, an interesting phenomenon has been predicted to occur – polar runaway. This effect is expected to arise because the scattering rate (to be obtained below) varies (for phonon emission) as $1/\sqrt{E}$, so that it becomes less significant at higher energies. This is an unstable situation for which no steady state exists, and the carriers run off to very high energies (Stratton, 1958). Equations (4.1) and (4.2) assume that a stable steady state exists, although the existence of this steady state has not been fully investigated for all scattering processes, and certainly not for all semiconductors.

In the case of hot carriers, the considerable power given to the carriers by the field and relaxed to the lattice can drive the phonon distribution out of equilibrium, so that a hot-phonon distribution can occur. At room temperatures, the phonon lifetimes are sufficiently short that such an effect

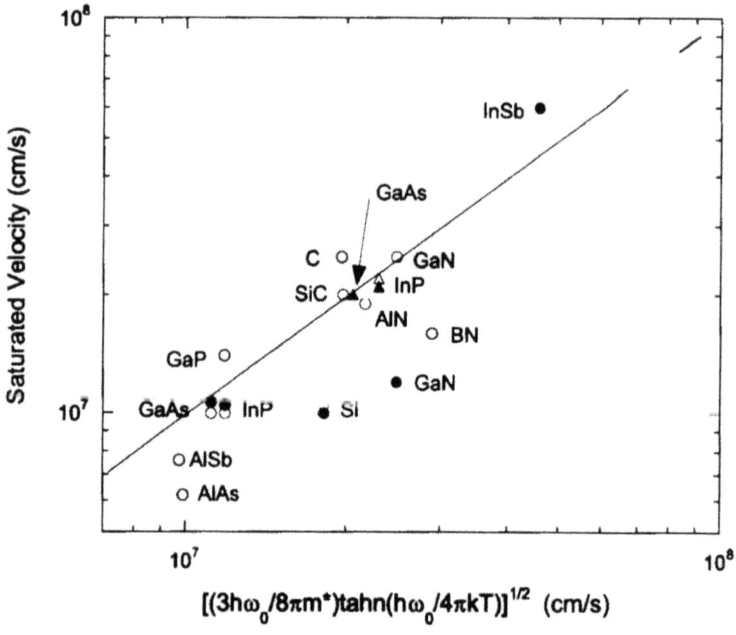

Figure 4.1 The saturated drift velocity for electrons in comparison with the scaling theory of equation (4.3). The data is determined either from experiment (solid circles) or from theoretical estimates (open circles). Also shown is the peak velocity (triangles) for those materials in which intervalley transfer plays a major role. In this case, the peak is plotted using the polar optical phonon in the Γ valley, while the saturated value is plotted using the intervalley phonon that dominates scattering in the steady state.

is unlikely to occur in normal circumstances. At low temperatures, however, it can be expected that the situation is different. The long-wavelength acoustic phonons can achieve lifetimes as long as a few nanoseconds, so that this distribution could reasonably be expected to be driven out of equilibrium (Ferry, 1974). This is the case in acousto-electric effects. The optical phonons, though, have a shorter lifetime, so that it is unlikely that this distribution will be driven out of equilibrium except, for example, under intense laser irradiation, where carriers are created high in the energy band and subsequently emit a shower of optical phonons (Kash *et al.*, 1985). The optical phonon lifetime, in fact, does not vary by more than a factor of 2 to 4 with temperature. However, the difference between room temperature and low temperature (4.2 K) lies in the number of optical phonons that are thermally excited. The process is analogous to excess carrier generation. If a number of phonons are emitted by the electrons (or holes), the importance of these excess phonons to the carrier-phonon scattering processes depends on how large the excess density is in comparison with the background. At room temperature, the background density of optical phonons is generally thought to be sufficient to wash out the hot-phonon effect, but this is not the case at low temperature. If the phonons are driven out of equilibrium, the saturation velocity can be expected to reflect this effect. We will return to this discussion in Chapter 6.

4.1.2 Transient transport

In high electric fields, we will find that there are at least two different relaxation times corresponding to the relaxation of momentum and energy. The momentum relaxation time τ_m describes the decay of the velocity (and the velocity fluctuations about a local near-equilibrium state) and is the relaxation time dealt with in the previous chapter, where the relaxation time approximation was discussed. However, the nonlinear transport in high electric fields arises primarily from the change in the distribution function in the presence of the high electric field, which leads to an increase in the average energy of the carriers. The response of the distribution function, which results in this increase in average energy, is characterized by its own relaxation time, which is referred to as the energy relaxation time τ_E, since the evolution of the distribution function represents the evolution of the average energy of the carrier ensemble.

If the energy relaxation process is slower than the momentum relaxation process, the velocity can overshoot its ultimate steady-state value (the saturation velocity) in high fields. This occurs because the distribution function first shifts (equivalent to the shift studied in previous chapters) in momentum space as the velocity rises to a value characterized mainly by its low-field mobility. As the distribution function then evolves to its non-equilibrium form, the mobility decreases to its ultimate high-field value, with a consequent decrease in the velocity. It can readily be shown that this "overshoot" behavior requires a more complicated behavior than was discussed in Chapter 3, where the momentum was assumed to obey the standard Langevin equation

$$\frac{dv}{dt} = \frac{eF}{m^*} - \frac{v}{\tau_m}. \tag{4.4}$$

When overshoot occurs, the left-hand side of (4.4) must have at least two zeros – one at a time corresponding to the steady state and one at a time corresponding to the peak velocity. However, the relaxation rate is an increasing function of energy (or velocity), so that the right-hand side has only a single zero. Thus a second time scale must be involved, which is the characteristic time of the energy relaxation. In the latter case, the motion of the particles is governed by a *retarded* Langevin equation, written as (Zwanzig, 1961)

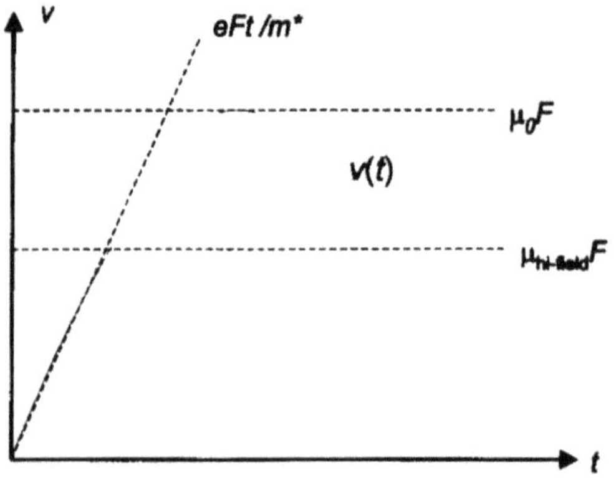

Figure 4.2 Schematic representation of the manner in which velocity overshoot is realized. Here, it is assumed that the energy relaxation time is longer than the momentum relaxation time.

$$m^*\frac{dv}{dt} = -\int_0^t \gamma(t-u)v(u)du + R(t) + eF. \tag{4.5}$$

The function $\gamma(t)$ is a "memory function" for the non-equilibrium system, and it will be related to the correlation function below. The other terms in (4.5) have the same meaning as used in Chapter 3. Equation (4.5) is a non-Markovian form of the Langevin equation, since the rate of change of the velocity at time t depends not only on the present time but also on all past time. We will return to this discussion below, but now we have a simple interpretation of the velocity overshoot, which was shown in Figure 1.3. As mentioned above, the velocity tries to rise initially to the value given by the mobility appropriate to the low field case. However, as the average temperature (energy) of the distribution increases, the mobility decreases so that the velocity also decreases toward this value, as depicted in Figure 4.2.

4.2 The Shockley model

At this point it is desirable to use a relatively simple model to illustrate the nature of the non-equilibrium distribution function and its behavior in external electric fields. The model that will be introduced here is an adaptation of one first introduced by Shockley (1951) to investigate velocity saturation in semiconductors, but it incorporates the full behavior of both energy and momentum relaxation. The approach we follow is slightly different from that of Shockley, although the rationale is quite the same, and the simplifications are made to make the model more transparent. To begin, the Boltzmann transport equation is written in the form (which differs from that in Chapter 3 by introducing the momentum by $m^*\mathrm{v}$) for a spatially homogeneous system (the high-field effects are of primary interest here, so the spatial gradients are not considered):

$$\frac{\partial f(\mathrm{v})}{\partial t} + \frac{eF}{m^*}\frac{\partial f(\mathrm{v})}{\partial \mathrm{v}} = \left.\frac{\partial f(\mathrm{v})}{\partial t}\right|_{scattering} . \tag{4.6}$$

The right-hand side is the scattering (or relaxation) term and will be treated here by two processes. The first process is elastic scattering, such as introduced in the last chapter and which is taken to be isotropic on the energy surface (even though a one-dimensional model is used here for simplicity). This may be expressed as

$$\left.\frac{\partial f(\mathrm{v})}{\partial t}\right|_{el} = -\frac{1}{\tau_m}[f(\mathrm{v}) - f(-\mathrm{v})]. \tag{4.7}$$

It may readily be established that this term vanishes in equilibrium where the distribution function is symmetric in velocity space. In fact, it is just this term that forces the distribution function to become symmetric in the three-dimensional equilibrium situation, so that no average current flows in the system. This point is not usually recognized, but (4.7) is an alternative form of the relaxation-time approximation. Here, states of opposite momentum are equilibrated with one another.

The second scattering process introduces the inelastic processes, such as those due to optical phonons. For this process it may be assumed that any electron that reaches the velocity v_0 (or $-\mathrm{v}_0$), which characterizes the energy of the optical phonon through $m^*\mathrm{v}_0^2/2 = \hbar\omega_0$, is immediately scattered back to zero velocity (zero energy) by the emission of a phonon. The scattering rate is taken to be infinite at this single value of energy, and this assures that no carriers can be accelerated beyond v_0. While this is a drastic assumption, and ignores the specific energy dependence of the phonon scattering, it provides an adequate treatment at this point to illustrate the role of the inelastic scattering. This scattering rate may be written as

$$\left.\frac{\partial f(\mathrm{v})}{\partial t}\right|_{inel} = -af(\mathrm{v}_0)[\delta(\mathrm{v} - \mathrm{v}_0) + \delta(\mathrm{v} + \mathrm{v}_0) - 2\delta(\mathrm{v})], \tag{4.8}$$

where a and $f(\mathrm{v}_0)$ are constants that will be evaluated later (two are used to simplify the mathematics below). With these definitions, the Boltzmann equation now becomes

$$\frac{\partial f(\mathrm{v})}{\partial t} + \frac{eF}{m^*}\frac{\partial f(\mathrm{v})}{\partial \mathrm{v}} = -\frac{1}{\tau_m}[f(\mathrm{v}) - f(-\mathrm{v})] - af(\mathrm{v}_0)[\delta(\mathrm{v} - \mathrm{v}_0) + \delta(\mathrm{v} + \mathrm{v}_0) - 2\delta(\mathrm{v})]. \tag{4.9}$$

In the remainder of this section, the steady-state and transient solutions will be obtained for this equation.

4.2.1 The time-independent case

If the field is constant in time, it may be expected that a stable steady state can be achieved in which a constant, time-independent drift velocity is reached. For the steady state, (4.9) becomes

$$\frac{eF}{m^*}\frac{\partial f(\mathrm{v})}{\partial \mathrm{v}} = -\frac{1}{\tau_m}[f(\mathrm{v}) - f(-\mathrm{v})] - af(\mathrm{v}_0)[\delta(\mathrm{v} - \mathrm{v}_0) + \delta(\mathrm{v} + \mathrm{v}_0) - 2\delta(\mathrm{v})]. \tag{4.10}$$

The approach to be followed is to split the distribution function $f(\mathrm{v})$ into symmetric and anti-symmetric parts, defined through the relationships

$$f_s(\mathrm{v}) = \tfrac{1}{2}[f(\mathrm{v}) + f(-\mathrm{v})], \quad f_a(\mathrm{v}) = \tfrac{1}{2}[f(\mathrm{v}) - f(-\mathrm{v})], \tag{4.11}$$

or

$$f(\mathrm{v}) = f_s(\mathrm{v}) + f_a(\mathrm{v}), \quad f(-\mathrm{v}) = f_s(\mathrm{v}) - f_a(\mathrm{v}). \tag{4.12}$$

The equivalent Boltzmann equation for $f(-\mathrm{v})$ may be written as

$$\frac{eF}{m^*}\frac{\partial f(-\mathrm{v})}{\partial \mathrm{v}} = -\frac{1}{\tau_m}[f(\mathrm{v}) - f(-\mathrm{v})] + af(\mathrm{v}_0)[\delta(\mathrm{v} - \mathrm{v}_0) + \delta(\mathrm{v} + \mathrm{v}_0) - 2\delta(\mathrm{v})], \tag{4.13}$$

where we have replaced v by –v in (4.10) and then multiplied through by a negative sign to reach this result.

Equations (4.10) and (4.13) can now be combined through (4.11) to yield the equations for the symmetric and anti-symmetric parts of the distribution function as

$$\frac{eF}{m^*}\frac{\partial f_s(\mathrm{v})}{\partial \mathrm{v}} = -\frac{2f_a(\mathrm{v})}{\tau_m}$$

$$\frac{eF}{m^*}\frac{\partial f_a(\mathrm{v})}{\partial \mathrm{v}} = -af(\mathrm{v}_0)[\delta(\mathrm{v} - \mathrm{v}_0) + \delta(\mathrm{v} + \mathrm{v}_0) - 2\delta(\mathrm{v})]. \tag{4.14}$$

In the equilibrium case ($F = 0$), the distribution function must be symmetric so that all odd moments vanish. Thus $f_a(\mathrm{v}) = 0$ for this case. This yields two constraints: first, $f_a(\mathrm{v}_0) = 0$ when $F = 0$, and, secondly, that it is convenient to let $a = eF/m^*$, in order to balance the constants in the two equations. It may also be noted that the first of equations (4.14) is equivalent to the form found in the relaxation-time approximation in Chapter 3.

Using the foregoing value of a, the second of equations (4.14) may be integrated to yield the solution

$$f_a(\mathrm{v}) = f(\mathrm{v}_0)[2u_0(\mathrm{v}) - u_0(\mathrm{v} - \mathrm{v}_0) - u_0(\mathrm{v} + \mathrm{v}_0)], \tag{4.15}$$

where u_0 is the Heaviside step function (=1 for an argument >0 and zero otherwise). Since this anti-symmetric part of the distribution function must vanish as the electric field goes to zero, it is expected that $f(\mathrm{v}_0)$ is a constant that varies with the applied field. Equation (4.15) can be used in the first of equations (4.14) to yield the symmetric distribution

$$f_s(\mathrm{v}) = \frac{m^* f(\mathrm{v}_0)}{eF\tau_m} \cdot \begin{cases} 0, & |\mathrm{v}| > \mathrm{v}_0, \\ C_1 + 2(\mathrm{v} + \mathrm{v}_0), & -\mathrm{v}_0 < \mathrm{v} < 0, \\ C_1 - 2(\mathrm{v} - \mathrm{v}_0), & 0 < \mathrm{v} < \mathrm{v}_0. \end{cases} \tag{4.16}$$

The continuity of the symmetric function at $\mathrm{v} = 0$ has been used to evaluate some of the integration constants. The constant C_1 is found from the normalization of the distribution function. However, what normalization should be invoked? Normally, one integrates the *entire* distribution function over momentum and normalizes to the density. Here, $f(\mathrm{v}_0)$ is still undetermined, and the integration constraint will be used to set this value (it will be found that the distribution still remains normalized). Thus, the determination of C_1 remains unspecified. This constant will be evaluated by requiring $f(-\mathrm{v}_0)$ vanishes when approached from the positive side (this requires us always to create averages by dividing by the integrated distribution function), which invokes continuity of the symmetric function at this point. The justification for this lies in the fact that $f(\mathrm{v}_0)$ is zero in the equilibrium situation, since the scattering rate is infinite at these points. The field will accelerate carriers from the negative momentum side of the distribution to the positive momentum side. This will not affect the limiting value of $f(-\mathrm{v}_0)$, so this can be retained at a value of zero. To achieve this requires that

$$f_s(-\mathrm{v}_0) + f_a(-\mathrm{v}_0) = -f(\mathrm{v}_0) + \frac{C_1 m^*}{eF\tau_m} f(\mathrm{v}_0) = 0, \tag{4.17}$$

or

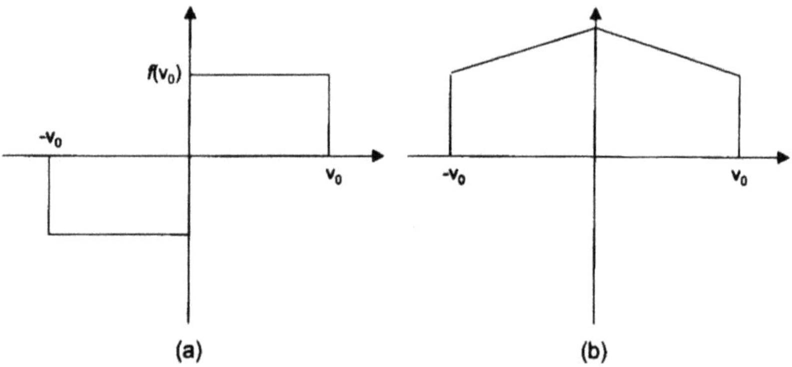

Figure 4.3 The anti-symmetric (a) and symmetric (b) parts of the distribution function for a non-zero electric field in the Shockley model.

$$C_1 = \frac{eF\tau_m}{m^*} = \mu_m F, \tag{4.18}$$

where $\mu_m = e\tau_m/m^*$. This choice of C_1, as mentioned, is also a choice of convenience, as there is now only a single remaining constant $f(v_0)$ to normalize in terms of the density. In Figure 4.3, the symmetric and anti-symmetric parts of the distribution function are shown. The final, total distribution function may now be found as

$$f(v) = f(v_0) \cdot \begin{cases} 0, & |v| > v_0, \\ 2\dfrac{v + v_0}{\mu_m F}, & -v_0 < v < 0, \\ 1 - 2\dfrac{v - v_0}{\mu_m F}, & 0 < v < v_0. \end{cases} \tag{4.19}$$

The drift velocity may now be calculated by integrating over the distribution function, given in (4.19), as

$$\langle v \rangle = \frac{1}{n}\int_{-\infty}^{\infty} vf(v)dv = \frac{1}{n}\int_{-v_0}^{v_0} vf_a(v)dv$$

$$= \frac{2}{n}\int_0^{v_0} vf_a(v)dv = \frac{2f(v_0)}{n}\int_0^{v_0} vdv = \frac{f(v_0)v_0^2}{n}, \tag{4.20}$$

where

$$n = \int_{-\infty}^{\infty} f(\mathrm{v})d\mathrm{v} = \int_{-\infty}^{\infty} f_s(\mathrm{v})d\mathrm{v} = 2\int_{0}^{\mathrm{v}_0} f_s(\mathrm{v})d\mathrm{v}$$

$$= 2f(\mathrm{v}_0)\left[\mathrm{v}_0\left(1 + 2\frac{\mathrm{v}_0}{\mu_m F}\right) - \frac{\mathrm{v}_0^2}{\mu_m F}\right] = 2f(\mathrm{v}_0)\mathrm{v}_0\left(1 + \frac{\mathrm{v}_0}{\mu_m F}\right) \tag{4.21}$$

is the required normalization in terms of the carrier density. The latter two equations may now be combined to yield

$$\mathrm{v}_d = \langle \mathrm{v} \rangle = \frac{(\mathrm{v}_0/2)}{1 + \dfrac{\mathrm{v}_0}{\mu_m F}} = \frac{\mu_m F}{1 + \dfrac{\mu_m F}{(\mathrm{v}_0/2)}} \to \frac{\mathrm{v}_0}{2}, \tag{4.22}$$

where the final value is that for extremely large fields. At low fields, the velocity rises linearly with an effective mobility of μ_m. There are no surprises here, and it is reassuring that the definition of the relaxation time in (4.7) yields the low field mobility. At high fields, the velocity saturates at a value of $\mathrm{v}_0/2$, as expected. The carrier continually accelerates in the region between $0 < \mathrm{v} < \mathrm{v}_0$, which accounts for the average found here. Thus, it is clear that this simple model shows the full nonlinear velocity saturation behavior, and that this behavior is characteristic of the drastic shift and restructuring of the distribution function. Further understanding can be obtained by looking at the overall distribution function, which is shown in Figure 4.4. As the field is increased, the left-hand ($\mathrm{v} < 0$) part of the distribution is reduced, while the right-hand ($\mathrm{v} > 0$) part is increased in population. However, the maximum of the distribution does not change, and this can be found from (4.19) and (4.21) to be

$$f(0^+) = f(\mathrm{v}_0)\left(1 + \frac{\mathrm{v}_0}{\mu_m F}\right) = \frac{n}{2\mathrm{v}_0}2\left(1 + \frac{\mathrm{v}_0}{\mu_m F}\right)^{-1}\left(1 + \frac{\mathrm{v}_0}{\mu_m F}\right) = \frac{n}{\mathrm{v}_0}. \tag{4.23}$$

Hence, there is no change in the peak height as the field increases from zero to large values. The slope of the distribution function, in both regions, is reduced $\mu_m F \gg \mathrm{v}_0$, for which the entire carrier distribution exists in the right-half plane $0 < \mathrm{v} < \mathrm{v}_0$. The shift of population from the left half to the right half is also shown in Figure 4.4. For the case of a high field, the distribution function is dominated by the optical phonon processes, and the density is constant over this velocity range, as the carriers continually sweep across the velocity space until reaching v_0, at which point they are scattered back to zero velocity. Any particular carrier exhibits a sawtooth waveform in time, being accelerated linearly in time up to v_0, then scattering

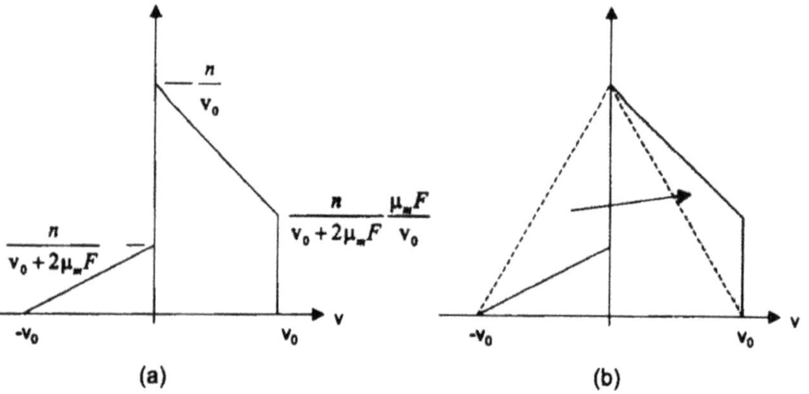

Figure 4.4 (a) The total distribution function, which comes from adding the two parts shown in Figure 4.3. (b) A schematic illustration of the shift of area from one part of the distribution function to the other, as the electric field is increased. The shift has been exaggerated, so that the areas do not appear to be equal. In fact, conservation of density will require that the total area moved from one part of the distribution to the other must be conserved.

back to zero and starting over. It is clear from this just why the average velocity is exactly $v_0/2$.

4.2.2 The time-dependent case

In a fashion similar to the above treatment, the transient response of the distribution function, and the average velocity, can be examined with this simple model. As the average velocity is zero in the equilibrium state (at $t = 0$), the function $f(v, t = 0)$ will be ignored in the following treatment. The time-dependent Boltzmann equation is given by

$$\frac{\partial f(v)}{\partial t}+\frac{eF}{m^*}\frac{\partial f(v)}{\partial v}=-\frac{1}{\tau_m}[f(v)-f(-v)]$$
$$-\frac{eF}{m^*}f(v_0)[\delta(v\pm v_0)-2\delta(v)], \tag{4.24}$$

where the inelastic scattering time and the equation both have been simplified. This equation will be solved by a Laplace transform technique. First, however, it must be noticed that $f(v_0)$, which is our normalization constant, may be a function of time in the transient case, as it is most definitely a function of the applied field. Therefore, it will be assumed that this is the case, and accordingly, the transform of (4.24) is

$$sf(\mathrm{v},s) + \frac{eF}{m^*}\frac{\partial f(\mathrm{v},s)}{\partial \mathrm{v}} = -\frac{1}{\tau_m}[f(\mathrm{v},s) - f(-\mathrm{v},s)]$$
$$-\frac{eF}{m^*}f(\mathrm{v}_0,s)[\delta(\mathrm{v} \pm \mathrm{v}_0) - 2\delta(\mathrm{v})]. \qquad (4.25)$$

This equation can be separated into its symmetric and anti-symmetric parts as

$$sf_s(\mathrm{v},s) + \frac{eF}{m^*}\frac{\partial f_a(\mathrm{v},s)}{\partial \mathrm{v}} = \frac{eF}{m^*}f(\mathrm{v}_0,s)[2\delta(\mathrm{v}) - \delta(\mathrm{v} \pm \mathrm{v}_0)]$$
$$sf_a(\mathrm{v},s) + \frac{eF}{m^*}\frac{\partial f_s(\mathrm{v},s)}{\partial \mathrm{v}} = -\frac{2}{\tau_m}f_a(\mathrm{v},s). \qquad (4.26)$$

As before, these can be solved in various regions and matched using boundary conditions. These boundary conditions, in the transformed case, are just

$$f(-\mathrm{v}_0,s) = 0,$$
$$\int_{\mathrm{v}'-\varepsilon}^{\mathrm{v}'+\varepsilon} \frac{\partial f(\mathrm{v},s)}{\partial \mathrm{v}}d\mathrm{v} = \begin{cases} 2f(\mathrm{v}_0,s), & \mathrm{v}' = 0, \\ -f(\mathrm{v}_0,s), & \mathrm{v}' = -\mathrm{v}_0, \\ -f(\mathrm{v}_0,s), & \mathrm{v}' = \mathrm{v}_0, \end{cases}$$
$$\left.\frac{\partial f(\mathrm{v},s)}{\partial \mathrm{v}}\right|_{\mathrm{v}\to-\mathrm{v}_0^+} = \frac{m^*}{eF}f(\mathrm{v}_0,s), \qquad (4.27)$$

and the derivative does not change elsewhere. Thus, even in the transient case, $F(\mathrm{v},s)$ is a smooth function except at the three points $\mathrm{v} = 0$ and $\mathrm{v} = \pm\mathrm{v}_0$. At these two points, step discontinuities occur, and these are primarily in the anti-symmetric part, as in Figures 4.3 and 4.4.

For the two active regions of interest, $-\mathrm{v}_0 < \mathrm{v} < 0$ and $0 < \mathrm{v} < \mathrm{v}_0$, the two equations (4.26) for the symmetric and the anti-symmetric parts of the distribution function can be combined to yield the general equation for the distribution function. The same equation results for either part of the distribution, so we can sum the symmetric and anti-symmetric parts in the regions away from the transition points into one equation for the total distribution function as

$$\left(\frac{eF}{m^*}\right)^2 \frac{\partial^2 f(\mathrm{v},s)}{\partial \mathrm{v}^2} - s\left(s + \frac{2}{\tau_m}\right)f(\mathrm{v},s) = 0, \qquad (4.28)$$

which has the general solution

$$f(v,s) = A\exp\left(\frac{Lv}{P}\right) + B\exp\left(-\frac{Lv}{P}\right) \tag{4.29}$$

with

$$L^2 = s\left(s + \frac{2}{\tau_m}\right), \quad P = \frac{eF}{m^*}. \tag{4.30}$$

The negative-velocity region ($v < 0$) is considered first. The boundary conditions of (4.27) give rise to the requirements

$$A\exp\left(-\frac{Lv_0}{P}\right) + B\exp\left(\frac{Lv_0}{P}\right) = 0$$

$$\frac{L}{P}\left[A\exp\left(-\frac{Lv_0}{P}\right) - B\exp\left(\frac{Lv_0}{P}\right)\right] = \frac{f(v_0,s)}{P\tau_m}. \tag{4.31}$$

These are readily solved to yield

$$f^<(v,s) = \frac{f(v_0,s)}{L\tau_m}\sinh\left(\frac{L(v + v_0)}{P}\right). \tag{4.32}$$

For the positive-velocity region, the constants A and B are replaced by C and D. The boundary conditions are then evaluated at $v = 0^+$ and at $v = v_0$, and these yield

$$C + D = f(v_0,s) + \frac{f(v_0,s)}{L\tau_m}\sinh\left(\frac{Lv_0}{P}\right)$$

$$C\exp\left(\frac{Lv_0}{P}\right) + D\exp\left(-\frac{Lv_0}{P}\right) = f(v_0,s). \tag{4.33}$$

These may then be used to give the distribution function in the positive-velocity region as

$$f^>(v,s) = f(v_0,s)\left[1 + \frac{1}{L\tau_m}\sinh\left(\frac{L(v_0 - v)}{P}\right)\right]. \tag{4.34}$$

The total number of particles is found by normalization (integrating over the two parts of the distribution function). This yields

$$N(s) = \frac{n}{s} = \int_{-v_0}^{0} F^{<}(v,s)dv + \int_{0}^{v_0} F^{>}(v,s)dv$$

$$= F(v_0,s)\left\{v_0 + \frac{2P}{L^2\tau_m}\left[\cosh\left(\frac{Lv_0}{P}\right) - 1\right]\right\}. \tag{4.35}$$

Here, the Laplace transform of the time-independent density has been used. Of course, (4.35) is used to evaluate the constant factor $f(v_0,s)$, which turns out not to be constant (as may be ascertained by examining this equation) due to the complicated s dependence of the quantity L. Similarly, the average velocity is found to be

$$\langle V(s)\rangle = \frac{1}{n}\left[\int_{-v_0}^{0} vf^{<}(v,s)dv + \int_{0}^{v_0} vf^{>}(v,s)dv\right] = \frac{v_0^2 f(v_0,s)}{n}$$

$$= \frac{v_0/2s}{1 + \frac{2P}{v_0 L^2\tau_m}\left[\cosh\left(\frac{Lv_0}{P}\right) - 1\right]}. \tag{4.36}$$

Here, (4.35) has been used. It is clear from this equation that the temporal behavior of the drift velocity has a quite complicated behavior. The long-time limit is found to be

$$v_d = \lim_{t\to\infty}\langle v(t)\rangle = \lim_{s\to 0} s\langle V(s)\rangle = \frac{v_0/2}{1 + \frac{v_0}{\mu_m F}}, \tag{4.37}$$

where the small argument expansion of the cosh function is used in the denominator prior to taking the limit, and the result is just (4.22), as expected. The general trend of the time response can be found for medium-time scales by including higher order terms in the cosh function expansion, so that

$$\langle V(s)\rangle = \frac{v_0}{s}\left\{1 + \frac{v_0}{\mu_m F}\left[1 + \frac{v_0^2\tau_m^2}{12\mu_m^2F^2}s\left(s + \frac{2}{\tau_m}\right)\right]\right\}^{-1}$$

$$= \frac{v_d}{s}\left\{1 + \frac{v_d}{\mu_m F}\frac{v_0^2\tau_m^2}{12\mu_m^2F^2}s\left(s + \frac{2}{\tau_m}\right)\right\}^{-1}. \tag{4.38}$$

The last form has been written with the long-time limit drift velocity. Clearly, the time behavior is very complicated. Depending on the relative size of the quantity $v_0/\mu_m F$, the response from the bracketed terms can be damped sinusoids or simple exponentials. To see this, the roots of the term in curly brackets define the characteristic equation, which can be rewritten as

$$\Lambda\tau_m^2 s^2 + \Lambda\tau_m s + 1 = 0, \quad \Lambda = \frac{v_d v_0^2}{12(\mu_m F)^3}. \tag{4.39}$$

Clearly, if $\Lambda < 1$ (large F), the roots are a complex conjugate pair and the velocity approaches steady state via a damped sinusoid. On the other hand, if $\Lambda > 1$ (small F), the response is simply two exponentials, which may or may not exhibit overshoot. In the latter case, the decay rates vary with the field and are not simply related to the momentum scattering rate. The Shockley model thus exhibits not only the saturation velocity effect, but also the full-velocity overshoot effect that is expected at high fields.

4.3 Inelastic phonon scattering

In the previous section, the importance of the inelastic scattering processes was made quite evident. We now want to turn to the details of these processes. In Section 3.3.1, acoustic-mode phonon scattering processes were treated. In the tetrahedrally coordinated semiconductors, there are two atoms per unit cell site and optical mode interactions are also allowed, where the two atoms vibrate relative to each other. These phonons are rather energetic, being of the order of 30 to 50 meV in energy, and lead to *inelastic scattering processes*, since there is a significant gain or loss of energy by the carrier during the scattering process. Although one normally thinks of scattering occurring just within a single minimum, or valley, of the band, these optical phonons can also cause *intervalley* or *interband* scattering. Such examples are scattering from the light-hole valence band to the heavy-hole valence band by a mid-zone phonon near the Γ point, or a Γ-to-L valley scattering in the conduction band by a zone-edge optical (or high-energy acoustic) phonon.

4.3.1 Zero-order deformation potential scattering

The matrix element for this scattering mechanism is generally found using a deformable ion model, in which the two sublattices are assumed to simply move relative to one another. Thus, the potential field of each ion is displaced slightly. This causes a resulting shift in the bond charges, which leaves a small excess positive charge where the ions have moved apart and a slight negative charge in the regions where they are closer together. This produces a macroscopic deformation field D, which is usually given in units of eV/cm. This scattering is a zero-order process, in that the resulting interaction potential is independent of the wave vector, or

$$\delta E = D u_{\mathbf{q}}, \tag{4.40}$$

so that

$$M(\mathbf{k},\mathbf{q}) = \left(\frac{\hbar D^2}{2V\rho\omega_{\mathbf{q}}}\right)^{\frac{1}{2}} \left[\sqrt{N_{\mathbf{q}}}\,\delta(E_{\mathbf{k}} - E_{\mathbf{k+q}} - \hbar\omega_{\mathbf{q}}) + \sqrt{N_{\mathbf{q}} + 1}\,\delta(E_{\mathbf{k}} - E_{\mathbf{k-q}} - \hbar\omega_{\mathbf{q}})\right]. \tag{4.41}$$

Here, ρ is the mass density, and N_q is the phonon occupation function. Although the delta function is shown in (4.41), it has already been incorporated into the integrals that appear in Section 3.3.1 (see, e.g., (3.116)).

In the case of optical phonon scattering within a single band (or valley) through the long-wavelength phonons near the zone center, the dispersion relation for the optical modes is quite flat, with very little dependence on the magnitude of the wave vector **q**. This implies that a reasonable approximation is to take $\omega_q = \omega_0$ to be constant in the integrations over the phonon wave vectors. For intervalley phonon scattering, or for scattering between different valence-band valleys, the dominant part of the phonon wave vector is quite large, so that no significant error is made by continuing to treat the frequency of the optical (or intervalley) phonon as a constant. Moreover, the scattering is isotropic; that is, there is no q dependence in the matrix element once ω_0 is taken as a constant. This means that one can use the density-of-states result (3.115), but for which the emission and absorption terms are separated, as follows:

$$\begin{aligned}\Gamma(k) &= \frac{2\pi}{\hbar}\frac{\hbar D^2}{2V\rho\omega_0}\left[\frac{V}{4\pi^2}\left(\frac{2m^*}{\hbar^2}\right)^{\frac{3}{2}}\right]\left[N_q\sqrt{E_k + \hbar\omega_0} + (N_q + 1)\sqrt{E_k - \hbar\omega_0}\,u_0(E_k - \hbar\omega_0)\right] \\ &= \sqrt{\frac{m^*}{2}}\frac{m^* D^2}{\pi\rho\hbar^3\omega_0}\left[N_q\sqrt{E_k + \hbar\omega_0} + (N_q + 1)\sqrt{E_k - \hbar\omega_0}\,u_0(E_k - \hbar\omega_0)\right].\end{aligned} \tag{4.42}$$

The Heaviside step function u_0 has been added to the last term to ensure that the argument of the square root is positive; i.e., that carriers with an energy $E_k < \hbar\omega_0$ cannot emit a phonon. For inter-valley scattering, (4.42) must still be multiplied by the number of final ellipsoids to which the carrier can scatter, although this factor can easily be included in the density-of-states effective mass m^* that appears in the equation. The optical phonon scattering rate calculated here is the mean free time for collisions, but because this process also relaxes both the energy and the momentum in a very efficient fashion, it is closely related to the relaxation times for the latter quantities. Other than the shift in the onset, the energy dependence of the optical scattering is quite similar to that for the acoustic modes, but it is much more temperature dependent because of the complete form of the Bose–Einstein distribution N_q that is retained here.

4.3.2 Selection rules

When scattering occurs either within a single valley or band minimum, or between different valleys, whether equivalent or not, it is not always the case that just any old phonon will couple properly to move the carrier from the initial state to the final state. For example, the top of the valence band is predominantly formed from the anion p states, while the bottom of the conduction band at the Γ point is predominantly formed from the cation s states, as discussed in Chapter 2. If an electron is going to scatter from the Γ point to the L point in the conduction band, for example, it is necessary that the cation atom be in motion (due to the phonon wave) in order to couple to the electron. We know that the cation motion for the L-point phonon mode is the LO mode if the cation is the lighter of the two atoms and the LA mode if it is the heavier of the two atoms (Ferry, 1991). Although this is a hand-waving argument, it can be placed on quite firm ground through group theory.

Space group selection rules are usually calculated by group-theoretical techniques. It is beyond the scope of this book to go through these techniques, so we merely summarize the macroscopic features of the arguments. If a given set of M physical quantities such as the matrix elements coupling the carriers in different valleys by the phonons are to be calculated, the selection rules determine the number of n_M independent matrix elements in the set M. For example, consider the required selection rule for an electron in Si, in which the transition is made from the valley located at $(k_0,0,0)$ to the valley at $(-k_0,0,0)$, where $k_0 = 0.857\pi/a$ (this is the point at which the minimum of the conduction band appears in Figure 2.10). This transition has been termed a g-phonon (the details of the phonon scattering in Si are discussed later), and the selection rule can be written as

$$\Delta_1(k_0) \otimes \Delta_1(-k_0) = \Delta_1(2k_0), \tag{4.43}$$

where Δ_1 represents the required symmetry for the electron wave function in the appropriate minimum of the conduction band and $\otimes$ represents a group-theoretical convolution operation. In short, the two wave functions on the left can only be coupled by a phonon with the wave function symmetry appearing on the right-hand side. The problem is that the wave vector on the right extends beyond the edge of the Brillouin zone and is therefore termed an *umklapp process*, as the wave vector must be reduced by a reciprocal lattice vector – but, which reciprocal lattice vector? The point $2k_0$ lies on the prolongation of the (100) direction (Δ direction) beyond the X point into the second Brillouin zone (see Figure 2.7). The symmetry Δ_1 passes over into a symmetry function $\Delta_{2'}$ as q passes the X point. Thus the desired phonon must have a wave vector along the (100) axis and have the symmetry $\Delta_{2'}$ for it to couple the two valleys discussed above. If there were no phonons of this symmetry, the transition would be forbidden to zero order, which is the coupling calculated in the previous section. Fortunately, the LO phonon branch has just this symmetry in Si, so that the desired phonon has $q = 0.3\pi/a$ and is an LO mode. As a second example, consider the scattering from the central Γ-point minimum in GaAs to the L valleys. The latter valleys lie some 0.29 eV above the central Γ minimum (see Figure 2.9), and scattering to these valleys is the process by which the Gunn effect (discussed in the next chapter) occurs in this material. An electron that gains sufficient energy in the central valley can be scattered to the satellite valleys, where the mass is heavier and the mobility is much lower. Thus the symmetry operation is given by

$$\Gamma_1(0) \otimes L_1(-\mathbf{k}_L) = L_1(\mathbf{k}_L), \tag{4.44}$$

where $\mathbf{k}_L = (\pi/2a, \pi/2a, \pi/2a)$ is the position of the L point in the Brillouin zone. This value of $\mathbf{k}_L$ is now the required phonon wave vector, and the required phonon must have the symmetry given by the right-hand side of (4.44). Not surprisingly, the branch with this symmetry is the LO branch if the cation atom is the lighter atom, and the LA branch if the cation atom is the heavier atom, just as the hand-waving argument above suggested. In Table 4.1, the allowed phonons for the materials of interest are delineated, based on the proper group-theoretical calculations (Birman, 1962; Birman and Lax, 1966; Lax and Birman, 1972).

4.3.3 *First-order scattering*

If the zero-order matrix element for the optical or inter-valley interaction vanishes, as is the case, for example, for the umklapp phonons via the acoustic modes in Si, it is expected that D is identically equal to zero. However, the general electron–phonon interaction is an expansion in powers of q, and the zero-order interaction is just the q^0 order term. Moreover, the

Table 4.1 Optical-mode selection rules

Material	*Intravalley*	*Intervalley*
Si	forbidden	*g*: $\Delta_{2'}$ (LO) *f*: Σ_1 (LA, TO)
Ge	Γ (LO)	X_1 (LA, LO)
$A^{III}B^{V}$	Γ (polar LO)	Γ → L: L[a] Γ → X: X[a] L → L: X[a]

Note:
[a] LO if $m_V > m_{III}$, otherwise LA mode

selection rules are strictly limiting only upon this zero-order interaction. In first-order interactions, a term arises that is of the form $\Xi_0 \mathbf{q}\cdot\mathbf{e}_q$. Here, Ξ_0 is the first-order optical coupling constant (in obvious agreement in notation with the acoustic deformation potential in Section 3.3.1). In fact, this approach yields a form exactly like the acoustic deformation potential approach (Ferry, 1976; Siegel *et al.*, 1976), because the latter approach is also a first-order scattering process. It turns out that such an approach can also occur for the optical modes. To proceed, one can use directly (3.119), with the change in notation of the deformation potential and the constant frequency, as

$$|M(k,q)|^2 = \frac{\hbar\Xi_0^2 q^2}{2\rho V\omega_0}(N_q + 1), \tag{4.45}$$

and an equivalent term for the absorption term. We use (3.116), due to the q dependence of the matrix element, which must be generalized before we insert the matrix element, and

$$\Gamma(k) = \frac{2\pi}{\hbar}\frac{V}{(2\pi)^2}\int_0^\pi d\vartheta\sin\vartheta\int_0^\infty q^2 dq\,|M(k,q)|^2\delta(E_k - E_{k\pm q} \pm \hbar\omega_0). \tag{4.46}$$

The integration over the azimuthal angle has already been performed. The integration over the polar angle involves the argument of the δ-function, as

$$\begin{aligned} E_k - E_{k\pm q} \pm \hbar\omega_0 &= \frac{\hbar^2 k^2}{2m^*} - \frac{\hbar^2(\mathbf{k}\pm\mathbf{q})^2}{2m^*} \pm \hbar\omega_0 \\ &= -\frac{\hbar^2 q^2}{2m^*} \mp \frac{\hbar^2 kq}{m^*}\cos\vartheta \pm \hbar\omega_0. \end{aligned} \tag{4.47}$$

Hence, the integration over the δ-function provides a factor of

$$\frac{m^*}{\hbar^2 kq}$$

plus a restriction on the q-integration to the range

$$k - \sqrt{k^2 - \frac{2m^*\omega_0}{\hbar}} < q < k + \sqrt{k^2 - \frac{2m^*\omega_0}{\hbar}} \tag{4.48}$$

for emission of a phonon, and

$$\sqrt{k^2 + \frac{2m^*\omega_0}{\hbar}} - k < q < \sqrt{k^2 + \frac{2m^*\omega_0}{\hbar}} + k \tag{4.49}$$

for absorption of a phonon. By denoting these respective limits as q_- and q_+, we can write (4.46) as

$$\Gamma(k) = \frac{m^* V}{2\pi\hbar^3 k} \int_{q_-}^{q_+} |M(k,q)|^2 q\,dq. \tag{4.50}$$

We may now use this result with the matrix element (4.45) to give the first-order optical phonon scattering as

$$\Gamma(k) = \frac{m^*\Xi_0^2}{4\pi\rho\hbar^2 k\omega_0} \left\{ (N_q + 1) \int_{q_-^e}^{q_+^e} q^3 dq + N_q \int_{q_-^a}^{q_+^a} q^3 dq \right\}. \tag{4.51}$$

The integrations are straightforward, and the final scattering rate is just

$$\Gamma(k) = \frac{\sqrt{2}(m^*)^{\frac{5}{2}}\Xi_0^2}{\pi\rho\hbar^5\omega_0} \left\{ N_q(2E_k + \hbar\omega_0)\sqrt{E_k + \hbar\omega_0} \right.$$

$$\left. + (N_q + 1)(2E_k - \hbar\omega_0)\sqrt{E_k - \hbar\omega_0}\, u_0(E_k - \hbar\omega_0) \right\}, \tag{4.52}$$

where the Heaviside step function has been added to the emission term to assure that the argument of the square root is positive. The first-order process has a much smaller magnitude at low energies, but a much stronger energy dependence than the zero-order optical and intervalley process. Thus it is much weaker in normal situations, but can become the dominant process for energetic carriers at high electric fields.

4.3.4 Polar-optical phonon scattering

The non-polar optical phonon interactions discussed in the previous sections arose through the deformation of the energy bands. This led to a macroscopic deformation potential or field. In compound semiconductors, the two atoms per unit cell have differing charges, and the optical phonon interaction involving the relative motion of these two atoms has a strong Coulomb potential contribution to the interaction. We will see in a later chapter that this Coulomb interaction modifies the dielectric function by the dispersion of the long-wavelength LO mode near the zone center. This, in turn, is due to the interaction of the effective charges on the atoms of the lattice in polar semiconductors. Of interest here, however, is the fact that this mode of lattice vibration is a very effective scattering mechanism for electrons in the central valley of the group III–V and II–VI semiconductors. Particularly, in the central Γ conduction band valley, the non-polar interaction is generally weak, and the polar interaction can be dominant. It can also be effective for holes, although the TO non-polar interaction can be quite effective and compete with the polar interaction. In terms of the expansion in orders of q mentioned above, the polar interaction is q^{-1} which arises from its Coulombic nature.

The polarization of the dipole field, that accompanies the vibration of the polar mode is given essentially by the effective charge times the displacement. The latter is just the phonon mode amplitude u_q, which has been used previously. Hence, we can write the polarization as

$$P_q = \sqrt{\frac{\hbar}{2\gamma V\omega_0}}\, \mathbf{e}_q (a_q^{\dagger} e^{-i\mathbf{q}\cdot\mathbf{r}} + a_q e^{i\mathbf{q}\cdot\mathbf{r}}), \tag{4.53}$$

where $\mathbf{e}_q$ is the polarization unit vector for the mode vibration, a^{+} and a are creation and annihilation operators for mode q, and the effective interaction parameter (which is related to the effective charge) is

$$\frac{1}{\gamma} = \omega_0^2 \left(\frac{1}{\varepsilon_\infty} - \frac{1}{\varepsilon(0)} \right). \tag{4.54}$$

Here, ε_∞ and $\varepsilon(0)$ are the high-frequency and low-frequency dielectric permittivities, respectively (this difference gives the strength of the polar interaction, and vanishes in non-polar materials where these two values are equal). Comparing (4.53) with (4.41), we see that (4.54) replaces the value D^2/ρ. This polarization leads to a local electric field, which is a longitudinal field in the direction of propagation of the phonon wave, and it is this field that scatters the carriers. The interaction energy is given by (3.129), just as for the piezoelectric interaction (which is the acoustic mode corresponding to this Coulomb interaction) in terms of the polarization and the interaction field. These lead to a screened version of the polar interaction, in which the perturbing energy is given as

$$\delta E = \left(\frac{\hbar e^2}{2\gamma V\omega_0}\right)^{\frac{1}{2}} \frac{q}{q^2 + q_D^2}(a_q^{+}e^{-i\mathbf{q}\cdot\mathbf{r}} - a_q e^{i\mathbf{q}\cdot\mathbf{r}})e^{-i\omega t} \tag{4.55}$$

In keeping with the use of a simple screening (discussed in the piezoelectric scattering), the harmonic motion of the phonon can lead to a reduction of the screening (we will discuss this in a later chapter), so that q_D would be smaller than the Debye screening length. In this case, the phonon energy is often comparable to the electron energy. A good approximation, however, is to ignore this and use the Debye screening value. It should be emphasized that this is a very simple approximation to the full dynamic screening, and its validity has not been tested. Use of (4.55) leads to the matrix element

$$|M(k,q)|^2 = \left(\frac{\hbar e^2}{2\gamma V\omega_0}\right)\frac{q^2}{(q^2 + q_D^2)^2}[(N_q + 1)\delta(E_{\mathbf{k}} - E_{\mathbf{k-q}} - \hbar\omega_0)$$
$$+ N_q\delta(E_{\mathbf{k}} - E_{\mathbf{k+q}} + \hbar\omega_0)], \tag{4.56}$$

where, again, the delta functions have been included, although they are already taken into account in the derivations of Section 3.3.1. This result is now inserted into (4.50) to give the scattering rate as

$$\Gamma(k) = \frac{m^* e^2}{4\pi\hbar^2 k\gamma\omega_0}\left[(N_q + 1)\int_{q_-^e}^{q_+^e} \frac{q^3 dq}{(q^2 + q_D^2)^2} + N_q \int_{q_-^a}^{q_+^a} \frac{q^3 dq}{(q^2 + q_D^2)^2}\right]. \tag{4.57}$$

The limits for the emission and absorption terms are given by (4.48) and (4.49), respectively. The final result for the screened interaction is

$$\Gamma(k) = \frac{m^* e^2\omega_0}{4\pi\hbar^2 k}\left(\frac{1}{\varepsilon_\infty} - \frac{1}{\varepsilon(0)}\right)\left[G(k) - \frac{q_D^2}{2}H(k)\right], \tag{4.58}$$

where

$$G(k) = (N_q + 1)\ln\left[\frac{\left(k + \sqrt{k^2 - q_0^2}\right)^2 + q_D^2}{\left(k - \sqrt{k^2 - q_0^2}\right)^2 + q_D^2}\right]^{\frac{1}{2}}$$

$$+ N_q \ln\left[\frac{\left(\sqrt{k^2 + q_0^2} + k\right)^2 + q_D^2}{\left(\sqrt{k^2 + q_0^2} - k\right)^2 + q_D^2}\right]^{\frac{1}{2}}, \tag{4.58a}$$

$$H(k) = (N_q + 1)\frac{4\sqrt{k^2 - q_0^2}}{(q_0^2 - q_D^2)^2 + 4k^2 q_D^2}$$

$$+ N_q \frac{4\sqrt{k^2 + q_0^2}}{(q_0^2 + q_D^2)^2 + 4k^2 q_D^2}, \tag{4.58b}$$

$$q_0^2 = \frac{2m^*\omega_0}{\hbar} = \frac{\hbar\omega_0}{E_k}, \tag{4.58c}$$

and the emission terms should be multiplied by the Heaviside function to assure that they occur only when the carrier energy is larger than the phonon energy. This value of q_0 is the so-called "dominant phonon" wave vector, and can be used to estimate the reduction in screening that can occur. If this is done, the Debye wave vector q_D is reduced at most by a factor of $2^{\frac{1}{2}}$.

Screening plays a significant role in the scattering of carriers by the polar optical phonon interaction. In both terms, the screening wave vector acts to reduce the amount of scattering that occurs. In the first term, the screening wave vector works to reduce the magnitude of the ratio of terms that occurs inside the logarithm arguments, hence reducing the scattering strength. The second term is negative, which also reduces the strength. In the absence of screening, where $q_D \sim 0$ (which occurs at very low densities of free carriers), the equation above reduces to the more normal form

$$\Gamma(k) = \frac{m^* e^2 \omega_0}{4\pi\hbar^2 k}\left(\frac{1}{\varepsilon_\infty} - \frac{1}{\varepsilon(0)}\right)\left[(N_q + 1)\ln\left(\frac{k + \sqrt{k^2 - q_0^2}}{k - \sqrt{k^2 - q_0^2}}\right)u_0(E_k - \hbar\omega_0)\right.$$

$$\left. + N_q \ln\left(\frac{\sqrt{k^2 + q_0^2} + k}{\sqrt{k^2 + q_0^2} - k}\right)\right]. \tag{4.59}$$

It is assumed here that spin degeneracy of the final states has not been taken into account in the prefactors.

4.3.5 Scattering in semiconductors

Before continuing, it is perhaps fruitful to review the nature of the understanding of the various scattering processes that are important in typical semiconductors. Here, only Si, Ge, and a few of the group III–V materials are reviewed, primarily because a full understanding of transport in nearly all semiconductors is still lacking. What is presented here is the state of understanding that currently exists, with a few of the speculations that appear in the literature.

4.3.5.1 Silicon

The conduction band of Si has six equivalent ellipsoids located along the Δ [these are the (100) axes] lines about 85 percent of the way to the zone edge at X. Scattering within each ellipsoid is limited to acoustic phonons and impurities (discussed earlier), as the intra-valley optical processes are forbidden, as indicated in Table 4.1. Acoustic mode scattering, by way of the deformation potential, is characterized by two constants Ξ_u and Ξ_d, which are thought to have values of 9 eV and −6 eV, respectively (Ridley, 1982). The effective deformation potential is then the sum of these, or about 3 eV. Non-polar optical scattering occurs for scattering between the equivalent ellipsoids. There are two possible phonons that can be involved in this process. One, referred to as the *g*-phonon, couples the two valleys along opposite ends of the same (100) axis. This is the umklapp process discussed previously, and has a net phonon wave vector of $0.37\pi/a$. The symmetry allows only the LO mode to contribute to this scattering. At the same time, *f*-phonons couple the (100) valley to the (010) and (001) valleys, and so on. The wave vector has a magnitude of $2^{\frac{1}{2}}(0.85)\pi/a = 1.2\pi/a$, which lies in the square face of the Brillouin zone (Figure 2.7) along the extension of the (110) line into the second Brillouin zone. The phonons here are near the X-point phonons in value but have a different symmetry. Nevertheless, Table 4.1 illustrates that both the LA and TO modes can contribute to the equivalent inter-valley scattering. Note that the energies of the LO *g*-phonon and the LA and TO *f*-phonons are all nearly the same value, while the low-energy inter-valley phonons are forbidden. Long (1960), however, has found from careful analysis of the experimental mobility versus temperature that a weak low-energy inter-valley phonon is required to fit the data. In fact, he treats the allowed high-energy phonons by a single equivalent inter-valley phonon of 64.3 meV, but must introduce a low-energy inter-valley phonon with an energy of 16.4 meV. The presence of the low-energy phonons is also confirmed by studies of magnetophonon resonance (where the phonon frequency is equal to a multiple of the cyclotron frequency) in Si inversion layers, which indicates that scattering by the low energy phonons is a weak contributor to the transport (Eaves *et al.*, 1975). The low-energy phonon is certainly forbidden and Long treats it with a very weak coupling constant. Ferry (1976) points out that the forbidden low-energy inter-valley phonon must be treated by the first-order interaction and fits the data with a coupling constant of $\Xi_0 = 5.6$ eV, while the allowed transition is treated with a coupling constant of $D = 9 \times 10^8$ eV/cm. There are few experimental data to confirm these values directly, so they must be taken merely as an indication of the order of magnitude to be expected for these interactions. However, when used in Monte Carlo simulations, they fit quite closely to results computed with a full-band structure used in the calculations (discussed further below), although a value of Ξ_0 closer to 6 eV seems to give better behavior.

The valence band has considerable anisotropy, and the degeneracy of the bands at the zone center can be lifted by strain. Nevertheless, the acoustic deformation potential is thought to have an effective value of about 2.5 eV. Optical modes can couple holes from one valence band to the other, but there is little information on the strength of this coupling.

4.3.5.2 Germanium

The conduction band of germanium has four equivalent ellipsoids located at the zone edges along the (111) directions – the L points. The acoustic mode is characterized by the two deformation potentials, Ξ_u and Ξ_d, which are thought to have values of about 16 and –9 eV, respectively. These lead to an effective coupling constant of about 9 eV. Optical intra-valley scattering is allowed by the LO mode. Equivalent intervalley scattering is also allowed by the X-point LA and LO phonons (which are degenerate). The coupling constant for these phonons is fairly well established at 7×10^8 eV/cm from studies of the transport at both low fields (as a function of temperature) and high fields (Paige, 1969).

The holes in Ge are characterized by the anisotropic valence bands, just as in Si, and the acoustic deformation potentials are very close to the values for Si. Again, little is known about the coupling constants for inter-valence band scattering by optical modes.

4.3.5.3 Group III–V Compounds

In GaAs, InP, and InSb, the acoustic deformation potential is about 7 eV, although many other values have been postulated in the literature. The conduction band is characterized by Γ, *L*, *X* ordering of the various minima. The transport in the Γ valley is dominated by the polar LO mode scattering, while at sufficiently high energy, the carriers can scatter to the *L* and *X* minima through nonequivalent inter-valley scattering. The deformation fields for these two processes are fairly well established in GaAs to be 7×10^8 and 1×10^9 eV/cm, respectively (Kash *et al.*, 1983), through both experimental measurements and theoretical calculations (even though debate has not subsided in the literature). InP is thought to have the same values (Shah *et al.*, 1987). The L valleys are similar to those of Ge, so the L–L scattering should be given by the Ge values. L–X scattering is thought to have a deformation field of 5×10^8 eV/cm, although there is no real experimental evidence to support this. The Γ–L scattering rate in InSb is thought to be somewhat stronger, on the order of 1×10^9 eV/cm (Fawcett and Ruch, 1969; Curby and Ferry, 1973).

Again, the holes are characterized by anisotropic valence bands, but in GaAs it is thought that the dominant acoustic deformation potential is about 9 eV, while inter-valence band scattering has been treated through both the polar LO mode and the non-polar TO mode. The latter is thought to have a deformation field of about 1×10^9 eV/cm, and this value has been used in some discussions of transport (Osman and Ferry, 1987).

4.4 The Boltzmann equation once again

It is now apparent from the discussions above that it is necessary to find the proper form for the distribution function in order to solve the transport problem completely. While simple models give qualitative agreement with the expected results, full understanding requires the full and complete details of the carrier–lattice interactions. In this section, several analytical approaches to achieve closed-form (or nearly closed-form) solutions are discussed. These are followed in a later section by a discussion of the full numerical approaches that have become prevalent in recent years. The treatment is limited to a discussion of electrons.

In trying to understand the relaxation-time approximation of the previous chapter, it was necessary to go beyond the simple approach. This is particularly true in situations in which an applied field may drive the distribution function further from equilibrium than assumed. In the present section, we will examine a more useful and intuitive approach in which the distribution function is expanded in a series of Legendre polynomials $P_m(\cos\theta)$, where θ is the angle between the electric field and the velocity (or momentum) of the carriers. The rationale for this choice of expansion is that the electric field is a symmetry-breaking operator that introduces a preferred axis along which a shift of the distribution function occurs. On the other hand, there is no breaking of symmetry in the azimuthal plane around the electric field vector (in the absence of a magnetic field), so that the polar angle becomes a good expansion function for the cylindrical symmetry of the problem. This approach will in turn lead to the useful iterative approach introduced by Rode (1975), which is based on the Legendre expansion.

The general expansion of the distribution function takes the polar axis along the electric field vector *F*, with the polar angle being θ. Then the distribution function can be expanded as

$$f(E,\cos\theta) = \sum_{m=0}^{\infty} f_m(E)P_m(\cos\theta). \tag{4.60}$$

In general, for isotropic and elastic scattering, it is completely reasonable to terminate this series at $m = 1$, which is the basis of the relaxation-time approximation. If, however, the scattering is not predominantly elastic or isotropic, such as occurs for optical phonon scattering, further terms need to be kept in the series. In fact, in the case of polar optical phonon scattering, the scattering is so anisotropic that nearly all terms in the expansion must be retained, and this approach is too complicated for any case other than near equilibrium.

In the following, it will be assumed that the energy surfaces are spherical and parabolic, although the results can easily be extended to more complicated surfaces. Here, the steady-state, homogeneous semiconductor will be treated, so that the results are those for electrical conductivity in a homogeneous material. The force term of the Boltzmann equation then takes the form (Paige, 1964)

$$eF \cdot \frac{\partial f}{\partial k} = \frac{eF}{\hbar}\sum_{m=0}^{\infty}\left[\frac{mk^{m-1}}{(2m-1)}\frac{\partial}{\partial k}(k^{1-m}f_{m-1}) + \frac{m+1}{k^{m+2}(2m+3)}\frac{\partial}{\partial k}(k^{m+2}f_{m+1})\right]P_m(\cos\theta)$$

$$= \frac{eF}{\hbar}\left[\frac{2}{3k}f_1(E) + \frac{\hbar v}{3}\frac{\partial f_1}{\partial E} + \hbar v\frac{\partial f_0}{\partial E}\cos\theta\right] + \ldots. \tag{4.61}$$

The collision term may be expressed, for non-degenerate material, as

$$\left.\frac{\partial f}{\partial t}\right|_{coll.} = \left.\frac{\partial f_0}{\partial t}\right|_{coll.} + \left.\frac{\partial f_1}{\partial t}\right|_{coll.} + \ldots \tag{4.62}$$

where the second term, in order to simplify the equations somewhat, will be written as

$$f_1(E)\cos\theta = \mathbf{k}\cdot\mathbf{F}g(E) = kFg(E)\cos\theta$$

$$= -g(E)\int_{S_k}\mathbf{k}\cdot\mathbf{F}(1-\cos\theta')P(k,k')k'^2\sin\theta' d\theta' d\phi = -\frac{f_1}{\tau_m}. \tag{4.63}$$

Here, θ' is the angle between **k** and **k**′. Only the elastic scattering has been retained since this term primarily involves the relaxation of momentum and the inelastic effects will be included in the scattering term for f_0. It may be noted from (4.63) that this term defines a relaxation-time approximation, since it may readily be recognized to have the same form as (3.52). Thus, the momentum relaxation time has been introduced formally.

Before proceeding to compute the scattering term for the zero-order term in the Legendre expansion, a detour will be made. To proceed, equations (4.63), and the equivalent term for f_0, must be equated to (4.61) to establish the Boltzmann equation. This leads to

$$\left.\frac{\partial f_0}{\partial t}\right|_{coll.} - \frac{f_1}{\tau_m}P_1 = \frac{eF}{\hbar}\left[\frac{2}{3k}\left(f_1 + E\frac{\partial f_0}{\partial E}\right) + \hbar \mathrm{v}\frac{\partial f_0}{\partial E}P_1\right] = 0 \tag{4.64}$$

for the two lowest-order terms. As in any expansion in a generalized Fourier series, the coefficients of each of the Legendre polynomials must balance separately, so that two equations result:

$$f_1(E) = -e\tau_m \mathrm{v}F\frac{\partial f_0(E)}{\partial E},$$

$$\left.\frac{\partial f_0}{\partial t}\right|_{coll.} = \frac{2eF}{3\hbar k}\left(f_1 + E\frac{\partial f_0}{\partial E}\right). \tag{4.65}$$

The first of equations (4.65) is exactly the relaxation-time approximation (3.51). However, the relaxation-time approximation has not been assumed for this equation. Rather, the relaxation time approximation arises by the disregard of the second equation in (4.65). In equilibrium, it is usually assumed that the distribution function is the equilibrium function, which means that the second equation may be disregarded. However, when inelastic and anisotropic scatterers are present, this is not the case, no matter how small the electric field. Thus the second equation, with f_1 replaced by the first equation, gives a differential equation for the zero-order distribution function. The term $f_0(E)$ is often called the *energy distribution function.*

To formulate the collision operator for the energy distribution function, it must be recognized that all four processes described by the inelastic scattering interaction can be involved in the variation of this distribution function. This involves both emission and absorption from the state $f_0(E)$, which are "outgoing" processes, as well as emission into this state from that at $f_0(E + \hbar\omega_0)$ and by absorption from that at $f_0(E - \hbar\omega_0)$. The latter two are "incoming" processes. Thus, the collision operator can be written quite generally in the descriptive notation

$$\left.\frac{\partial f_0}{\partial t}\right|_{coll.} = \frac{2\pi}{\hbar}\sum_{\mathbf{q}}\{|M_{em}(q,\mathbf{k}+\mathbf{q})|^2 f_0(E+\hbar\omega_0)\delta(E_{k+q}-E_k+\hbar\omega_0)$$
$$-|M_{em}(q,k)|^2 f_0(E)\delta(E_k-E_{k-q}+\hbar\omega_0)$$
$$+|M_{ab}(q,\mathbf{k}-\mathbf{q})|^2 f_0(E-\hbar\omega_0)\delta(E_k-E_{k-q}+\hbar\omega_0)$$
$$-|M_{ab}(q,\mathbf{k}-\mathbf{q})|^2 f_0(E)\delta(E_k-E_{k+q}-\hbar\omega_0)\}. \qquad (4.66)$$

This can now be combined with equations (4.65) to provide a (truncated) second-order differential equation for the energy distribution function, once the inelastic processes have been identified. In general, however, this differential equation is quite complicated to solve. Very careful iterative techniques must be used to assure the calculation of the stable solution. In fact, useful results are obtained only in the situation in which $k_BT_e \gg \hbar\omega_0$, where T_e is an effective electron temperature, describing the average energy of the distribution (we will see more of this in a later section). For this situation, the distribution functions in (4.66) can be expanded in a Taylor series around the energy E, and the resulting differential equation becomes somewhat easier to solve (Yamashita and Watanabe, 1954; Reik and Risken, 1961). However, this approach has not been used in recent years, because of the rise of the Monte Carlo technique (discussed below), although recent work of the Bologna group (Gnudi *et al.*, 1993; Reggiani, 1998) has refocused on the solution of these equations by computational techniques.

4.5 Rode's method

In the presence of inelastic scattering processes, we have already pointed out that the relaxation-time approximation usually fails to provide adequate accuracy for calculating the real mobility in semiconductors. An approach like the Legendre expansion can give enhanced accuracy, but at relatively great computational and analytical expense. On the other hand, a technique introduced by Rode (1975) takes the spirit of the Legendre expansion and provides a natural iterative procedure to calculate the actual distribution function. From this distribution function, any of the various transport coefficients can be calculated. The heart of the approach lies in a generalized expansion, such as used in the Legendre polynomial expansion, except that only two terms are retained:

$$f(\mathbf{k}) = f_0(E) + g(\mathbf{k})\cos\theta. \qquad (4.67)$$

In form, this looks exactly like the first two terms of the Legendre expansion. The difference here is that $g(\mathbf{k})$ is not guaranteed to be a function only of the energy, but contains exact variations with k so that the expression (4.67) is exact in small electric fields where linear response is valid (it is assumed that f_0 retains the equilibrium form, which is why this method is limited to low electric fields).

The Boltzmann equation, in the case where the material may in fact be degenerate, is given by the combination of (3.43) and (3.46). Since f_0 is the equilibrium part of the distribution function, by definition it satisfies the collision integral exactly in order to satisfy detailed balance. To proceed, only the steady-state solution will be obtained, although the technique may easily be extended to treat the transient case. Moreover, it will be assumed that both the electric field and the spatial gradient are in the z direction only. Then (4.67) is inserted into the Boltzmann equation, and the coefficients of the cosθ terms are grouped together to yield the equation

$$\mathrm{v}\frac{\partial f_0}{\partial z} + \frac{eF}{\hbar}\frac{\partial f_0}{\partial k} = -g(\mathbf{k})\int\{P(\mathbf{k},\mathbf{k}')[1 - f_0(\mathbf{k}')] + P(\mathbf{k}',\mathbf{k})f_0(\mathbf{k}')\}d^3\mathbf{k}'$$
$$+ \tfrac{3}{2}\int_0^{\pi}\cos\theta_k\int_{\mathbf{k}'}\cos\theta_{k'}g(\mathbf{k}')\{P(\mathbf{k}',\mathbf{k})[1 - f_0(\mathbf{k})]$$
$$+ P(\mathbf{k},\mathbf{k}')f_0(\mathbf{k})\}d^3\mathbf{k}'\sin\theta_k d\theta_k, \tag{4.68}$$

where the integration over θ_k picks out the cosθ term of the second integral, and θ_k and $\theta_{k'}$ are the polar angles between the two wave vectors **k** and **k**′, respectively, and the z axis. In general, the scattering processes are actually functions of the angle between **k** and **k**′, rather than the polar angles. This angle may be denoted as θ_0, and a general property of isotropic and spherical energy surfaces is that

$$\int\cos\theta_{k'}A(\cos\theta_0)d^3\mathbf{k}' = \cos\theta_k\int\cos\theta_0 A(\cos\theta_0)d^3\mathbf{k}'. \tag{4.69}$$

This follows from the fact that the angles satisfy certain relationships, such as $\cos\theta_{k'} = \cos\theta_k\cos\theta_0 + \sin\theta_k\sin\theta_0\cos\phi = \cos\theta_k\cos\theta_0$, since the last term averages to 0 in the ϕ integration. This now allows us to simplify the expression (4.68) to

$$\mathrm{v}\frac{\partial f_0}{\partial z} + \frac{eF}{\hbar}\frac{\partial f_0}{\partial k} = -g(\mathbf{k})\int\{P(\mathbf{k},\mathbf{k}')[1 - f_0(E')] + P(\mathbf{k}',\mathbf{k})f_0(E')\}d^3\mathbf{k}'$$
$$+\int\cos\theta_0 g(\mathbf{k}')\{P(\mathbf{k}',\mathbf{k})[1 - f_0(E)]$$
$$+ P(\mathbf{k},\mathbf{k}')f_0(E)\}d^3\mathbf{k}'. \tag{4.70}$$

As remarked above, this result is exact for small electric fields where linear response theory is valid. After integration over $\mathbf{k}'$, the result is a function only of the single magnitude k, as there are no angular parts left in the equation and g is assumed to be a function of the magnitude of k alone. Thus $g(k)$ is isotropic for isotropic energy bands. The exact form of the Fermi–Dirac or Maxwellian distribution is maintained by the inelastic processes, as the shape of the distribution is unaffected by the elastic scatterers. This was evident for the relaxation-time approximation and will be reinforced in the next paragraph. Since all transport coefficients are given by integrals over g, all of the integrals can be performed, even with non-parabolic bands.

In the following, it is desirable to split the scattering function $P(\mathbf{k},\mathbf{k}')$ (and its reciprocal process) into separate elastic and inelastic parts. This is because the equilibrium distribution function is then evaluated on the same energy shell for the elastic processes and these all reduce to the single term

$$g(k)\int[1-\cos\theta_0]P(\mathbf{k},\mathbf{k}')d^3\mathbf{k}' \equiv \frac{g(k)}{\tau_m}, \tag{4.71}$$

which is precisely the relaxation-time approximation result, where the definition of the momentum relaxation time is obvious in light of previous discussions. Equation (4.70) can now be rewritten as

$$\begin{aligned} \mathrm{v}\frac{\partial f_0}{\partial z}+\frac{eF}{\hbar}\frac{\partial f_0}{\partial k} = &-g(\mathrm{k})\int\{P_{in}(\mathbf{k},\mathbf{k}')[1-f_0(E')]+P_{in}(\mathbf{k}',\mathbf{k})f_0(E')\}d^3\mathbf{k}' \\ &-\frac{g(\mathrm{k})}{\tau_m}+\int\cos\theta_0 g(\mathbf{k}')\{P_{in}(\mathbf{k}',\mathbf{k})[1-f_0(E)] \\ &+P_{in}(\mathbf{k},\mathbf{k}')f_0(E)\}d^3\mathbf{k}'. \end{aligned} \tag{4.72}$$

From this expression, the iterative nature of the process begins to appear. Indeed, in the case for which there is no spatial variation and no inelastic scattering, (4.72) yields exactly the relaxation-time approximation. In general, however, the first term on the right-hand side provides the basis for the iterative procedure, as the new iterate is formed from this basic beginning.

In order to simplify the equations somewhat in the following, a reduced notation will be introduced. Two functionals will be defined through

$$S_{out}=\int\{P_{in}(\mathbf{k},\mathbf{k}')[1-f_0(E')]+P_{in}(\mathbf{k}',\mathbf{k})f_0(E')\}d^3\mathbf{k}',$$

$$S_{in}(g')=\int\cos\theta_0 g(k')\{P_{in}(\mathbf{k}',\mathbf{k})[1-f_0(E)]+P_{in}(\mathbf{k},\mathbf{k}')f_0(E)\}d^3\mathbf{k}'. \tag{4.73}$$

These two terms can obviously be related to the scattering out of the state k and the scattering into the state k. With these definitions, (4.72) can then be written as

$$g(k) = \frac{S_{in}(g') - v\frac{\partial f_0}{\partial z} - \frac{eF}{\hbar}\frac{\partial f_0}{\partial k}}{S_{out}(k) + \frac{1}{\tau_m}}. \tag{4.74}$$

This result is an integral equation for the linear deviation from the equilibrium distribution function $g(k)$, and may be solved iteratively. The iterative process begins by assuming that the zero iterate is identically $g_0(k) = 0$. In this case, S_{in} is also zero, and the next iterate is quite simple to determine once S_{out} has been determined. The next iterate can then be formed from the sequence for the $(r + 1)$ term in the interation as

$$g_{r+1}(k) = \frac{S_{in}(g'_r) - v\frac{\partial f_0}{\partial z} - eFv\frac{\partial f_0}{\partial E}}{S_{out}(k) + \frac{1}{\tau_m}}. \tag{4.75}$$

Note that the only term which changes from one iteration to the next is S_{in}. The energy derivative has also been introduced here, as it is more commonly used, and has been utilized almost exclusively in earlier sections of this chapter. There are only two steps to the iteration: (i) compute the change in S_{in} for a given order g_r and (ii) use (4.75) to compute the new g_{r+1}. This sequence can be carried out until the function $g(k)$ converges to a steady solution. Transport coefficients can be computed from integrals introduced in earlier sections. For example, in the absence of any spatial gradients, the mobility may be found from

$$\mu = \frac{1}{3F}\frac{\int vk^2 g(k)dk}{\int k^2 f_0(k)dk}. \tag{4.76}$$

The Rode method has been used for non-parabolic bands, and with nonlinear screening of the impurity interaction. It has also been extended to the presence of magnetic fields and for multi-valley semiconductors. It is the most usable, accurate method for obtaining low-field transport coefficients in materials with multiple, complicated scattering processes.

4.6 The hydrodynamic equations

The treatment of high-electric-field effects can generally be broken into two distinct regions. These two can be differentiated by the concept of the electron temperature, which differs from the lattice temperature introduced previously. The concept of such a temperature is valid only when the spherically symmetric part of the distribution function, which we have called the *energy* distribution $f_0(E)$ above, remains in a Maxwellian form as $\exp(-E/k_BT_e)$. If the distribution function differs from this form, particularly in the case of high electric fields, the concept of the electron temperature becomes quite vague. However, Fröhlich and Paranjape (1956) showed many years ago that if the density of electrons is sufficiently high, energy and momentum exchanges are dominated primarily by inter-electronic collisions. These collisions provide the fast time scale that dominates the distribution function and forces it into a quasi-equilibrium form, as described by Bogoliubov (1946). If this assumption is valid, the electron distribution function in momentum space will be in internal equilibrium at an electron temperature T_e, which is a function of the electric field, although it may also be necessary to include other parameters, which themselves are related to constants of the motion. In general, the electron temperature is greater than the lattice temperature. The distribution function can then be thought of as shifted in momentum space, as in the near-equilibrium case, but the spreading of the distribution is fully characterized by the electron temperature. The shift in momentum space is described by the drift velocity (or drift momentum, as complications due to non-parabolic bands will not be included).

In a later chapter, the electron–electron interaction is treated in some detail. Here, a simpler argument will be presented, which illustrates the size of the carrier density needed to achieve the quasi-equilibrium distribution function. The energy loss due to energy exchange among carriers is characterized by the rate at which individual electrons are scattered. The energy loss may be approximated by the product of the inter-electronic Coulomb energy and the plasma frequency. The latter characterizes the time scale of the inter-electronic interactions. This gives

$$\left.\frac{dE}{dt}\right|_{e-e} \sim -\omega_p \frac{e^2}{4\pi\varepsilon_s \langle r\rangle} \sim -\omega_p \frac{e^2}{4\pi\varepsilon_s (v/\omega_p)} = -\frac{e^2\omega_p^2}{4\pi\varepsilon_s v} = -\frac{ne^4}{4\pi\varepsilon_s^2 m^* v}. \quad (4.77)$$

It should be remarked that carriers which are moving faster than the average velocity generally lose energy by this process, whereas those that are traveling less than the average velocity gain energy by this process. In general, averaging (4.77) over a Maxwellian produces a zero result, since inter-electronic collisions conserve total momentum, energy, and the number of carriers in the system. For this reason, it is usually assumed that, if a parameterized Maxwellian distribution is taken for the carriers, the details of the inter-electronic interactions can be neglected under the premise that they are already included *de facto* within the distribution function.

In order for inter-electronic collisions to dominate the energy relaxation, it is necessary for the energy relaxation rate described above to dominate the lattice scattering rate, which we may approximate within this simple approach as

$$\left.\frac{dE}{dt}\right|_{phonon} \sim \frac{\hbar\omega_0}{\tau_{phonon}}. \tag{4.78}$$

If this quantity is smaller than (4.77), the assumption of an electron temperature is valid. If, on the other hand, (4.78) is not smaller than (4.77), the distribution function can deviate markedly from a Maxwellian form. For the latter situation, an electron temperature cannot be formally defined, and the form of the distribution function must be found from a full solution of the Boltzmann equation, best achieved by e.g. the ensemble Monte Carlo technique discussed below. For most semiconductors, the critical density for which carrier–carrier scattering dominates may be estimated by the requirement that

$$\frac{ne^4}{4\pi\varepsilon_s^2 m^* v} > \frac{\hbar\omega_0}{\tau_{phonon}}, \tag{4.79a}$$

or

$$n > \frac{\hbar\omega_0 4\pi\varepsilon_s^2 m^* v}{e^4 \tau_{phonon}} \sim 5\times 10^{17}\ \mathrm{cm}^{-3}, \tag{4.79b}$$

where numbers appropriate to Si have been used.

The general approach followed, and that which will be described in detail in the following, for the case where the distribution function can be expressed as a generalized Maxwellian is to multiply the Boltzmann equation with powers of the momentum $p = m^* v = \hbar k$, and then integrate over the momentum itself. This produces a series of equations that balance the various moments of the Boltzmann equation and can be used to evaluate the electron temperature and drift velocity, which have been included in the distribution function as parameters, so that this drifted Maxwellian distribution is expressed as

$$f(\mathbf{p}, \mathbf{v}_d, T_e) \equiv f(\mathbf{p}) = C\exp\left[-\frac{(\mathbf{p} - m^*\mathbf{v}_d)^2}{2m^* k_B T_e}\right]. \tag{4.80}$$

Here, we have inserted the average momentum $m^*\mathbf{v}_d$ and the electron temperature T_e. C is a constant for normalization purposes. The first two terms of this generalized distribution function may be found by expanding the exponential factor as

$$f(\mathbf{p},\mathbf{v}_d,T_e) = C\exp\left[-\frac{p^2}{2m^*k_BT_e} + \frac{\mathbf{p}\cdot\mathbf{v}_d}{k_BT_e} - \frac{v_d^2}{2m^*k_BT_e}\right]$$

$$\approx C\exp\left[-\frac{p^2}{2m^*k_BT_e}\right]\left[1 + \frac{\mathbf{p}\cdot\mathbf{v}_d}{k_BT_e} + \ldots\right]$$

$$= C\exp\left[-\frac{E}{k_BT_e}\right] + C\left(\frac{\mathbf{p}\cdot\mathbf{v}_d}{k_BT_e}\right)\exp\left[-\frac{E}{k_BT_e}\right] + \ldots$$

$$\equiv f_0(E) + f_1(E) + \ldots. \tag{4.81}$$

It is clear from this that

$$f_1(E) = \left(\frac{\mathbf{p}\cdot\mathbf{v}_d}{k_BT_e}\right)f_0(E). \tag{4.82}$$

The actual moments of the Boltzmann equation are complicated by the fact that the semiconductor is often describable by multiple valleys and hence is a coupled, and complicated, system. Indeed, there will be conditions in which even sets of valleys, which are equivalent in equilibrium, can become inequivalent in a high electric field. Thus full account for transfer between inequivalent valleys will be considered as the hydrodynamic equations are developed. The starting point is, of course, the Boltzmann transport equation

$$\frac{\partial f}{\partial t} + \mathbf{v}\cdot\frac{\partial f}{\partial \mathbf{r}} + e\mathbf{F}\cdot\frac{\partial f}{\partial \mathbf{p}} = \sum_{\mathbf{p}'}[P(\mathbf{p},\mathbf{p}')f(\mathbf{p}') - P(\mathbf{p}',\mathbf{p})f(\mathbf{p})]. \tag{4.83}$$

Note that we are using the momentum here, rather than the crystal wave vector **k**, which has been used previously. A function $\phi(\mathbf{p})$ is assumed to be a simple power of the momentum p (e.g., p^n), and its average is described by

$$\langle\phi(\mathbf{p})\rangle = \frac{1}{n}\int \phi(\mathbf{p})f(\mathbf{p})d^3\mathbf{p}. \tag{4.84}$$

The Boltzmann equation is now multiplied by $\phi(\mathbf{p})$ and then integrated (or summed as the case may be) over the momentum. This results in the form, for carriers in valley i,

$$\frac{\partial(n_i\langle\phi\rangle)}{\partial t} = n_i e\mathbf{F}\cdot\left\langle\frac{\partial\phi_i}{\partial\mathbf{p}_i}\right\rangle - \frac{1}{m^*}\nabla\cdot(n_i\langle\mathbf{p}_i\phi_i\rangle)$$

$$+ \sum_{\mathbf{p}_i,\mathbf{p}_i'}[P(\mathbf{p}_i,\mathbf{p}_i')f_i(\mathbf{p}_i') - P(\mathbf{p}_i',\mathbf{p}_i)f_i(\mathbf{p}_i)]\phi_i(\mathbf{p}_i)$$

$$+ \sum_j\sum_{\mathbf{p}_i,\mathbf{p}_j'}[P(\mathbf{p}_i,\mathbf{p}_j')f_j(\mathbf{p}_j') - P(\mathbf{p}_j',\mathbf{p}_i)f_i(\mathbf{p}_i)]\phi_i(\mathbf{p}_i). \tag{4.85}$$

The first term on the right-hand side has been integrated by parts. Different sums have been provided for scattering within valley i and to/from valley j, which is assumed to be part of a different set of equivalent valleys (which must be summed over). The second term on the right-hand side accounts for diffusive forces. When the initial and final states lie in the same valley (or an equivalent valley), they are describable by the same distribution function (which itself is denoted by the subscript according to the valley index), and the various momenta are specific functions for the valleys in which they are described. However, when the initial and final states lie in non-equivalent valleys, the two distribution functions are, in general, not the same quasi-equilibrium distribution; for example, a family of distributions may be required to describe all of the valleys that are involved in the far-from-equilibrium transport problem. With this in mind, the intra-valley (or intra-equivalent valleys) terms on the right-hand side of (4.85) (the first scattering integrals) can be treated by a simple change of variables, since the summation is over both momenta, and this term becomes

$$n_i\langle\Gamma_\phi(\mathbf{p}_i)\rangle = n_i\sum_{\mathbf{p}'}P(\mathbf{p}_i',\mathbf{p}_i)[\phi(\mathbf{p}_i') - \phi(\mathbf{p}_i)]. \tag{4.86}$$

If $\phi = C_0$, this term vanishes and intra-valley terms do not contribute to the lowest order (density) balance equation. The remaining scattering term, for scattering between nonequivalent valleys, can be written as

$$\left.\frac{d(n_i\langle\phi\rangle)}{dt}\right|_{intervalley} = \sum_{\mathbf{p}_i}\phi(\mathbf{p}_i)\sum_{\mathbf{p}_j'}[P(\mathbf{p}_i,\mathbf{p}_j')f_j(\mathbf{p}_j') - P(\mathbf{p}_j',\mathbf{p}_i)f_i(\mathbf{p}_i)]$$

$$= \sum_{\mathbf{p}_j'}f_j(\mathbf{p}_j')\phi(\xi_j)\sum_{\mathbf{p}_i}P(\mathbf{p}_i,\mathbf{p}_j')$$

$$- \sum_{\mathbf{p}_i}f_i(\mathbf{p}_i)\phi(\mathbf{p}_i)\sum_{\mathbf{p}_j'}P(\mathbf{p}_j',\mathbf{p}_i), \tag{4.87}$$

where the energy-conserving delta function inherent in the scattering rates has been used to introduce ξ_j as a shifted momentum defined by

$$E_j(\xi_j) = E_i(\mathbf{p}_i) \pm \hbar\omega_0 + \Delta_{ji}. \tag{4.88}$$

The shift between the energy references has been included via Δ_{ji} to account for any energy differences between non-equivalent valley sets. Thus the momentum function $\phi(\mathbf{p})$ has now been shifted to the other summation. Introducing the inter-valley scattering rate as

$$\Gamma_{ij}(\mathbf{p}) = \sum_{\mathbf{p}'} P(\mathbf{p}'_j, \mathbf{p}_i), \tag{4.89}$$

(4.87) now becomes

$$\left.\frac{d\langle n_i \phi\rangle}{dt}\right|_{intervalley} = n_j \langle \phi(\xi_j)\Gamma_{ji}(\mathbf{p}'_j)\rangle_j - n_i \langle \phi(\xi_i)\Gamma_{ij}(\mathbf{p}_i)\rangle_i, \tag{4.90}$$

and the subscript on the average refers to the particular distribution function (for non-equivalent sets of valleys) used for the average. By inserting the details of the various scattering processes, the individual equations can be readily set up for any situation. The first moment becomes the carrier density balance equation, where $\phi(\mathbf{p}) = C_0$,

$$\frac{\partial n_i}{\partial t} + \nabla \cdot (n_i \mathbf{v}_{di}) = \sum_j [-\langle \Gamma_{ij}\rangle_i + \langle \Gamma_{ji}\rangle_j]. \tag{4.91}$$

The first two terms (on the left-hand side) lead to the continuity equation and the right-hand side corresponds to a generation and recombination due to inter-valley scattering. The momentum balance equation, where $\phi(\mathbf{p}) = \mathbf{p}$, is found to be

$$\frac{\partial(n_i m^* \mathbf{v}_{di})}{\partial t} = -\nabla \cdot (n_i m^* \langle \mathbf{v}\mathbf{v}\rangle_i) - n_i e\mathbf{F} - n_i \langle \Gamma(\mathbf{p})\rangle_i - n_i \sum_j \langle \mathbf{p}\Gamma_{ij}\rangle_i$$
$$+ \sum_j n_j \langle (\mathbf{p}_j - \mathbf{p}_{0,j})\Gamma_{ji}\rangle_j, \tag{4.92}$$

where $\mathbf{p}_{0,j}$ describes the effective minimum of the conduction band in the other valleys. The energy balance equation is found by using $\phi(\mathbf{p}) = p^2/2m^* = E$, and this is found to be

$$\frac{\partial(n_i \langle E\rangle_i)}{\partial t} = -\nabla \cdot (n_i \langle E\mathbf{v}\rangle_i) - n_i e\mathbf{F} \cdot \mathbf{v}_{di} - n_i \langle \Gamma_E(\mathbf{p})\rangle_i - n_i \sum_j \langle E\Gamma_{ij}\rangle_i$$
$$+ \sum_j n_j \langle (E - \Delta_{ji})\Gamma_{ji}\rangle_j. \tag{4.93}$$

In these equations, the quantity Δ_{ji} represents the shift between the bottom of valley set j and valley set i (this is a positive quantity if the bottom of valley j lies above that of valley i, and conversely). In each case, the last two terms on the right-hand side represent the transfer of particles (or momentum, or energy) in the inter-valley scattering process between non-equivalent sets of valleys. The last term is that for "in" scattering, and the next-to-last term is for "out" scattering processes for the set of valleys i. In the case of a single set of equivalent valleys, these two terms cancel each other.

As an example of the hydrodynamic equation approach, transport in a polar material such as GaAs will be considered. In these materials, electronic motion in the central valley of the conduction band is dominated by the polar mode of the optical phonons. The example considered is to calculate the limiting velocity of the carriers in the Γ valley (if such exists). In this single valley, the steady-state balance equations for the energy and momentum are

$$-e\mathbf{F}\cdot\mathbf{v}_d = \frac{1}{n}\langle\Gamma_E\rangle, \tag{4.94}$$

and

$$-e\mathbf{F} = \frac{1}{n}\langle\Gamma_p\rangle, \tag{4.95}$$

and both relaxation averages are only for the polar-optical phonon modes. The polar-mode scattering process for the optical phonons leads to the scattering rate (4.59) for the unscreened interaction, which we will use here. In order to apply the balance equations, the average rates of energy and momentum loss due to the scattering process must be known. In the case of the optical phonons, the energy loss (or gain) with each scattering process is a constant, since the optical modes have been taken to be non-dispersive in the discussion earlier in this chapter. Thus the energy loss rate can be expressed simply as

$$\begin{aligned}\Gamma_E &= -\hbar\omega_0\left(\frac{1}{\tau_{em}} - \frac{1}{\tau_{abs}}\right)\\ &= -\frac{m^*e^2\omega_0^2}{4\pi\hbar k}\left(\frac{1}{\varepsilon_\infty} - \frac{1}{\varepsilon(0)}\right)\left[(N_q+1)\ln\left(\frac{k+\sqrt{k^2-q_0^2}}{k-\sqrt{k^2-q_0^2}}\right)u_0(E_k-\hbar\omega_0)\right.\\ &\quad\left. - N_q\ln\left(\frac{\sqrt{k^2+q_0^2}+k}{\sqrt{k^2+q_0^2}-k}\right)\right].\end{aligned} \tag{4.96}$$

The general forms (4.87) can now be used to compute the average energy loss rate by incorporating an average of the distribution function, as represented by f_0, as

$$-\langle\Gamma_E\rangle = -\frac{m^* e^2\omega_0^2}{2\pi(e^x - 1)}\left(\frac{1}{\varepsilon_\infty} - \frac{1}{\varepsilon(0)}\right)\int_0^\infty \frac{\rho(E)e^{-E/k_BT_e}}{\sqrt{2m^*E}}\left[\sinh^{-1}\left(\frac{E}{\hbar\omega_0}\right)^{\frac{1}{2}}\right.$$

$$\left. - e^x \sinh^{-1}\left(\frac{E}{\hbar\omega_0} - 1\right)^{\frac{1}{2}} u_0(E_k - \hbar\omega_0)\right] dE. \qquad (4.97)$$

Only the spherically symmetric part of the distribution function has been used in the integrals, as the energy-loss function is an even function of the momentum (it only depends on the energy). In the above equation, we have introduced the density of states $\rho(E)$ and the short-hand notation $x = \hbar\omega_0/k_BT_L$, where T_L is the lattice temperature. We will later use an equivalent x_e, in which the lattice temperature is replaced by the electron temperature. The quantity $E - \hbar\omega_0$ is replaced by E' in the second term in brackets, and the integration modified by this change of variables, so that the total function can be written as

$$-\langle\Gamma_E\rangle = \frac{m^* e^2\omega_0^2}{4\pi^3(e^x - 1)}\left(\frac{1}{\varepsilon_\infty} - \frac{1}{\varepsilon(0)}\right)\frac{2m^*}{\hbar^3}(e^{x-x_e} - 1)\int_0^\infty \sinh^{-1}\left(\frac{E}{\hbar\omega_0}\right)^{\frac{1}{2}} e^{-E/k_BT_e}\, dE. \quad (4.98)$$

The integral can be evaluated, with the change of variables $y = \sinh(2u)$, with $y = E/\hbar\omega_0$, so that

$$-\langle\Gamma_E\rangle = \frac{m^{*2}e^2\omega_0^3}{2\pi^3\hbar^2(e^x - 1)}\left(\frac{1}{\varepsilon_\infty} - \frac{1}{\varepsilon(0)}\right)(e^{x-x_e} - 1)\int_0^\infty \sinh^{-1}\sqrt{y}\, e^{-x_e y} dy$$

$$= \frac{m^{*2}e^2\omega_0^2}{8\pi^3\hbar^3(e^x - 1)}\left(\frac{1}{\varepsilon_\infty} - \frac{1}{\varepsilon(0)}\right)(e^{x-x_e} - 1)k_BT_e e^{-x_e/2}K_0(x_e/2). \qquad (4.99)$$

Thus, a relatively straightforward closed-form solution has been achieved. Here, K_0 is a modified Bessel function of the second kind. Similarly, the density may be evaluated from f_0 (the average of the first-order term vanishes, of course) as

$$n = \int_0^\infty \rho(E)e^{-E/k_BT_e}\, dE = \frac{1}{4}\left[\frac{2m^*k_BT_e}{\pi\hbar^2}\right]^{\frac{3}{2}}. \qquad (4.100)$$

An arbitrary normalization with the Fermi energy level has been omitted, as it will divide out of both numerator and denominator for these equations. This finally leads to the energy balance equation

$$eF \cdot v_d = \frac{m^* e^2 \omega_0^2 (e^{x - x_e} - 1)}{2\pi\sqrt{2\pi m^* k_B T_e}(e^{x} - 1)} \left(\frac{1}{\varepsilon_\infty} - \frac{1}{\varepsilon(0)} \right) e^{-x_e/2} K_0(x_e/2). \tag{4.101}$$

Calculating the momentum loss rate is somewhat more difficult, as the momentum exchanged in a collision is not a constant that can be removed from the summation over final states. Hence, we must return to the use of equations such as (4.50) and incorporate directly the momentum exchanged in the relaxing collision. The quantity of interest is just

$$\left\langle \frac{dp}{dt} \right\rangle_{polar} = \left\langle \frac{dp_F}{dt} \right\rangle_{polar} \sim -\left\langle \frac{m^* v_d}{\tau_{PO}} \right\rangle, \tag{4.102}$$

according to the format of (4.87). However, the momentum exchange involves the phonon wave vector and this must be included in the summation over the latter quantity, which is a modification of (4.50). The quantity that needs to be determined is just

$$\frac{dp_F}{dt} = \frac{2\pi}{\hbar} \sum_{q} \{ \hbar q_F |M_{abs}(k,q)|^2 \delta(E_{k+q} - E_k - \hbar\omega_0)$$

$$- \hbar q_F |M_{em}(k,q)|^2 \delta(E_{k-q} - E_k + \hbar\omega_0) \}, \tag{4.103}$$

where p_F and $\hbar q_F$ are the momentum along the field direction. This can be found from the relationship

$$q \cdot F = qF[\cos\theta\cos\theta_F + \sin\theta\sin\theta_F\cos(\phi - \phi_F)], \tag{4.104}$$

where θ and θ_F are the angles that q and F make with the direction of **k**, and the latter is taken as the polar direction. The second term vanishes under the integration over the azimuthal ϕ angle in (4.46), and the value of $\cos\theta$ can be taken from the argument of the delta function by the prescription of (4.47) as

$$\cos\theta = \mp\frac{q}{2k} \pm \frac{m^*\omega_0}{\hbar k q}. \tag{4.105}$$

This result can be used to incorporate qF into (4.103), and with the matrix element from (4.56) (setting $q_D = 0$ for no screening), we can write the momentum exchange as

$$\frac{dp_F}{dt}=\frac{m^*e^2\omega_0}{4\pi\hbar k(e^x-1)}\left(\frac{1}{\varepsilon_\infty}-\frac{1}{\varepsilon(0)}\right)\left\{\int_{q_a^-}^{q_a^+}\left(\frac{m^*\omega_0}{\hbar kq}-\frac{q}{2k}\right)dq\right.$$

$$\left.-e^x\int_{q_e^-}^{q_e^+}\left(\frac{m^*\omega_0}{\hbar kq}+\frac{q}{2k}\right)dq\right\}\cos\theta_p. \qquad (4.106)$$

The second term in each case can be integrated trivially, while the first term is just the original integral encountered in Section 4.3.4. The cosine term is just the expansion term used to create the various moments of the distribution function and used in the heirarchy of (4.85). Recognizing that this is just the contribution to the momentum balance equation, we can ignore this cosine term. The result is

$$\frac{dp_F}{dt}=-\frac{m^*e^2\omega_0}{4\pi\hbar k(e^x-1)}\left(\frac{1}{\varepsilon_\infty}-\frac{1}{\varepsilon(0)}\right)\left\{k\sqrt{1+\frac{\hbar\omega_0}{E}}-\frac{2m^*\omega_0}{\hbar k}\sinh^{-1}\left(\frac{E}{\hbar\omega_0}\right)^{\frac{1}{2}}\right.$$

$$\left.+e^xu_0(E-\hbar\omega_0)\left[k\sqrt{1-\frac{\hbar\omega_0}{E}}+\frac{2m^*\omega_0}{\hbar k}\sinh^{-1}\left(\frac{E}{\hbar\omega_0}-1\right)^{\frac{1}{2}}\right]\right\}. \qquad (4.107)$$

This must still be averaged over the distribution function.

The averaging procedure outlined above can now be used to compute the average momentum loss rate. This gives

$$\frac{1}{n}\left\langle\frac{dp_F}{dt}\right\rangle=\frac{m^*v_d}{nk_BT_e}\int_0^\infty v(E)\rho(E)\left(\frac{dp_F}{dt}\right)e^{-E/k_BT_e}dE$$

$$=-\frac{2e^2v_d\omega_0}{3\hbar k_BT_e(e^x-1)}\left(\frac{1}{\varepsilon_\infty}-\frac{1}{\varepsilon(0)}\right)\left(\frac{m^*}{2\pi k_BT_e}\right)^{\frac{3}{2}}\int_0^\infty G(E)e^{-E/k_BT_e}dE,$$

$$G(E)=\sqrt{E(E+\hbar\omega_0)}-\hbar\omega_0\sinh^{-1}\left(\frac{E}{\hbar\omega_0}\right)^{\frac{1}{2}}+e^x\left[\sqrt{E(E-\hbar\omega_0)}\right.$$

$$\left.-\hbar\omega_0\sinh^{-1}\left(\frac{E}{\hbar\omega_0}-1\right)^{\frac{1}{2}}\right]u_0(E-\hbar\omega_0). \qquad (4.108)$$

As previously, we can shift the energy range of the integration in the two terms involving the emission process, and this results in the same integrals as for the absorption terms, but with a prefactor that involves both the lattice temperature (e^x) and the electron temperature (e^{-x_e}). The two integrals are now

$$
\begin{aligned}
I_1 &= \int_0^\infty \sqrt{E(E+\hbar\omega_0)}\,e^{-E/k_BT_e}\,dE \\
&= (\hbar\omega_0)^2 \int_0^\infty u^{\frac{1}{2}}(u+1)^{\frac{1}{2}} e^{-x_e u}\,du \\
&= \frac{(\hbar\omega_0)^2}{x_e} e^{x_e/2} K_1(x_e/2),
\end{aligned} \tag{4.109}
$$

and

$$
\begin{aligned}
I_2 &= \hbar\omega_0 \int_0^\infty \sinh^{-1}\left(\frac{E}{\hbar\omega_0}\right)^{\frac{1}{2}} e^{-E/k_BT_e}\,dE \\
&= (\hbar\omega_0)^2 \int_0^\infty \sinh^{-1}(u^{\frac{1}{2}}) e^{-x_e u}\,du \\
&= \frac{(\hbar\omega_0)^2}{x_e} e^{x_e/2} K_0(x_e/2).
\end{aligned} \tag{4.110}
$$

As before, K_0 and K_1 are modified Bessel functions of the second kind. Finally, the relations above can be combined to give the net momentum loss rate per electron

$$
\begin{aligned}
\frac{1}{n}\left\langle \frac{dp_F}{dt} \right\rangle &= \frac{m^* e^2 v_d}{3\hbar(e^x - 1)} \left(\frac{1}{\varepsilon_\infty} - \frac{1}{\varepsilon(0)}\right) \left(\frac{2m^*\omega_0}{\pi\hbar}\right)^{\frac{1}{2}} x_e^{\frac{3}{2}} e^{x_e/2} \\
&\quad \times \left[(e^{x-x_e}+1)K_1(x_e/2) + (e^{x-x_e}-1)K_0(x_e/2)\right].
\end{aligned} \tag{4.111}
$$

This can now be equated to the field term, which gives the momentum balance equation:

$$
\begin{aligned}
eF &= m^* v_d \frac{e^2 x_e^{\frac{3}{2}} e^{x_e/2}}{3\hbar(e^x - 1)} \left(\frac{1}{\varepsilon_\infty} - \frac{1}{\varepsilon(0)}\right) \left(\frac{2m^*\omega_0}{\pi\hbar}\right)^{\frac{1}{2}} \\
&\quad \times \left[(e^{x-x_e}+1)K_1(x_e/2) + (e^{x-x_e}-1)K_0(x_e/2)\right].
\end{aligned} \tag{4.112}
$$

Equations (4.101) and (4.112) can now be solved simultaneously to find the drift velocity and electron temperature at a given electric field F. It must be pointed out that this is an exceedingly simplified model. Transport in GaAs is more complicated and factors such as inter-valley scattering and band non-parabolicity should be considered. However, the results are useful as an illustrative example of the application of the drifted Maxwellian technique. It should be noticed that the velocity does not really saturate for polar-optical-mode scattering, because if one solves the foregoing equations for the drift velocity, it is found that this quantity continues to depend on the electron temperature. This is found for most polar materials; the drift velocity is not found to exhibit the hard saturation found in silicon and germanium. The non-saturation can be demonstrated quite easily. Equations (4.101) and (4.112) can be combined to solve for the drift velocity as

$$v_d^2 = \frac{3k_BT_e}{2m^*}\left[1+\coth\left(\frac{x-x_e}{2}\right)\frac{K_1(x_e/2)}{K_0(x_e/2)}\right]^{-1}$$

$$\sim \frac{3\hbar\omega_0}{4m^*}\tanh\left(\frac{\hbar\omega_0}{2k_BT_e}\right)\ln\left(\frac{2k_BT_e}{\hbar\omega_0}\right), \qquad (4.113)$$

where the latter form arises for the limiting case of $x_e \ll 1, x$. The difference between this and (4.3) is the logarithm term. The temperature continues to increase with increasing electric field, as it does for nearly all forms of scattering, and the logarithmic term here makes the argument of the latter term continue to increase. This leads to a non-saturating drift velocity. This is a weak increase, and can be characterized as a "soft" saturation. A further problem arises due to the fact that the temperature will begin to increase almost exponentially with the field at very high field values (typically on the order of a few hundred kilovolts per centimeter). In the latter case, the velocity will begin to increase rapidly with the field. This is the onset of "runaway," which was studied by Stratton (1958) in his early uses of the drifted Maxwellian approach. Generally, approaches based on more exact dynamics, such as the ensemble Monte Carlo technique, also show the presence of runaway, but it normally occurs at higher fields than those found using the hydrodynamic equations. Although it is not clear that the method of solution has failed, it is also not clear that it applies very well to this situation, and this incongruity has never been resolved. The feature of runaway, mentioned earlier, appears due to the fact that the scattering rate for polar scattering has a weak decrease as the energy increases (for non-parabolic bands), which arises primarily from the prefactor in (4.58) varying as $1/k \sim 1/\sqrt{E}$.

The accuracy of the hydrodynamic equations obtained above ranges, in its applications to solving transport problems, from very good when the carrier density is relatively high, to exceedingly poor at lower carrier densities. The latter results arise in cases in which the assumptions about the form of the distribution function are just not valid, such as polar-mode scattering when the carrier density is low so that the distribution function is highly asymmetric with strong peaking along the field direction. It should be pointed out, however, that the validity of the hydrodynamic equation approach goes beyond the assumption of the distribution function, and can be used to evaluate any parameterized distribution function if the integrals can be evaluated. It turns out that the drifted Maxwellian is the general form of the distribution function if only two adiabatic invariants are introduced – the drift velocity and the electron temperature.

4.7 A return to electron kinetics

In Section 4.1.2, we discussed the transient velocity (the overshoot velocity) of the carriers. When overshoot occurs, the left-hand side of (4.4) must have at least two zeros – one at a time corresponding to the steady state and one at a time corresponding to the peak velocity. However, the relaxation rate is an increasing function of energy (or velocity), so that the right-hand side has only a single zero. Thus a second time scale must be involved, which is the characteristic time of the energy relaxation. With the Shockley model, it was found that the second time scale was related to the inelastic scattering, and, in the last section, we found that this could actually be treated through the energy balance equation, one of the hydrodynamic equations. The introduction of the second time scale leads to a description of the kinetic motion of the particles through a *retarded* Langevin equation, written as (Zwanzig, 1961)

$$m^* \frac{dv}{dt} = -\int_0^t \gamma(t-u)v(u)du + R(t) + eF. \tag{4.5}$$

The function $\gamma(t)$ is a "memory function" for the non-equilibrium system, and this will be related to the correlation function in this section. The other terms in (4.5) have the same meaning as used in Chapter 3. Equation (4.5) is a non-Markovian form of the Langevin equation, since the rate of change of the velocity at time t depends not only on the present time but also on all past time.

4.7.1 Retardation and the correlation function

We now want to explore what this equation tells us about the kinetic motion. This equation can easily be solved by a Laplace transform technique. Through this, we will generate the characteristic function $X(s)$, which will be related to the correlation function. We begin by Laplace transforming (4.5) to give (we also reinsert the particle index as a subscript, in keeping with the notation of the last chapter)

$$sV_i(s) - v_i(0) = \frac{eF}{sm^*} - V_i(s)\Gamma(s) + \frac{1}{m^*}R_m(s). \tag{4.114}$$

This may be rewritten, in analogy with (3.8), as

$$V_i(s) = v_i(0)X(s) + \frac{eF}{m^*}\frac{X(s)}{s} + \frac{1}{m^*}R_m(s)X(s), \tag{4.115}$$

where

$$X(s) = \frac{1}{s + \Gamma(s)}. \tag{4.116}$$

The function $\Gamma(s)$ is the Laplace transform of the kernel $\gamma(t)$ in (4.5). This leads to the solution

$$v_i(t) = v_i(0)x(t) + \frac{eF}{m^*}\int_0^t x(t')dt' + \frac{1}{m^*}\int_0^t R_m(t')x(t-t')dt'. \tag{4.117}$$

This is a general expression of the velocity of each carrier under the influence of the external field and the collisions. The function $x(t)$ may be found by averaging (4.117) over the entire ensemble, as discussed in Chapter 3, with the assumption that the random force has a zero average for the ensemble, and is also uncorrelated with $x(t)$. This leads to

$$v_d(t) = \frac{1}{N}\sum_i v_i(t) = \frac{eF}{m^*}\int_0^t x(t')dt', \tag{4.118}$$

which tells us that $x(t)$ is the relaxation function discussed by Kubo (1957). (In fact, this equation is a form of the Kubo formula.) To understand this function, consider the two-time correlation function

$$\begin{aligned}\varphi_v(t,t') &= \langle v_i(t)v_i(t')\rangle - \langle v_i(t)\rangle\langle v_i(t')\rangle \\ &= \langle v_i(t)v_i(t')\rangle - v_d(t)v_d(t').\end{aligned} \tag{4.119}$$

(This definition differs also from that of Chapter 3 in that it is not normalized by the mean-square velocity.) The two-time behavior arises because the distribution function is non-stationary. That is, in our far-from-equilibrium system, the distribution function is evolving with time, as are the average values of the carrier temperature and the drift velocity. These latter two quantities are termed integral constraints in thermodynamics, and no system in which these integral constraints are varying with time can be ergodic. (Ergodicity normally is taken to mean a system in which the time average is equal to the ensemble average, but this also requires that the ensemble average is *not* varying with time.) Hence, our correlation function depends not only on the observation time, but also upon the time at which the correlation was initialized. We can evaluate at least one form of (4.119) if both sides of (4.117) are multiplied by $v_i(0)$ and then the ensemble average is taken, under the assumptions that $\langle v_i(0)\rangle = 0$ (as above) and that the initial velocity is uncorrelated with the random force. The result is

$$\varphi_v(0,t) = \langle v_i^2(0)\rangle x(t). \tag{4.120}$$

Hence, x(t) is readily recognized as the normalized velocity auto-correlation function beginning at the initial time at which the field is applied (here, $t = 0$).

In (3.12), it was found that $x(t)$ was simply e^{-t/τ_m}, where τ_m was the momentum relaxation time. The function is more complicated here, because of the retardation in the Langevin equation. However, we can set some limits. For example, in the long-time limit for stationary fields, where the drift velocity becomes constant, the latter can be removed from the integral in (4.118), and we find that

$$\int_0^\infty x(t)dt = \tau_m. \tag{4.121}$$

Hence, we would recover this result if $\Gamma(s) = 1/\tau_m$, that is, $\gamma(t) = (1/\tau_m)\delta(t)$. But, this would not yield the required retardation. Hence, we need to seek another expression for the retardation function $\gamma(t)$. Since we desire to incorporate the temporal behavior of the velocity, which is to exhibit an overshoot behavior, the correlation function must be a non-monotonic function, which is quite different from the behavior found in equilibrium systems. The important addition is that of energy relaxation, which we examined in Section 4.1.2 and found that it led to a time variation of the relaxation time. This concept can be included, at least phenomenologically, by expanding the δ-function as

$$\gamma(t) = \frac{1}{\tau_m\tau_E}e^{-t/\tau_E}, \tag{4.122}$$

where the coefficient has been chosen to assure compatibility with (4.121) (this assures that the integral of $\gamma(t)$ is the same as that found in the equilibrium case, at least in form). This leads to

$$X(s) = \frac{\tau_m(1 + s\tau_E)}{s\tau_m(1 + s\tau_E) + 1}, \tag{4.123}$$

which clearly satisfies the long-time limit of (4.121). The Laplace transform of the drift velocity is then readily found to be

$$V_d(s) = \frac{eF}{m^*} \frac{s + \dfrac{1}{\tau_E}}{s\left[s\left(s + \dfrac{1}{\tau_E}\right) + \dfrac{1}{\tau_E \tau_m}\right]}, \tag{4.124}$$

which will generally lead to the overshoot behavior. In fact, a comparison with (4.38) for the Shockley model shows the same general form for the characteristic equation (denominator), and if we set $\tau_E = \tau_m/3$, the denominator differs only in the constant within the parentheses.

Clearly, the result (4.118) also tells us that the velocity overshoot remains closely connected to the velocity auto-correlation function. In fact, it is just the integral of this normalized quantity. This is illustrated in Figure 4.5. These results demonstrate a number of important aspects of hot-carrier behavior. First, the dynamics become retarded with a memory effect [the function $\gamma(t)$ is the memory function] because of the extra time behavior corresponding to the evolution of the distribution function. This, in turn, opens the door for velocity overshoot to occur. Moreover, this process, when coupled with the velocity saturation effect, clearly indicates the far-from-equilibrium nature of this nonlinear transport. In the field range where velocity saturation and velocity overshoot can occur, the distribution function is determined by carrier–lattice interactions and by boundary conditions in the form, for example, of applied fields, and is not simply related to the equilibrium form. There is one caveat, however, and that is that the transient velocity (which may be calculated in an ensemble Monte Carlo process) will be a result that depends upon all the scattering processes. On the other hand, the Kubo relationship obtained above is for the total correlation function, but we have assumed that there is no correlation between the random force and the velocity in the derivation. However, in the case of strong carrier–carrier interactions, this is not the case, and great care must be used in computing the actual correlation function in the many-body situation.

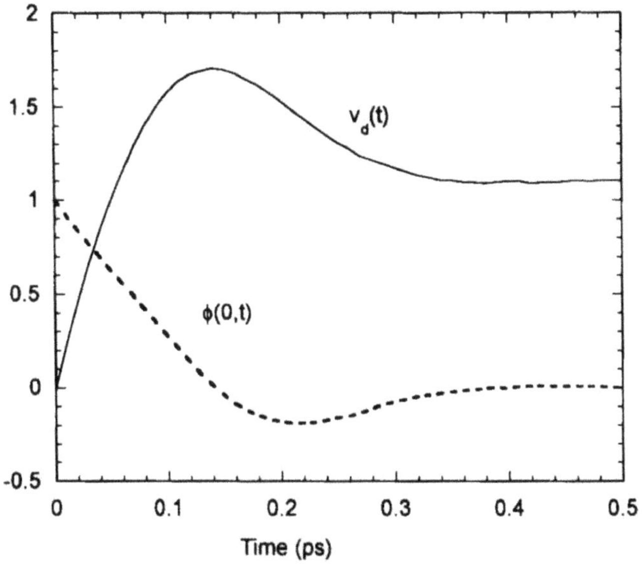

Figure 4.5 The drift velocity is compared with the normalized velocity auto-correlation function to show the intimate integral relationship between these two quantities. These have been calculated with an ensemble Monte Carlo technique. The velocity is in units of 10^7 cm/s, while the correlation function is normalized to unity at $t = 0$. The plot is for a field of 40 kV/cm in Si and a temperature of 300 K.

4.7.2 Energy relaxation and the memory function

In the above discussion, it was found that the retarded Langevin equation led to a situation in which the relaxation rate was actually varying with time, which is expected for velocity overshoot. Let us now examine this process a little further. The assumed time variation of the relaxation function is given by (4.122), and this leads to

$$\frac{d\gamma}{dt} = -\frac{1}{\tau_E}\gamma(t). \tag{4.125}$$

Indeed, $\gamma(t)$ evolves with the energy relaxation time. However, what does this say about the scattering rate $\Gamma_m = 1/\tau_m$? Certainly, the variation of the memory function with time is a representation of the variation of the scattering rate with time. However, the experience with the hydrodynamic equations of the previous section tells us that it is the energy that predominantly is the variable with time, and which responds with the energy relaxation time. On the other hand, the scattering processes are energy dependent, so if we write the general relaxation rate as

$$\Gamma_m = \Gamma_{m0}\left(\frac{E}{E_0}\right) = \frac{1}{\tau_m}\left(\frac{E}{E_0}\right), \tag{4.126}$$

where E_0 is the equilibrium thermal energy, then we can generally say

$$\frac{d\Gamma_m}{dt} \sim \frac{\Gamma_{m0}}{E_0}\frac{dE}{dt} \sim \frac{\Gamma_{m0}}{E_0}\left[eF\mathrm{v}_d(t) - \frac{E - E_0}{\tau_E}\right], \tag{4.127}$$

where we have introduced a simple relaxation time for the change in energy with collisions in (4.93). This may be Laplace transformed to give

$$\Gamma_m(s) = \frac{\Gamma_{m0}}{s} + \frac{eF\Gamma_{m0}}{E_0}\frac{V(s)}{s + \frac{1}{\tau_E}}$$

$$= \frac{\Gamma_{m0}}{s} + \frac{e^2F^2\Gamma_{m0}}{m^*E_0}\frac{X(s)}{s\left(s + \frac{1}{\tau_E}\right)}. \tag{4.128}$$

The time evolution of the relaxation process is clearly governed by both the energy relaxation time and the momentum relaxation time, the latter through the relaxation function $X(s)$. The characteristic function (denominator) of the second term is identical to that for the drift velocity in (4.124), so it is possible to also observe energy overshoot in the approach to the far-from-equilibrium steady state at high fields, when velocity overshoot is observed. However, the numerator is different, so that the details will differ as well. We can examine the long-time limit using (4.128), and find that

$$\lim_{t\to\infty}\Gamma_m(t) = \lim_{s\to 0} s\Gamma_m(s) = \Gamma_{m0}\left(1 + \frac{e^2F^2\tau_E\tau_m}{m^*E_0}\right). \tag{4.129}$$

Thus, we can see that the average energy gained in the field is a balanced process between the driving force of the field and the dissipative processes of the energy and momentum relaxation times.

4.7.3 *Far-from-equilibrium diffusion*

The treatment of diffusion in Chapter 3 can also be extended to the far-from-equilibrium situation, and we will discover an important new point. If we integrate the velocity, we find that

$$r_i(t) - \langle r_i(t)\rangle = \int_0^t \mathrm{v}_i(t_1)dt_1 - \int_0^t \mathrm{v}_d(t_1)dt_1. \tag{4.130}$$

We are now interested in the mean-square fluctuation in this position, so that

$$\begin{aligned}\langle [r_i(t) - \langle r_i(t)\rangle][r_i(t') - \langle r_i(t')\rangle]\rangle &= \int_0^t\int_0^{t'} \langle \mathrm{v}_i(t_2)\mathrm{v}_i(t_1)\rangle dt_1 dt_2 \\ &\quad - \int_0^t\int_0^{t'} \mathrm{v}_d(t_2)\mathrm{v}_d(t_1)dt_1 dt_2 \\ &= \int_0^t\int_0^{t'} \varphi_v(t_1,t_2)dt_1 dt_2.\end{aligned} \tag{4.131}$$

The definition of the diffusion coefficient can be found from (3.23) to give

$$D(t) = \frac{1}{2}\frac{d\langle \delta r_i^2(t)\rangle}{dt} = \int_0^t \varphi_v(t',t)dt'. \tag{4.132}$$

This integral is clearly different from the one in (4.118), which may be rewritten as

$$\mu(t) = \frac{e}{m^*\langle \mathrm{v}_i^2(0)\rangle}\int_0^t \varphi_v(0,t')dt'. \tag{4.133}$$

The former integral sums the value of the correlation function from a variable time up to the present time, while the latter integral sums the value of the final states of the correlation function. That is the mobility is essentially a sum over a single correlation function, while the diffusion coefficient sums a number of correlation functions, each beginning at a different time. Only in equilibrium can these two be shown to be equal. This is a general failing of the Einstein relationship, or more commonly known as a fluctuation–dissipation theorem, in the far-from-equilibrium situation which has been known for a great many years (Price, 1979). Much work has been expended in trying to find a new expression for this theorem, but the resolution has not yet occurred.

4.8 The ensemble Monte Carlo technique

Most of the analytical methods described above, and others found in use by various groups, are quite difficult to evaluate carefully in the real situation of a semiconductor with non-parabolic energy bands and complicated scattering processes. An alternative approach is to use the computer to completely solve the transport problem with a stochastic methodology fully in keeping with the retarded Langevin equation for carrier dynamics within an ensemble of carriers. The ensemble Monte Carlo (EMC) technique has been used now for more than three decades as a numerical method to simulate far-from-equilibrium transport in semiconductor materials and devices. It has been the subject of many reviews (Jacoboni and Reggiani, 1983; Jacoboni and Lugli, 1989; Hess, 1991). The approach taken here is to introduce the methodology, and how it is implemented, as well as to point out the differences between the EMC approach and a numerical solution of the Boltzmann equation. Indeed, many people believe that the EMC approach actually solves the Boltzmann equation, but this is true only in the long-time limit. For short times, the EMC is actually a more exact approach to the problem.

The EMC is built around the general Monte Carlo technique, in which a random walk is generated to simulate the stochastic motion of particles subject to collision processes. These collisions provide both the momentum relaxation process and the random force that appears in e.g. (4.5). Random walks, and stochastic techniques, may be used to evaluate complicated multiple-dimensional integrals (Kalos and Whitlock, 1986). In the Monte Carlo transport approach, we simulate the basic free flight of a carrier, and randomly interrupt this flight with instantaneous scattering events, which shift the momentum (and energy) of the carrier. Here, the length of each free flight, and the selection of the appropriate scattering process, are selected by weighted probabilities, with the weights adjusted according to the physics of the transport process. In this way, very complicated physics can be introduced without any additional complexity of the formulation (albeit at much more extensive computer time in most cases). At appropriate times through the simulation, averages are computed to determine quantities of interest, such as the drift velocity, average energy, and so forth. By simulating an ensemble of carriers, rather than the single carrier normally used in a Monte Carlo procedure, the non-stationary time-dependent evolution of the carrier distribution, and the appropriate ensemble averages, can be determined quite easily without resorting to any need for time averages. In the following sections, we will outline the general approach for the EMC procedure, and finally turn to its connection with the Boltzmann equation.

4.8.1 Free flight generation

As mentioned above, the dynamics of the particle motion is assumed to consist of free flights interrupted by instantaneous scattering events. The latter change the momentum and energy of the particle according to the physics of the particular scattering process. Of course, we cannot know precisely how long a carrier will drift before scattering, as it continuously interacts with the lattice and we only approximate this process with a scattering rate determined by first-order time-dependent perturbation theory, as in Section 4.3 above. Within our approximations, we may simulate the actual transport by introducing a probability density $P(t)$, where $P(t)dt$ is the joint probability that a carrier will both arrive at time t without scattering (after its last scattering event at $t = 0$), and then will actually suffer a scattering event at this time (i.e., within a time interval dt centered at t). The probability of actually scattering within this small time interval at time t may be written as $\Gamma[\mathbf{k}(t)]dt$, where $\Gamma[\mathbf{k}(t)]$ is the total scattering rate of a carrier of wave vector $\mathbf{k}(t)$ (we use almost exclusively the wave vector, rather than the velocity, in this section). This scattering rate represents the sum of the contributions of each scattering process that can occur for a carrier of this wave vector (and energy). The explicit time dependence indicated is a result of the evolution of the wave vector under any accelerating electric (and magnetic) fields. In terms of this total scattering rate, the probability that a carrier has not suffered a collision after time t is given by

$$\exp\left(-\int_0^t \Gamma[\mathbf{k}(t')]dt'\right). \tag{4.134}$$

Thus, the probability of scattering within the time interval dt after a free flight time t, measured since the last scattering event, may be written as the joint probability

$$P(t)dt = \Gamma[\mathbf{k}(t)]\exp\left(-\int_0^t \Gamma[\mathbf{k}(t')]dt'\right)dt. \tag{4.135}$$

Random flight times may now be generated according to the probability density $P(t)$ by using, for example, the pseudo-random number generator available on nearly all modern computers and which yields random numbers in the range [0,1]. Using a simple, direct methodology, the random flight time is sampled from $P(t)$ according to the random number r as

$$r = \int_0^t P(t')dt'. \qquad (4.136)$$

For this approach, it is essential that r is uniformly distributed through the unit interval, and the result t is the desired flight time. Using (4.135) in (4.136) yields

$$r = 1 - \exp\left(-\int_0^t \Gamma[\mathbf{k}(t')]dt'\right). \qquad (4.137a)$$

Since $1 - r$ is statistically the same as r, this latter expression may be simplified as

$$-\ln(r) = \int_0^t \Gamma[\mathbf{k}(t')]dt'. \qquad (4.137b)$$

The set of equations (4.137) are the fundamental equations used to generate the random free flight for each carrier in the ensemble. If there is no accelerating field, the time dependence of the wave vector vanishes, and the integral is trivially evaluated. In the general case, however, this simplification is not possible, and it is expedient to resort to another *trick*. Here, we will introduce a fictitious scattering process that has no effect on the carrier. This process is called *self-scattering*, and the energy and momentum of the carrier are unchanged under this process (Rees, 1969). However, we will assign an energy dependence to this process in just such a manner that the total scattering rate is a constant, as

$$\Gamma_{self}[\mathbf{k}(t)] = \Gamma_0 - \Gamma[\mathbf{k}(t)] = \Gamma_0 - \sum_i \Gamma_i[\mathbf{k}(t)], \qquad (4.138)$$

and the summation runs over all real scattering processes. Since the self-scattering process has no effect upon the carrier, it will not change the observable transport properties at all, but its introduction eases the evaluation of the free flight times, as now

$$t = -\frac{1}{\Gamma_0}\ln(r). \qquad (4.139)$$

The constant total scattering rate Γ_0 is chosen a priori so that it is larger than the maximum scattering encountered during the simulation interval.

In the simplest case, a single constant is used globally through the simulation (constant gamma method), although other schemes have been suggested that modify the value of Γ_0 at fixed time increments in order to become more computationally efficient.

4.8.2 *Final state after scattering*

We now consider the next step of the simulation in some detail. Consider that a typical electron arrives at time t (arbitrarily selected by the methods of the previous paragraph) in a state characterized by momentum $\mathbf{p}_a$, position $\mathbf{x}_a$, and energy E. At this time, the duration of the accelerated flight has been determined from the probability of not being scattered, given above with a random number r_1, which lies in the interval [0,1]. At this time, the energy, momentum, and position are updated according to the energy gained from the field during the accelerative period to the values mentioned above – that is, they gain a momentum and energy according to their acceleration in the applied field during the time t. Once these new dynamical variables are known, the various scattering rates can now be evaluated for this particle (in practice, these rates are usually stored as a table to enhance computational speed). A particular rate is selected as the germane scattering process according to a second random number r_2, which is used in the following approach: All scattering processes are ordered in a sequence with process 1, process 2, . . . , process $n - 1$, and finally the self-scattering process. The ordering of these processes does not change during the entire simulation. Hence, at time t, we can use this new random number r_2 to select the process according to

$$\sum_{i=1}^{s-1}\Gamma_i[E(t)] < r_2\Gamma_0 < \sum_{i=1}^{s}\Gamma_i[E(t)]. \qquad (4.140)$$

In this way, process s is selected. Then, the energy and momentum conservation relations are used to determine the post-scattering momentum and energy $\mathbf{p}_2$ and E_2 (that is, $E_2 = E \pm \hbar\omega_0$, depending upon whether the process is absorption or emission, respectively, and the momentum is suitably adjusted to account for the phonon momentum).

Additional random numbers are used to evaluate any individual parts of the momentum that are not well defined by the scattering process, such as the angles θ, ϕ associated with the process. For example, in polar scattering the polar angle is well defined by the $1/q$ variation of the matrix element. On the other hand, the azimuthal angle ϕ change is not specified by the matrix element, so that ϕ is randomly selected by a third random number as $2\pi r_3$. In isotropic scattering processes such as non-polar optical and acoustic scattering, both angles are randomly selected. A fourth random number is now used to select the polar angle, according to the distribution of these angles. Let us consider once more the polar optical scattering as an illustration. The probability of scattering through a polar angle θ is provided by the square of the matrix element weighted delta function, which gives the angular probability to be proportional to $1/q^2 = 1/|\mathbf{k} - \mathbf{k}'|^2$. This is just the un-normalized function

$$P(\theta) = \frac{\sin\theta}{2E \pm \hbar\omega_0 - 2\sqrt{E(E \pm \hbar\omega_0)}\cos\theta}. \tag{4.141}$$

This distribution function is then used to select the scattering angle θ with the random number r_4 through the equation

$$r_4 = \frac{\int_0^\theta P(\theta')d\theta'}{\int_0^\pi P(\theta')d\theta'} = \frac{\ln[(1-\xi\cos\theta)/(1-\xi)]}{\ln[(1+\xi)/(1-\xi)]},$$

$$\xi = \frac{2\sqrt{E(E \pm \hbar\omega_0)}}{\left(\sqrt{E} - \sqrt{E \pm \hbar\omega_0}\right)^2} \tag{4.142}$$

Finally, this last expression can be inverted to yield the actual scattering angle selected by this random number as

$$\theta = \cos^{-1}\left[\frac{(1+\xi) - (1+\xi)^{r_4}}{\xi}\right]. \tag{4.143}$$

This approach is easily extended to non-parabolic bands.

The final set of dynamical variables obtained after completing the scattering process are now used as the initial set for the next iteration, and the process is continued for several hundred thousand cycles. This particular algorithm is one that is amenable to full vectorization and is relatively computationally efficient, in that it can utilize the hardware "scatter-gather" routines available on large vector computers. On high speed work stations, though, such subtleties are not necessary and the program is quite efficient on the pipelined architecture of most PCs. In one general variant, the program begins by creating the large scattering matrix in which all of the various scattering processes are stored as a function of the energy; that is, this scattering table may be set up with 1 meV increments in the energy.

This includes the self-scattering process. The energy is discretized, and the size of each elemental step in energy is set by the dictates of the physical situation that is being investigated. The initial distribution function is then established – the N electrons actually being simulated are given initial values of energy and momentum corresponding to the equilibrium ensemble, and they are given initial values of position and other possible variables corresponding to the physical structure being simulated. At this point, $t = 0$. If the inter-carrier forces are being computed in real space by a molecular dynamics interaction (we come to this in a later chapter), the initial values of these forces, corresponding to the initial distributions in space are also computed. Part of the initialization process is also to assign to each of the N electrons a t_i according to (4.137b), which is its individual time at which it ends its free flight and undergoes scattering. Then each electron undergoes its free flight and a scattering process, which may be self-scattering. New times are selected for each particle and the process is repeated as long as desired.

4.8.3 *Time synchronization*

The key problem in treating an ensemble of particles is that each particle has its unique time scale. However, we want to compute ensemble averages for such quantities as the drift velocity and average energy, with the former defined as

$$\mathrm{v}_d(t) = \frac{1}{N}\sum_{i=1}^{N}\mathrm{v}_i(t). \tag{4.144}$$

For the best accuracy, all the particles need to be aligned at the same time t, which here runs from the beginning of the simulation. Thus, we need to overlay the system with a global time scale, with which each local particle time scale can be synchronized. In practice, this is achieved by introducing a global time variable T, which is discretized into steps as $n\Delta T$. Then at integer multiples of this time step, all particles are stopped in their free flights, and ensemble averages are computed. This is shown schematically in Figure 4.6. Each horizontal line is a particle time path. It is composed of accelerations (straight lines) and scattering processes (X's). There are N of these time lines all running horizontal in the figure. Vertically, we have indicated the time "pauses" $n\Delta T$, at which the entire ensemble is halted, and averages computed. In general, these pauses will always occur during the free flights, but it means that we have to keep our accounting carefully in order to properly align these two time scales.

If one is incorporating non-linear effects, such as molecular dynamics (Chapter 7), or non-equilibrium phonons or degeneracy induced filling of the final states after scattering (next section), then these processes are updated

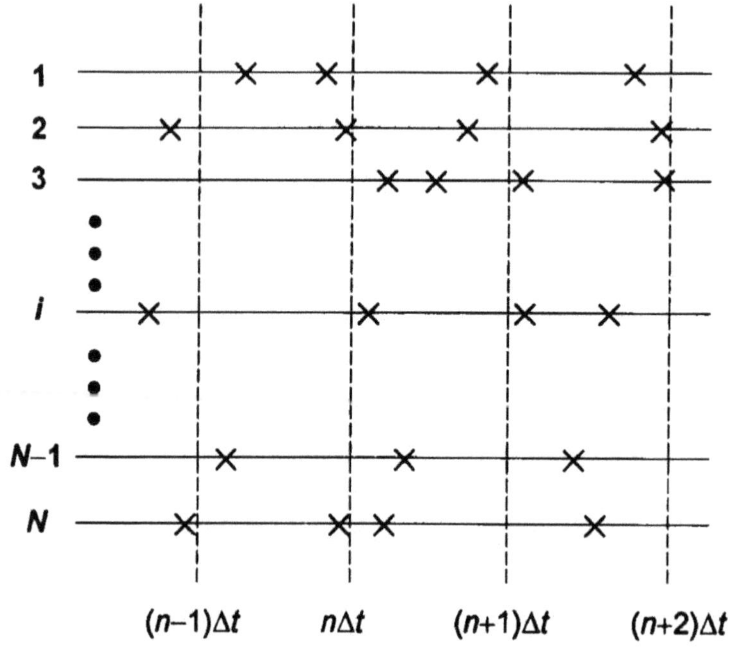

Figure 4.6 Representation of the two time scales for each particle. One is the "local" time scale corresponding to the horizontal line for each particle. This measures the time along each path, which is a sequence of accelerations and scattering events (x's). The second, or "global" time scale is the measuring time scale, which is represented by the vertical dashed lines. At sequential times $n\Delta t$, the entire simulation is "stopped" and ensemble averages are computed. Nonlinear processes are also updated at these time iterates.

on the pauses of the T time scale as well. In this sense, the imposition of the second time scale synchronizes the distribution and gives the global, or laboratory, time scale of interest in experiments.

4.8.4 Rejection techniques for nonlinear processes

In the case of polar-optical phonon scattering, it was possible to actually integrate the angular probability function (4.141). This is not always the case, and one has therefore to resort to other statistical methods. One of these is the so-called rejection technique. Suppose the probability density function for the process, such as (4.141), is quite nonlinear and not easily integrated to get the total probability. Then one can use a pair of random numbers (r_1, r_2) to evaluate the angle. Consider Figure 4.7, in which we plot a complicated probability density function. Here, it is assumed that the maximum coordinate x is unity, so that the range of the function's

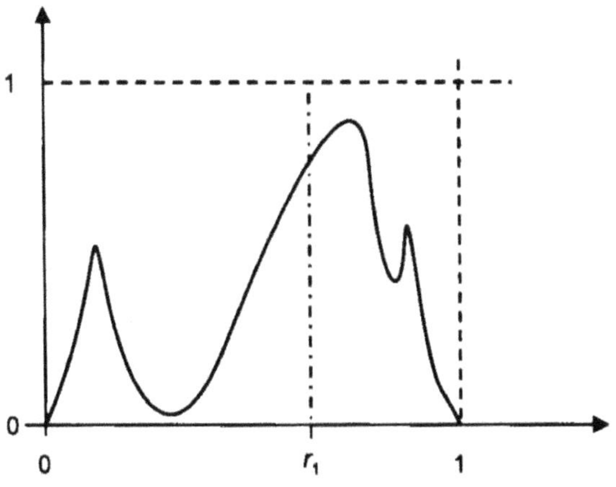

Figure 4.7 Evaluation of a probability density (curved line) by a rejection technique. If necessary, one random number selects the position along the x axis, denoted by r_1. This value is accepted if $f(r_1) > r_2$, a second random number.

argument is from zero to one (one can easily use other values, such as π, by proper normalization). The maximum value of the function is also set near unity, and one can always renormalize this to the span of the random numbers. Now, the first random number is taken to correspond to the span of the function (the x axis). This determines the argument of the function that is to be evaluated. For example, let us assume that this is $r_1 = 0.65$ (this is indicated by the dashed line in the figure). We now use the second random number to determine whether $f(r_1) > r_2$. If this relationship holds, then the value r_1 is accepted for the argument of the function, and the scattering process proceeds with this value. Certainly, values of r_1 for which the function is large are more heavily weighted in this rejection process. We consider this in more detail for two important processes: (1) state filling due to the degeneracy of the electron gas, and (2) non-equilibrium phonons.

Degeneracy and Fermi–Dirac statistics have been introduced through the concept of a secondary self-scattering process (Bosi and Jacoboni, 1976; Lugli and Ferry, 1985) based upon the rejection technique. We call it secondary self-scattering, because if the condition $f(r_1) > r_2$ is not satisfied, we treat the rejection exactly as a self-scattering process, which was introduced earlier. Each of the scattering processes must include a factor of $[1 - f(E)]$, where $f(E)$ is the dynamic distribution function, and represents the probability that the final state after scattering is empty. Rather than recompute the scattering rates as the distribution function evolves in order to incorporate the degeneracy, all scattering rates are computed as if the final states were always empty. A grid in momentum space is maintained and the number of particles in each state is tracked (each cell of this grid has its population divided by the total number of states in the cell, which depends on the cell size, to provide the value of the distribution function in that cell). The scattering processes themselves are evaluated, but the acceptance of the process depends on a rejection technique. That is, an additional random number is used to accept the process if

$$r < 1 - f(\mathbf{p}_{final}, t). \tag{4.145}$$

Thus, as the state fills, most scattering events into that state are rejected and treated as a self-scattering process.

The most delicate point of the degeneracy method involves the normalization of the distribution function $f(\mathbf{p})$. The extension of the secondary self-scattering method to the ensemble Monte Carlo algorithm involves the fact that there are N electrons in the simulation ensemble, which represent an electron density of n. The effective volume V of "real space" being simulated is N/n. The density of allowed wave vectors of a single spin in k-space is just $V/(2\pi)^3$. In setting up the grid in the three-dimensional wave vector space, the elementary cell volume is given by $\Omega_k = \Delta k_x \Delta k_y \Delta k_z$. Every cell can accommodate at most N_c electrons, with $N_c = 2\Omega_k V/(2\pi)^3$, where the factor of 2 accounts for the electron spin. For example, if the density is taken to be 10^{17} cm^{-3}, $N = 10^4$, and $\Delta k_x \Delta k_y \Delta k_z = (2 \times 10^5 \text{ cm}^{-1})^3$ ($k_F = 2.4 \times 10^6$ cm^{-1} at 77 K), then $V = 10^{-13}$ cm^3 and $N_c = 6.45$. N_c constitutes the maximum occupancy of a cell in the momentum space grid. (Obviously, a more careful choice of parameters would have N_c come out to be an integer, for convenience.) A distribution function is defined over the grid in momentum space by counting the number of electrons in each cell. The distribution function is normalized to unity by dividing the number in each cell by N_c for use in the rejection technique. It should be noted that N_c should be sufficiently large that round-off to an integer (if the numbers do not work out properly, as in the case above) does not create a significant statistical error.

A second usage is for the consideration of non-equilibrium phonon distributions (Lugli *et al.*, 1987). In the derivations presented earlier in this chapter, the assumption was made that the phonons are in equilibrium and characterized by N_q. However, under a number of circumstances, such as the excitation of the semiconductor by an intense laser pulse, the carriers are created high in the energy band, and then decay by a cascade of phonon emission processes (we will treat this in detail in Chapter 6). As a result of this cascade the phonon distribution is driven out of equilibrium, and this affects both the emission and absorption processes by which the carriers interact with the phonons. As before, we use q, rather than k, for the momentum of the phonons. Once again, the momentum space is discretized for the phonon distribution, so that an individual cell in this discretized space has volume $\Delta q_x \Delta q_y \Delta q_z$. This small volume has available a number of states given by $V/(2\pi)^3$, where V is determined by the effective simulation volume N/n, as previously. The difference between state filling for carrier degeneracy and phonon state filling is that there is no limit to the number of phonons that can exist within the state. The basic approach assumes that the phonons are out of equilibrium, and the carrier scattering processes are evaluated with an assumed $N_{max}(\mathbf{q})$. Then, within the simulation, the number of phonons emitted, or absorbed, with wave vector q is carefully monitored. At the synchronization times of the global time scale, the phonon population in each cell of momentum space is updated from the emission/absorption statistics that have been gathered during that time step. One must also include phonon decay, which is through a 3-phonon process to other modes of the lattice vibrations, so that the update algorithm is simply

$$N(\mathbf{q}, t + \Delta t) = N(\mathbf{q}, t) + G_{net,\Delta t}(\mathbf{q}) - \left[\frac{N(\mathbf{q}, t) - N_{q0}}{\tau_{phonon}}\right]\Delta t, \tag{4.146}$$

where N_{q0} is the equilibrium distribution, $G_{net,\Delta t}(\mathbf{q})$ is the net (emission minus absorption) generation of phonons in the particular cell *during the time step*, and τ_{phonon} is the phonon lifetime. During the simulation, each phonon scattering process is evaluated as if the maximum assumed phonon population were present. Then, a rejection technique is used, by which the phonon scattering process is rejected (and assumed to be a secondary self-scattering process) if

$$r_{test} > \frac{N(\mathbf{q}, t)}{N_{max}}. \tag{4.147}$$

Here, N_{max} is the peak value that was assumed in setting up the scattering matrices. While this is assumed here to be a constant for all phonon wave vectors, this is not required. A more sophisticated approach would use a momentum-dependent peak occupation.

4.9 Relationship between the Boltzmann equation and EMC

There has been considerable discussion in the literature about the connection between the Boltzmann equation and the EMC technique. Most of this discussion relates to whether or not they yield the same results, and if so upon what time scale. In fact, it was easily pointed out many years ago that the MC procedure only approached the Boltzmann result in the long-time limit (Rees, 1969; Boardman *et al.*, 1970). Yet, there are still efforts to put more significance into the Boltzmann equation on the short time scale. The problem, as we shall see, is that the Boltzmann equation is Markovian in its scattering integrals. Just as we had to move to a *retarded* Langevin equation, a retarded, or non-Markovian, form of the Boltzmann equation is required for short time scales. Here, we shall develop the path integrals from which the Monte Carlo procedure is begun, and show where this important difference arises.

4.9.1 The Boltzmann path integral

To develop this approach, the Boltzmann equation will be written in terms of a path integral. In this, the streaming terms on the left-hand side will be written as partial derivatives of a general derivative of the time motion along a "path" in a 6-dimensional phase space; this is then used to develop a closed-form integral equation for the distribution function. This integral has itself been used to develop an iterative technique, but provides one basis of the connection between the Monte Carlo procedure and the Boltzmann equation. To begin, the Boltzmann equation is written as

$$\left(\frac{\partial}{\partial t} + \mathbf{v} \cdot \nabla + eF \cdot \frac{\partial}{\partial \mathbf{p}}\right) f(\mathbf{p},\mathbf{r},t) = -\Gamma_0 f(\mathbf{p},\mathbf{r},t) + \int d^3\mathbf{p}' W(\mathbf{p},\mathbf{p}') f(\mathbf{p}',\mathbf{r},t), \tag{4.148}$$

where

$$\Gamma_0 = \int d^3\mathbf{p}' W(\mathbf{p}',\mathbf{p}) \tag{4.149}$$

is the total out-scattering rate, including self-scattering. That is, (4.149) provides the entire rate of decrease of population $f(\mathbf{p},\mathbf{r},t)$ due to scattering of particles out of this state. The remaining scattering term provides the complementary scattering of particles into the state.

At this point it is convenient to transform to a variable that describes the motion of the distribution function along a trajectory in phase space. It is usually difficult to think of the motion of the distribution function, but perhaps easier to think of the motion of a typical particle that characterizes the distribution function. For this, the motion is described in a six-dimensional phase space, which is sufficient for the one-particle distribution function being considered here (Budd, 1966). The coordinate along this trajectory is taken to be s, and the trajectory is rigorously defined by the semiclassical trajectory, which can be found by any of the techniques of classical mechanics (i.e., it corresponds to that path which is an extremum of the action). It is as easy to remember, however, that it follows Newton's laws, where the forces arise from all possible potentials – induced and self-consistent ones in device simulations. Each normal coordinate can be parameterized as a function of this variable as

$$\mathbf{r} \to \mathbf{x}^*(s), \quad \mathbf{p} = \hbar\mathbf{k} \to \mathbf{p}^*(s), \quad t \to s, \tag{4.150}$$

and the partial derivatives are constrained by the relationships

$$\frac{d\mathbf{x}^*}{ds} = \mathbf{v}, \quad \mathbf{x}^*(t) = \mathbf{r}, \quad \frac{d\mathbf{p}^*}{ds} = e\mathbf{F}, \quad \mathbf{p}^*(t) = \mathbf{p}. \tag{4.151}$$

With these changes, the Boltzmann equation becomes simply

$$\frac{df}{ds} + \Gamma_0 f = \int d^3\mathbf{p}^{*\prime} W(\mathbf{p}^*, \mathbf{p}^{*\prime}) f(\mathbf{p}^{*\prime}, \mathbf{x}^*, s). \tag{4.152}$$

This is now a relatively simple equation to solve. It should be recalled at this point that $W(\mathbf{p}^*,\mathbf{p}^{*\prime})$ is the probability per unit time that a collision scatters a carrier from state $\mathbf{p}^{*\prime}$ to $\mathbf{p}^*$, and these variables will be retarded due to the phase-space variations described above. The form (4.152) immediately suggests the use of an integrating factor $\exp(\Gamma_0 s)$, so that this equation becomes

$$\frac{d}{ds}(f(\mathbf{p}^*)e^{\Gamma_0 s}) = \int d^3\mathbf{p}^{*\prime} W(\mathbf{p}^*, \mathbf{p}^{*\prime}) f(\mathbf{p}^{*\prime}, \mathbf{x}^*, s) e^{\Gamma_0 s}, \tag{4.153}$$

where the momenta evolve in time as the energy increases in time along the path s due to the acceleration of the external fields. In fact, on the phase space path defined by s, the energy does not increase, but as the "laboratory" coordinates are restored, this energy increase will appear (this is just a choice of gauge for the field and momentum). Indeed, the major time variation lies in the momenta themselves. The Boltzmann equation can now be rewritten as

$$f(\mathbf{p}^*, t) = f(\mathbf{p}^*, 0)e^{-\Gamma_0 t} + \int_0^t ds \int d^3\mathbf{p}^{*\prime} W(\mathbf{p}^*, \mathbf{p}^{*\prime}) f(\mathbf{p}^{*\prime}, \mathbf{x}^*, s) e^{-\Gamma_0 (t-s)} \tag{4.154}$$

and, if we restore the time variables appropriate to the laboratory space, we arrive at

$$f(\mathbf{p}, t) = f(\mathbf{p}, 0)e^{-\Gamma_0 t} + \int_0^t dt' \int d^3\mathbf{p}' W(\mathbf{p}, \mathbf{p}' - e\mathbf{F}t') f(\mathbf{p}' - e\mathbf{F}t', t') e^{-\Gamma_0 (t-t')}. \tag{4.155}$$

This last form is often referred to as the Chambers–Rees path integral (Rees, 1972), and is the form from which an iterative solution can be developed.

The integral (4.155) has two major components. The first is the scattering process by which the carriers described by $f(p')$ are scattered (by the processes within W). The second is the following ballistic drift under the influence of the field, with a probability of the drift time given by $\exp[-\Gamma_0(t-t')]$. These are the two parts of the Monte Carlo algorithm, and it is from such an integral that we recognize that the Monte Carlo method is merely evaluating the integral stochastically. The problem with it is that there is no retardation in the scattering process, so that the scattering rate and energy are supposed to respond instantaneously to changes in the momentum along the path $\mathbf{p}' - e\mathbf{F}t'$. That is, the number of particles represented by the distribution function within the integral instantaneously responds during the previous drift. In essence, this is the Markovian assumption, and is true only in the long-time limit.

4.9.2 The EMC procedure: beyond the Boltzmann equation

When a distribution of carriers in a semiconductor is subjected to a high, spatially homogeneous electric field that is initiated at $t = 0$, the ensemble accelerates to a new, far-from-equilibrium distribution function that is characteristic of the nonlinear, high-electric-field transport problem. In the above section, the case for the Boltzmann equation was developed, and it was shown how the path integral could be solved by a Monte Carlo procedure. However, the *ensemble* Monte Carlo technique is a random-walk process in the $6N$-dimensional phase space and the general probability integrals reduce to the path integral form of the Boltzmann equation only in the limit $t \to \infty$. Yet, the Boltzmann transport equation itself is only valid as a transport equation in this long-time limit (Zwanzig, 1961), and it has been known for some time that more generalized transport equations must be used on the short-time scale. Indeed, these generalized transport equations are amenable to a path integral solution that agrees with the ensemble Monte Carlo technique on a short-time scale, and reduce to the Boltzmann transport equation on the longer time scale.

In the past few years, the EMC techniques have been developed to a form particularly suited for calculation of the transient processes. As discussed above, the EMC technique is a hybrid method in which an ensemble of electrons is adopted. This ensemble is composed of N carriers with variables $\{R_i\}$, $i = 1, 2, 3, \ldots, N$, and $R_i = \{\mathbf{p}_i, \mathbf{r}_i, \ldots\}$ includes all necessary descriptors of each carrier's state. At each time step, all R_i are calculated by the Monte Carlo process, and the set $\{R_i\}$ is treated as an ensemble evolving in time. The EMC technique has an advantage over the normal Monte Carlo technique in that an ensemble distribution function exists and evolves with its parameters. Thus, all dynamic transport variables are computed by ensemble averages, and there is no need to rely on the *possible* existence of the ergodic theorem. The variance in any ensemble average is controlled by the use of a sufficiently large ensemble N, and a variety of sophisticated variance reduction techniques exist to improve these estimates. Of even more importance is the fact that the ensemble is constructed in such a way that the ensemble averages evolve according to a generalized retarded transport equation rather than the Boltzmann transport equation. In this regard, the EMC process is a *many-electron* distribution function that contains the full information describing the phase-space evolution of the ensemble. In construing any averaging process to map the dynamic variables onto those obtained from a one-electron distribution function, the full retardation of the process is incorporated, so that proper non-Markovian response results. The approach followed here is styled closely after that of Krafter and Silbey (1980) and Kenkre *et al.* (1981).

Consider the single-particle Monte Carlo process. Here one defines $S_n(\mathbf{p}_0,\mathbf{p},t)$ to be the probability that an electron, initially at $\mathbf{p}_0$ at $t = 0$, passes through the state $\mathbf{p}$ at time t during the nth free flight. (Here S is not the scattering matrix but is just a probability transition function.) The explicit time dependence of this definition must be maintained since the electron can pass through $\mathbf{p}$ at any time $0 < t < \infty$. Then the probability function S_n satisfies the iterative relationship (Boardman *et al.*, 1970)

$$S_n(\mathbf{p}_0,\mathbf{p},t) = \sum_{\mathbf{p}'}\int_0^t dt' S_{n-1}(\mathbf{p}_0,\mathbf{p}',t-t')W(\mathbf{p}',\mathbf{p}-e\mathbf{F}t')e^{-\Gamma_0 t'}, \tag{4.156}$$

where $W(\mathbf{p}',\mathbf{p}-e\mathbf{F}t')$ is the total scattering rate summed over all possible processes (including self-scattering), and the exponential decay term accounts for weighting of the ballistic trajectory in the second argument of W. This term is readily recognized as part of the path integral that was obtained from the Boltzmann equation. If the operator $\hat{W}$ is defined by

$$\hat{W}S = \sum_{\mathbf{p}'}\int_0^t dt' S(\mathbf{p}_0,\mathbf{p}',t-t')W(\mathbf{p}',\mathbf{p}-e\mathbf{F}t')e^{-\Gamma_0 t'}, \tag{4.157}$$

then (4.156) is the nth term in the perturbation series expansion

$$S = S_0 + \hat{W}S_0 + \hat{W}\hat{W}S_0 + \hat{W}\hat{W}\hat{W}S_0 + \ldots, \tag{4.158}$$

where S_0 is the initial state $S_0(\mathbf{p}_0,\mathbf{p},0)$. The expansion (4.158) is the iterative solution of the integral equation

$$S(\mathbf{p}_0,\mathbf{p},t) = \sum_{\mathbf{p}'} \int_0^t dt' S(\mathbf{p}_0,\mathbf{p}',t-t') W(\mathbf{p}',\mathbf{p}-e\mathbf{F}t') e^{-\Gamma_0 t'} \quad (4.159)$$

along the appropriate trajectory, and corresponds to the general expansions found in quantum mechanics when the interaction representation is used. If we neglect the time dependence within W (we return to this later), replace the sum over the initial state with an integral, and change the variables within the integral, so that the $t - t'$ term is in the exponent, this equation is the same as (4.155). Therefore, it may be asserted under these conditions that (4.159) is the solution of the differential equation

$$\frac{dS}{dt} + \Gamma_0 S = \sum_{\mathbf{p}'} W(\mathbf{p}',\mathbf{p}) S(\mathbf{p}',t). \quad (4.160)$$

In (4.160), it may be recognized that Γ_0 includes all relevant out-scattering processes, including self-scattering. However, the connection to the Boltzmann equation is recognized only when the time derivative (first term on the left-hand side) is interpreted to be a coalescence of all of the streaming terms into a path-variable form, as done previously. Hence, a connection can be made to the Boltzmann equation by defining an initial state distribution function (which can be the equilibrium distribution) of the N electrons used in the ensemble Monte Carlo technique as

$$f(\mathbf{p},0) = \sum_{\mathbf{p}_0} n(\mathbf{p}_0) S(\mathbf{p}_0,\mathbf{p},0), \quad (4.161)$$

where $n(\mathbf{p}_0)$ represents the fraction of electrons in state $\mathbf{p}_0$ and is normalized to the total number of electrons per unit volume, as

$$\sum_{\mathbf{p}_0} n(\mathbf{p}_0) = n. \quad (4.162)$$

Here $n(\mathbf{p}_0)$ is not a probability function but the actual distribution of carriers at the initial time.

While the formal evolution of (4.161) is the same as that obtained for the Boltzmann transport equation, the similarity ends there. We must note that we made a crucial assumption above: the function W is independent of time. If we make the connection with the Boltzmann equation, then this W is related to the scattering functions. But, as the energy evolves under the field (and diffusive) streaming, the scattering function does change as the individual probabilities of various scattering events are very energy dependent. Hence, the detailed form of W is time evolving. However, it is important to note that the ensemble Monte Carlo, and its representative distribution function, are many-electron distribution functions rather than the one-electron function of the Boltzmann equation. In this regard, the momentum $\mathbf{p}$ in (4.161) and (4.162) are in fact the ensemble of momentum values for the N carriers. Hence, $n(\mathbf{p}_0)$ the one-electron equilibrium distribution must be projected from the ensemble of particles. Within this many-body phase space, the scattering function is a function only of the explicit momentum variables, and not an explicit function of time. Its time dependence arises only in its projection into an equation for the subsequent one-electron distribution function. To reduce the many-electron function to the corresponding one-electron function, it is necessary to perform some type of condensation process. Here, this reduction will be achieved by introducing projection operators. One typical projection operation (and the one of interest to us) at time t can be defined on any arbitrary function F by the operation

$$PF = f_1(\bar{\mathbf{p}},t)\int d^{3N}\mathbf{p}'F(\mathbf{p}',t), \tag{4.163}$$

with the integral running over all the momenta coordinates of the various electrons and normalized properly to the density, and $\bar{\mathbf{p}}$ is an effective one-electron momentum. Here, f_1 is the normalized, single-electron distribution function that is parameterized in such a way that all central moments are equal to those in the actual distribution (this may well be a drifted Maxwellian, as used in the hydrodynamic equations). Such an approach can be used to calculate the equations of motion for various correlation functions (Zwanzig, 1961), but the interest here is in the dynamic variables necessary for transport in semiconductors. The methodology of the ensemble Monte Carlo process produces an averaging procedure over the possible Monte Carlo evolutions that satisfies (4.163).

To proceed, it is first simpler if (4.160) is rewritten with the introduction of the generalized collision operator (where it is assumed that summations will be made over the intermediate states, and the path variable has been reintroduced in keeping with the streaming nature of the "time" derivative)

$$\frac{dS}{ds} = \hat{C}S \equiv \sum_{\mathbf{p}'}[W(\mathbf{p}',\mathbf{p}) - \Gamma_0\delta(\mathbf{p}-\mathbf{p}')]S. \tag{4.164}$$

Since S is a many-electron distribution, the scattering process W maps all of these particles into the post-scattering particle set. Hence, W is no longer a simple one-particle scattering process. In general, this has the operator solution

$$S(s) = \exp(\hat{C}s)S(0) \equiv G(s)S(0), \tag{4.165}$$

with

$$\frac{dS(0)}{ds} = \hat{C}S(0) = 0, \tag{4.166}$$

as this state is in equilibrium. In (4.165), $G(s)$ is termed the propagator for $S(s)$ and defines how states in the initial-time $S(0)$ evolve forward in time to give $S(s)$.

At this point it is fruitful to make a number of general remarks of an informative nature on projection operators. In general, $P(PF) = PF$, or $P^2 = P$, since the projection of an already projected state is just itself. Similarly, it is possible to define a *residual* operator $Q = 1 - P$, with $QP = PQ = 0$. Using (4.165) in (4.164), we can operate first with P and then with Q to achieve the two equations

$$\frac{d(PG)}{ds} = P\hat{C}PG + P\hat{C}QG,$$
$$\frac{d(QG)}{ds} = Q\hat{C}PG + Q\hat{C}QG. \tag{4.167}$$

The second of these equations is first solved for QG, and this is then used in the first equation to give the result

$$\frac{d(PG)}{ds} = P\hat{C}PG + \int_0^s ds' P\hat{C}Qe^{Q\hat{C}(s-s')}Q\hat{C}PG(s')$$
$$+ P\hat{C}Qe^{Q\hat{C}s}QS(0). \tag{4.168}$$

It is clear that the projected equation (4.168) already has the form of a retarded non-Markovian equation. The last term has a number of interpretations, but clearly retains information about the initial state of the ensemble and how it was "created." It is just as clear that this term will decay away with the transient behavior and represents the need to break up initial correlations among the electrons (with their initial configuration) in evolving to the far-from-equilibrium steady state.

The form of (4.168) is, in fact, a standard one in transport theory and was initially obtained by Zwanzig (1961). Generally, we are interested in a specific average of the evolving system, such as the drift velocity or the diffusivity. Thus, one may look just for an equation of motion for this quantity. More generally, a projection operator is useful and leaves us with an equation just such as (4.168) (Mori, 1965). In fact, using a projection operator consists of rejecting most of the information contained in the initial equation of the entire system, and this has often been regarded as introducing irreversibility into the equation of motion for the quantity of interest, as well as retardation and memory terms. Choosing a finite set of parameters, such as momentum, energy, and so on, each having a measurable average, allows a micro-canonical distribution that is parameterized with this parameter set to be defined on a subspace in phase space. This in turn satisfies a generalized transport equation containing memory functions in the convolution integrals. Any equation of motion obtained for the averaged values of the set of parameters exhibits retardation, and the memory function appearing in each of these equations can be related to a correlation function, just as was done in the last chapter and in preceding sections of this chapter.

Let us now investigate the collision term in more detail. The Laplace transform is introduced (as the normal Laplace transform variable has already been used as the path variable; z is taken as the transform variable), so (4.168) becomes

$$PG(z) = \frac{PG(0)}{z - \langle\hat{C}\rangle - \langle(\delta\hat{C})^2\rangle(z - Q\hat{C})^{-1}}, \tag{4.169}$$

where we have written

$$P\hat{C} = \langle\hat{C}\rangle, \quad Q\hat{C} = (1 - P)\hat{C} = (\hat{C} - \langle\hat{C}\rangle) = \delta\hat{C} \tag{4.170}$$

everywhere except for the term arising from the exponential within the integral (we note that there are two factors of $Q\hat{C}$ in the integral), and the initial state memory term has been ignored for the moment. From (4.164), one can also obtain

$$G(s) = \frac{G(0)}{z - \hat{C}}, \tag{4.171}$$

which may be operated upon by the projection operator P to yield

$$PG(s) = \left\langle \frac{1}{z - \hat{C}} \right\rangle PG(0). \tag{4.172}$$

Equating (4.170) and (4.172) yields

$$\left\langle \frac{1}{z - \hat{C}} \right\rangle = \frac{1}{z - \langle\hat{C}\rangle - \langle(\delta\hat{C})^2\rangle(z - Q\hat{C})^{-1}} \equiv \frac{1}{z - \hat{M}(z)}$$
$$= \frac{1}{z}\left[1 + \frac{\hat{M}(z)}{z - \hat{M}(z)}\right]. \tag{4.173}$$

Here, we have introduced the retarded, memory function $\hat{M}(z)$. Combining this with (4.172), and the definition (4.163), we arrive at the final result for the projected one-particle distribution function as

$$zf_1(z) = f_1(0) + \hat{M}(z)f_1(z) + \left\langle \hat{C}Q\frac{1}{z - Q\hat{C}}Qf(0)\right\rangle, \tag{4.174}$$

where the initial-state memory term with $f(0)$ has been reintroduced. The connection between G and S needs to be stated a little more clearly at this point. In (4.165), $G(s)$ was defined to be the propagator for $S(s)$, describing how $S(0)$ evolved forward along the path variables. But the initial state is defined through $n(\mathbf{p}_0)$, so that the value of each component particle in $S(\mathbf{p}_0,0)$ can be thought of as unity, with the true initial value arising only from the integration of the product $nS(0)$ over $\mathbf{p}_0$. In this sense, $Gf(0) = S(s)n$ can be considered to be an identity. With this interpretation, the connection $G = S$ can be used immediately. The inverse transform of (4.174) can now be written immediately as

$$\frac{df_1(s)}{ds} = \int_0^s ds'\hat{M}(s - s')f_1(s') + P\hat{C}e^{Q\hat{C}s}Qf(0), \tag{4.175}$$

in path-variable form. The memory term is dominated by the averaged scattering process, and the self-scattering terms can be separated out if desired. By doing so it is possible to return to normal coordinates for (4.175). Then the equation is strongly connected not with the Boltzmann equation but with the Prigogine–Resibois equation, which is derived for quantum systems (Kreuzer, 1981). In the present case, of a homogeneous semiconductor in a high electric field, (4.175) may be unfolded to give

$$\frac{\partial f_1(t)}{\partial t} + e\mathbf{F}\cdot\frac{\partial f_1(t)}{\partial \mathbf{p}} = \int_0^t dt'\hat{M}(t - t')f_1(t') + \textit{initial memory terms}. \tag{4.176}$$

If we ignore the initial memory terms, and the terms in $\langle(\delta\hat{C})^2\rangle$ that contribute to $\hat{M}$ in (4.173), retaining only the average scattering process, then (4.176) can be reduced to

$$\frac{\partial f_1(t)}{\partial t} + e\mathbf{F}\cdot\frac{\partial f_1(t)}{\partial \mathbf{p}} = -\Gamma_0 f_1(\mathbf{p},t) + \int_0^t dt'\sum_{\mathbf{p}'} W^*[\mathbf{p},\mathbf{p}' - e\mathbf{F}(t - t'),$$
$$t - t']f_1(\mathbf{p} - e\mathbf{F}t',t'), \tag{4.177}$$

where we have re-introduced the terms in the scattering kernel from (4.164). The asterisk on the scattering function reminds us that this may not be the scattering function that appears in the Boltzmann equation, but is an "averaged" function for the many-particle distribution function. We have omitted the term in $\langle(\delta\hat{C})^2\rangle$, so that this is also a difference in the present case from the Boltzmann equation. Is the retardation in (4.177) important? One instance of its direct effect is that the presence of many electrons changes the screening in a dynamic manner for a number of scattering processes. Hence, the actual scattering rate evolves with time, just as the distribution function, and this is one source of the retardation and memory process. If we assume the long-time limit, then one may assume that the distribution function may be removed from the integral, or equivalently that the scattering function contains a δ-function for the third argument in its form in (4.177). Under this condition, the Boltzmann equation is recovered.

From (4.176), it is also possible to obtain the retarded Langevin equation. Suppose that interest is centered on the velocity of the individual carriers. Then the average velocity can be written as

$$\mathbf{v}_d(t) = \int d^3\mathbf{p}\mathbf{v}f_1(\mathbf{p},t). \tag{4.178}$$

Carrying out this process on (4.176), with the understanding that the equilibrium initial state gives no contribution to the drift velocity, we arrive at

$$\frac{\partial \mathbf{v}_d(t)}{\partial t} = \frac{e\mathbf{F}}{m^*} + \int_0^t \gamma(t-t')\mathbf{v}_d(t') + \left\langle \frac{\mathbf{R}(t)}{m^*} \right\rangle. \tag{4.179}$$

This, of course, is just the average of (4.5), the retarded Langevin equation. However, several steps have been rather "glossed over" along the way. First, the term in $\langle(\delta\hat{C})^2\rangle$ from (4.173) has been retained, as it is the random-force term for the Langevin equation. There may also be a contribution to the random force from the initial-state memory term, a point that has been suggested (Pottier, 1983), although we have set this term to zero as discussed above. Second, the steps from the memory function in the collision operator to the generalized memory function of (4.179) are not at all obvious. The recognition of the $\langle(\delta\hat{C})^2\rangle$ term as a contribution to the random force is behind a number of attempts to add such a force to the Boltzmann equation in order to more effectively discuss topics such as noise in the carrier system. While interesting, their use is still quite exploratory.

PROBLEMS

1. A particular semiconductor has a zero-field mobility composed of ionized impurity scattering of 3500 cm²/Vs and of acoustic scattering of 4500 cm²/Vs. Assume that the average energy of the carriers is given approximately by

$$\frac{3}{2}k_B(T_e - T) = e\mu F^2 \tau_E$$

with $\tau_E = 10^{-12}$ s. Plot μ as a function of the electric field at 22 K.

2. In the Shockley theory, the optical phonon "velocity" $v_0 = \sqrt{2\hbar\omega_0/m^*}$ plays a critical role in terms of the effective saturation velocity. Plot the actual saturation velocity observed experimentally versus v_0 for Si, Ge, GaAs, InP, InGaAs (lattice matched to InP), AlAs, PbTe, and SiC. Also give a table of values used for each material.
3. From the low-field mobility at 77 K (given by experiment) and the values found for the phonon energies, plot $v(t)$ for Si and Ge, for a step electric field applied at $t = 0$, with values of 30 kV/cm and 5 kV/cm, respectively, using the Shockley theory.
4. Plot the scattering rates for GaAs for the processes of acoustic phonons, piezoelectric, and polar-optical phonons at 77 K over the energy range 0 to 0.5 eV (for the central valley of the conduction band only).
5. For an electron at an energy of 0.45 eV in the central conduction band of GaAs, what are the relative strengths of intravalley scattering versus scattering to the L valleys?
6. Plot the scattering rates for electrons and holes in Si as a function of the free carrier energy (in the range $0 < E < 1.0$ eV) at 300 K. Include acoustic phonons, equivalent inter-valley optical phonons, and for the holes, intravalley optical phonon scattering. Treat the light holes and the heavy holes separately.
7. Using Rode's method, compute the mobility as a function of the impurity concentration for *n*-type GaAs at 300 K. Include scattering by acoustic and non-polar optical phonons and by impurities.
8. Using Rode's method extended to a quasi-two-dimensional semiconductor, calculate the mobility, as a function of temperature from 4.2 to 300 K, in the inversion channel of a GaAs/GaAlAs heterostructure. You may include just acoustic and non-polar optical phonon scattering, but should take account of the fact that the inversion channel density is degenerate over most of the range of temperature. Calculate for $n_s = 4 \times 10^{11}$ cm^{-2}. You may assume that the "infinite well" Airy function solutions can be used to calculate the sub-band energies and the layer widths, but should question the equi-partition approximation to the acoustic phonons.
9. Using Rode's method for a quasi-two-dimensional semiconductor, compute the mobility to be expected for a silicon inversion layer. Assume that the number of Coulomb scattering centers at the interface is 10^{11} cm^{-2} and that the surface roughness scattering can be characterized by $\Delta = 0.3$ nm and $L = 2$ nm. Evaluate the mobility at 300 K, and plot as a function of the surface normal electric field. Be sure to include both acoustic deformation potential and equivalent inter-valley scattering.

10. Through the balance equations with a drifted Maxwellian distribution function, determine the drift velocity and electron temperature as a function of electric field in a case for which only ionized impurity and acoustic phonon scattering are present. Use the properties and constants for germanium at 15 K, with $N_d - N_a = 10^{15}$ cm^{-3}, and 50 percent compensation.
11. Calculate the energy and momentum relaxation rates for non-polar-optical phonons.
12. Using a drifted Maxwellian and the hydrodynamic equations, calculate the velocity as a function of the electric field for InP at 300 K. Assume a field range up to 50 kV/cm, a donor concentration of 10^{17} cm^{-3}, and sufficient compensation that the low-field mobility is 4500 cm^2/Vs.
13. Using an ensemble Monte Carlo technique, calculate the velocity-field curve, the average energy-field curve, and the velocity-time curves (for three representative values of the field) for InSb and InAs at 77 K. Assume that the carrier doping is 10^{14} cm^{-3} in each case. Give a table of parameters used in the calculation. For simplicity, you may assume that the conduction band is represented by parabolic bands, and consider only acoustic and polar optical phonons and impurity scattering. (Note that in the impurity scattering the $1 - \cos\theta$ term must be left out of the computation of the scattering rate.)
14. Determine values for the energy and momentum relaxation time at a field of 40 kV/cm to fit the velocity auto-correlation function of Figure 4.5. From these values, determine the time-dependent diffusion constant.

REFERENCES

Birman, J. L., 1962, *Phys. Rev.*, 127, 1093.

Birman, J. L., and Lax, M., 1966, *Phys. Rev.*, 145, 620.

Boardman, A. D., Fawcett, W., and Swain, S., 1970, *J. Phys. Chem. Sol.*, 31, 1963.

Bogoliubov, N. N., 1962, in *Studies in Statistical Mechanics*, edited by de Boer, J. and Uhlenbeck, G. E. (Amsterdam: North-Holland).

Bosi, S., and Jacoboni, C., 1976, *J. Phys.*, C9, 315.

Budd, H., 1966, *J. Phys. Soc. Jpn. (Suppl.)*, 21, 424.

Conwell, E., 1967, *High Field Transport in Semiconductors* (New York: Academic Press).

Curby, R. C., and Ferry, D. K., 1973, *Phys. Stat. Sol. (a)*, 20, 569.

Eaves, L., Stradling, R. A., Tidey, R. J., Portal, J. C., and Askenazy, S., 1975, *J. Phys. C*, 8, 1975.

Fawcett, W., and Ruch, J. G., 1969, *Appl. Phys. Lett.*, 15, 368.

Ferry, D. K., 1974, *Phys. Rev.*, 90, 766.

Ferry, D. K., 1976, *Phys. Rev.*, B14, 1605.

Ferry, D. K., 1991, *Semiconductors* (New York: Macmillan).

Fröhlich, H. and Paranjape, V. V., 1956, *Proc. Phys. Soc. London*, B69, 21.

Fröhlich, H., and Seitz, F., 1950, *Phys. Rev.*, **79**, 526.
Gnudi, A., Ventura, D., Baccarani, G., and Odeh, F., 1993, *Sol.-State Electron.*, **36**, 575.
Grosvalet, J., Motsch, C., and Tribes, R., 1963, *Sol.-State Electron.*, **6**, 65.
Gunn, J. B., 1963, *Sol.-State Commun.*, **1**, 88.
Hess, K., 1991, Ed., *Monte Carlo Device Simulation: Full Band and Beyond* (Boston: Kluwer Academic Publishers).
Jacoboni, C., and Lugli, P., 1989, *The Monte Carlo Method for Semiconductor Device Simulation* (Vienna: Springer-Verlag).
Jacoboni, C., and Reggiani, L., 1983, *Rev. Mod. Phys.*, **65**, 645.
Kalos, M. H., and Whitlock, P. A., 1986, *Monte Carlo Methods* (New York: Wiley).
Kash, J., Tsang, J. C., and Hvam, J., 1985, *Phys. Rev. Lett.*, **24**, 2151.
Kash, K., Wolff, P. A., and Bonner, W. A., 1983, *Appl. Phys. Lett.*, **42**, 173.
Kenkre, V. M., Montroll, E. W., and Schlesinger, M. F., 1981, *J. Stat. Phys.*, **9**, 45.
Krafter, J., and Silbey, R., 1980, *Phys. Rev. Lett.*, **44**, 55.
Kreuzer, H. J., 1981, *Nonequilibrium Thermodynamics and Its Statistical Foundations* (London: Oxford University Press).
Kubo, R., 1957, *J. Phys. Soc. Jpn.*, **12**, 570.
Lax, M., and Birman, J. L., 1972, *Phys. Stat. Sol. (b)*, **49**, K153.
Long, D., 1960, *Phys. Rev.*, **120**, 2024.
Lugli, P., and Ferry, D. K., 1985, *IEEE Trans. Electron Dev.*, **32**, 2431.
Lugli, P., Jacoboni, C., Reggiani, L., and Kocevar, P., 1987, *Appl. Phys. Lett.*, **50**, 1251.
Mori, H., 1965, *Prog. Theor. Phys.*, **33**, 423.
Osman, M. A., and Ferry, D. K., 1987, *Phys. Rev. B*, **36**, 6018.
Paige, E. G. S., 1964, *The Electrical Conductivity of Germanium*, Vol. 8 of *Progress in Semiconductors*, edited by Gibson, A. F., and Burgess, R. E. (New York: John Wiley).
Paige, E. G. S., 1969, *IBM J. Res. Develop.*, **13**, 562.
Pottier, N., 1983, *Physica*, **A117**, 243.
Price, P. J., 1979, in *Semiconductors and Semimetals*, Vol. 14, edited by Willardson, R. K., and Beer, A. C. (New York: Academic Press), pp. 249–308.
Rees, H. D., 1969, *J. Phys. Chem. Sol.*, **30**, 643.
Rees, H. D., 1972, *J. Phys.*, **C5**, 64.
Reggiani, S., Vecchi, M. C., and Rudan, M., 1998, *VLSI Design*, **8**, 361.
Reik, H. G., and Risken, H., 1961, *Phys. Rev.*, **124**, 777.
Ridley, B. K., 1982, *Quantum Processes in Semiconductors* (Oxford: Clarendon Press).
Rode, D. L., 1975, in *Semiconductors and Semimetals*, Vol. 10, edited by Willardson, R. K., and Beer, A. C. (New York: Academic Press), pp. 1–89.
Shah, J., Deveaud, B., Damen, T. C., Tsang, W. T., Gossard, A. C., and Lugli, P., 1987, *Phys. Rev. Lett.*, **59**, 2222.
Shockley, W., 1951, *Bell. Sys. Tech. J.*, **30**, 990.
Siegel, W., Heinrich, A., and Ziegler, E., 1976, *Phys. Stat. Sol. (a)*, **35**, 269.
Stratton, R., 1958, *Proc. Roy. Soc. (London)*, **A246**, 406.
Trofimenkoff, F. N., 1965, *Proc. IEEE*, **53**, 1765.
Yamashita, J., and Watanabe, M., 1954, *Prog. Theor. Phys.*, **12**, 443.
Zwanzig, R. W., 1961, in *Lectures in Theoretical Physics*, edited by Brittin, W. E., Downs, B. W., and Downs, J. (New York: Interscience).

Chapter 5

Topics in high field transport

The earliest studies of high electric field effects in solids actually dealt with the breakdown properties of the dielectric materials. As a result, these studies were carried out at very high electric fields and generally resulted in the destructive breakdown of the sample being measured. With the studies of Ryder (1953) and Shockley (1951), though, emphasis shifted to a concentration on the transport properties themselves, especially in the new and interesting semiconductors germanium and silicon. It is interesting that once devices began to be made from semiconductor material, interest returned to the breakdown phenomena, in this case the avalanche breakdown in reverse-biased *p–n* junctions (McKay and McAfee, 1953). A critical parameter in these studies is the avalanche ionization rate, and its equivalent generation rate, for excess carriers. These, it turns out, are the results of just another scattering process – the impact ionizing collision, which occurs at very high electric fields. This scattering process leads to a "soft" breakdown, in that once a threshold electron energy is reached, the scattering process leads to a probability – not a certainty – that the ionizing process will occur, as this scattering still must compete with all other scattering processes.

In the early 1960s, attention was focused on the observation of negative differential conductivity (NDC), particularly in GaAs. This led to renewed interest in high field transport for the possibility of new sources of microwave power devices through the Gunn effect. However, the concepts of intervalley transfer are much older and were even observed in silicon.

In more recent years, the concept of velocity overshoot in short-gate field-effect transistors has arisen as the technology has advanced sufficiently far to fabricate real devices with gate lengths as short as 20 nm. This has, in turn, focused attention on the short-time behavior of carriers subject to high electric fields. Unfortunately, it was almost impossible to study the time-dependent dynamics of these carriers due to the sub-picosecond relaxation times involved. However, the advent of modern femtosecond pulse-length lasers has opened a new area in which the actual temporal behavior can be observed, and this will be discussed in the next chapter. In this chapter, the goal is to review the concepts of impact ionization and inter-valley transfer, and the manner in which they arise from high electric-field transport. Along the way, a treatment of the small-signal conductivity and its variation in high electric fields is introduced.

5.1 Impact ionization

As electrons (or holes) drift through a semiconductor under the influence of the electric field, they gain energy from the field and lose it through collisions with the lattice. At relatively high electric fields, though, a few electrons with energies in the tail of the distribution will have gained sufficient energy for an additional type of energy-dissipating collision – pair production through the *inverse Auger process*. An energetic electron (or hole) gives up its energy to a valence electron, raising the latter into the conduction band and leaving a hole in the valence band. This process is the reciprocal of Auger recombination, and this is the source of the name, but it is also called impact ionization, because the initial particle *knocks* the valence electron loose to create the electron–hole pair. The original particle has created two new particles (the new electron–hole pair). At low temperatures, collisions with neutral impurities can lead to the same effect, except that the hole is localized on the impurity so that only one new carrier is produced. The number of ionizing collisions produced will depend on the number of electrons (or holes) which have a sufficient energy for the pair production process and also will depend on the relative collision probabilities for other scattering processes.

The parameters that are desired for a discussion of the inverse Auger process are the ionization rate α (in units of cm^{-1}) and a related parameter, the generation rate $g(E)$ (in units of s^{-1}). The former can be expressed as

$$\alpha = \frac{1}{n_0 v_d} \sum_{E>E_T} \rho(E)\Gamma_{ion}(E)f(E), \tag{5.1}$$

where $\Gamma_{ion}(E) = v(E)/l_{ion}$, $v(E)$ is the velocity at energy E, $\rho(E)$ is the density of states, and l_{ion} is the mean free path for an ionizing collision. The latter quantity can roughly be calculated as $g = \alpha v$, although more properly, the multiplicative velocity term in g should be added to α prior to the summation over the energy, as

$$g(E) = \frac{1}{n_0 v_d} \sum_{E>E_T} \rho(E)v(E)\Gamma_{ion}(E)f(E). \tag{5.2}$$

In both of these equations, E_T is the threshold energy for the creation of an electron–hole pair.

Exact calculations for the ionization rate α, or generation rate g, must involve knowledge of the distribution function of the carriers, which is the major theoretical task involved. As observed in Chapter 4, solving for the distribution function can be a complicated task, since the added collision term for ionizing collisions must also be included. Most of the so-called "exact" theories for impact ionization, which have appeared over the years, are concerned with solving for the distribution function first. In this section, our concern is with impact ionization itself. A quite reasonable comprehension of the field dependence of the process can be obtained by considering two separate extreme cases. The first is for very high electric fields for which the distribution function is nearly spherically symmetric. The second is for lower values of the electric field, where concern is primarily with a few "lucky" electrons that are accelerated ballistically to a sufficient energy for ionization without having undergone any phonon collisions. After this brief introduction, attention will be focused on more general theories and the physical processes.

Wolff (1954), in one of the earliest theories applicable to semiconductors, assumed that a strong scattering interaction was always present in the semiconductor, and that this led to a quasi-Maxwellian distribution function, dominated by the spherically symmetric part. At very high electric fields, the diffusive effect (in energy space) dominates any anisotropy introduced by the scatterers, resulting in only the first two terms of the Legendre polynomial expansion being retained. In fact, the results of Chapter 4 lead us to recognize that the electron temperature, if one can be defined (or the average energy), rises very rapidly with electric field at fields on the order of the breakdown field. In turn, the scattering rates also rise very rapidly, which tends to increase the momentum randomization. Even polar scattering tends to be mainly back-scattering at very high energies, which reduces the asymmetry. The major effect of all of this, however, is that the symmetric part of the distribution spreads so rapidly with the rapidly increasing electron temperature that the major fraction of the carriers undergoing impact ionization comes from electrons in the tail of this part of the distribution. The diffusion approximation leads to a quasi-Maxwellian for $f_0(E)$ with an effective temperature related to the total mean free path L, as

$$T_e \sim \frac{(eFL)^2}{3\hbar\omega_0}, \tag{5.3}$$

and since α (or g) follows from an integration over f_0, one then obtains

$$\alpha(F) \approx A \exp\left(-\frac{K}{F^2}\right). \tag{5.4}$$

Shockley (1961), on the other hand, assumed a very weak interaction between the electron and the lattice, so that $f_0(E)$ retained a strongly peaked form, with the peak in the direction of the electric field F. In his model, impact ionization is due to a few "lucky" electrons in the tail of the asymmetric part of the distribution. This first ballistic transport model assumed that these few electrons are accelerated to the threshold energy E_T without collisions, so that the probability is proportional to

$$\alpha(F) \approx \mathbf{B}\exp\left(-\frac{E_T}{eFL}\right). \tag{5.5}$$

In general, the fields are neither so high as to validate Wolff's model nor so low as to validate Shockley's approach. As a consequence, a more detailed investigation is required.

5.1.1 A more general approach

Before proceeding to a discussion of the more exact solutions obtained by ensemble Monte Carlo approaches, it is worthwhile to justify the general results obtained by both Wolff and Shockley. It can be assumed that the total distribution function may be described as the sum of two parts. One part is spherically symmetric and obtainable, for example, by the hydrodynamic equation approach of Chapter 4. The second part is a delta function whose argument is the polar angle, so that it is directed along the electric field direction. This latter part contains all of the asymmetric terms of the distribution function. In essence, this approach will allow us to investigate the crossover between the two situations discussed in the previous section. The combined distribution function can be written as (Baraff, 1962)

$$f(\mathbf{p}, \cos\theta) = A(\mathbf{p}) + B(\mathbf{p})\delta(1 - \cos\theta). \tag{5.6}$$

To obtain the two functions A and B (which are different from the arbitrary constants of the earlier equations), the distribution function can be expanded in a Legendre polynomial series, as in Chapter 4. From this, for example, the general term can he written as

$$f_i(\mathbf{p}) = \frac{2i + 1}{2}\int_0^{\pi} f(\mathbf{p}, \cos\theta)P_i(\cos\theta)d(\cos\theta). \tag{5.7}$$

Using (5.6) in this equation gives the lowest order terms as

$$f_0(\mathbf{p}) = A(\mathbf{p}),$$
$$f_1(\mathbf{p}) = \tfrac{3}{2}B(\mathbf{p}),$$
$$f_2(\mathbf{p}) = \tfrac{5}{2}B(\mathbf{p}),$$
$$f_n(\mathbf{p}) = \frac{2n+1}{2}B(\mathbf{p}). \tag{5.8}$$

Baraff (1962) gives the solution for f_0, in analogy with the Druyvestyn distribution function, as

$$f_0(E) = \frac{1}{E^{\beta}}\exp\left(-\frac{E}{k_B T_e}\right) \tag{5.9}$$

with

$$\frac{1}{\beta} = \frac{2}{3} + \frac{eFL}{3\hbar\omega_0}, \quad T_e = \frac{(eFL)^2}{3\hbar\omega_0} + \frac{2}{3}(eFL). \tag{5.10}$$

For $eFL = \hbar\omega_0$, the result (5.9) is

$$f_0(E) = \frac{1}{E}\exp\left(-\frac{E}{eFL}\right), \tag{5.11}$$

which is precisely Shockley's result for his ballistic distribution. On the other hand, if $eFL \gg \hbar\omega_0$, only the leading term in the temperature remains, and

$$f_0(E) \sim \exp\left(-\frac{E\hbar\omega_0}{(eFL)^2}\right), \tag{5.12}$$

which is Wolff's form. Thus, for small fields the ionizing electrons come from ballistic behavior in the "spike" portion of the distribution function, while for large fields the bulk of the ionizing electrons come from the spherically symmetric part of the distribution.

It is worth remarking at this point that, in Shockley's regime, the band structure details are expected to be exceedingly important, since not only L but also E_T is orientation dependent (which will be seen to be the case below). Thus, conceptually one expects some slight orientation dependence for the lower-field ionization, but this is less likely to be the case in the diffusive regime of Wolff.

In general, the electric field is neither so large as to make Wolff's treatment proper, nor so small as to validate the Shockley treatment. Baraff's (1962) careful integral equation approach includes the energy dependence of the scatterers. In this approach, he considered only the acoustic and non-polar optical modes of the lattice, and used a spherically symmetric and parabolic energy band. These assumptions restrict the full validity of the results to almost no semiconductor in principle, yet the results have proven to be almost universally applicable to most semiconductors. In light of more detailed calculations that are available, it is not of interest to work completely through Baraff's calculation. Here the dimensionless ionization coefficient αL is plotted against the dimensionless reciprocal field E_T/eFL. Although it is unfortunate that Baraff's results (and indeed, any more exact results) do not yield a closed-form solution, a fairly simple analytic expression that relates exceptionally well to the numerical results has been obtained by Okuto and Crowell (1972). From this analytic fit, it is possible to express the ionization coefficient as

$$\alpha(F) = \frac{eF}{E_T}\exp\left(a - \sqrt{a^2 + x^2}\right), \tag{5.13}$$

with

$$x = \frac{E_T}{eFL}, \quad a = 0.217\left(\frac{E_T}{\hbar\omega_0}\right)^{1.14}. \tag{5.14}$$

This expression has generally good results for the region $a > 5$, although the results are acceptable over other regions.

In several applications, Baraff's theory has proven to be quite suitable for a variety of semiconductor materials. Even in cases where the band structure is highly non-parabolic, a reasonable fit with Baraff's theory can be obtained. Although it is not really expected that the fit to the ionization energy is very accurate, the value found for the mean free path is expected to be a reasonable value. Hence, the distribution function for the electrons can be expected to become quasi-Maxwellian for fields where ionization events are of considerable importance. In Figure 5.1, we plot some measured ionization rates for Si, along with two points obtained from an ensemble Monte Carlo simulation of high field transport in Si. For comparison, a fit with the Baraff theory is also shown. The parameters for the latter are an ionization energy of 1.18 eV, taken from Table 5.1 below, a dominant "phonon" energy of 74 meV, and an effective mean free path of 6.3 nm (at high fields). There is certainly much more data on the impact ionization in Si, but this data was chosen just to indicate that a good fit can be obtained with the Baraff theory, and with more detailed Monte Carlo simulations.

While Baraff's theory yields generally good results for ionization coefficients and for the generation rate $g(E)$, it is difficult to include, for example, strong anisotropy arising from the polar modes or non-parabolicity. For this

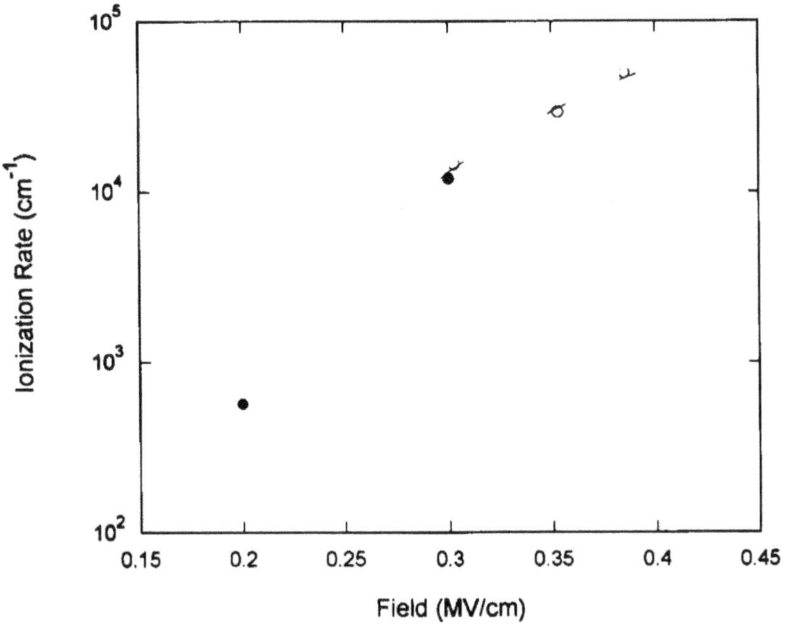

Figure 5.1 The ionization coefficient α for electrons in Si. The experimental data (open circles) is taken from Lee *et al.* (1964), and two ensemble Monte Carlo data points (solid circles) are shown for comparison. The solid curve is a fit to the data using the Baraff theory (the parameters are discussed in the text).

reason, more exact methods of solving the Boltzmann equation have been pursued. These include ensemble Monte Carlo solutions of the Boltzmann equation. In these approaches, the ionizing collision becomes just another scattering process.

5.1.2 Threshold energy

There are two major factors that will affect the shape exhibited by either the ionization rate or the generation rate when plotted against the electric field, in addition to that imparted by the distribution function. These are the ionization energy E_T and the energy dependence of the ionizing collision. In this section our attention turns to the calculation of the former parameter, the ionization energy E_T. The energy dependence of the ionizing collision is considered in the following section.

Consider a typical ionizing collision in which an electron in the conduction band interacts with a bound electron in the valence band. Prior to the collisions, the incident electron lies at a relatively high energy $E_c(\mathbf{p}_1)$ and the bound electron lies in the valence band at $E_v(\mathbf{p}_2)$. After the collision, the incident electron has created an electron–hole pair. The electrons now lie at $E_c(\mathbf{p}_3)$ and $E_c(\mathbf{p}_4)$, while the hole lies at $E_v(\mathbf{p}_2)$. Although this is an electron-initiated process, the hole-initiated process is quite similar. It differs only in the band in which the incident particle lies (there would be one incident hole and the result would be two holes and one electron rather than the two electrons and one hole of the process above). Several complications can occur. First, the incident and final electrons do not have to lie in the same conduction (or valence) band. Second, the process can involve phonons, just as in the indirect-absorption process.

During the ionizing collision, both the total energy and momentum must be conserved. The energy is expressible as

$$E_c(\mathbf{p}_1) = E_c(\mathbf{p}_3) + E_c(\mathbf{p}_4) - E_v(\mathbf{p}_2) + \sum_j \alpha_j \hbar\omega(\mathbf{q}), \tag{5.15}$$

where the latter sum runs over the phonons involved, and the sign on the valence band energy arises from the fact that a hole is left in state $\mathbf{p}_2$. By a similar argument, the momentum balance equation is

$$\mathbf{p}_1 = \mathbf{p}_4 + \mathbf{p}_3 - \mathbf{p}_2 + \sum_j \alpha_j \hbar \mathbf{q}_j. \tag{5.16}$$

In both of these equations, the coefficient $\alpha_j = \pm 1$ (times an integer, as several phonons in the same mode may be emitted). The threshold for the ionizing collision arises when the energy of the incident particle is a minimum (Anderson and Crowell, 1972) value that can still satisfy (5.15) and (5.16). Thus, only phonon absorption processes ($\alpha_j < 0$) are considered to be likely processes at threshold. The energy and momentum are minimized by differentiating the preceding equations as

$$d\mathbf{p}_1 = d\mathbf{p}_4 + d\mathbf{p}_3 - d\mathbf{p}_2 + \sum_j \alpha_j \hbar d\mathbf{q}_j,$$

$$d\mathbf{p}_1 \cdot \frac{\partial E_c(\mathbf{p}_1)}{\partial \mathbf{p}_1} = 0 = d\mathbf{p}_3 \cdot \frac{\partial E_c(\mathbf{p}_3)}{\partial \mathbf{p}_3} + d\mathbf{p}_4 \cdot \frac{\partial E_c(\mathbf{p}_4)}{\partial \mathbf{p}_4}$$

$$- d\mathbf{p}_2 \cdot \frac{\partial E_v(\mathbf{p}_2)}{\partial \mathbf{p}_2} + \sum_j \alpha_j \hbar d\mathbf{q}_j \cdot \frac{\partial \omega(\mathbf{q}_j)}{\partial \mathbf{q}_j}. \tag{5.17}$$

The last line of (5.17) can be re-expressed as

$$d\mathbf{p}_2 \cdot \mathbf{v}_2 = d\mathbf{p}_3 \cdot \mathbf{v}_3 + d\mathbf{p}_4 \cdot \mathbf{v}_4 + \sum_j \alpha_j \hbar d\mathbf{q}_j \cdot \mathbf{v}_j. \tag{5.18}$$

This last equation can now be combined with the first of equations (5.17), if a dot product of the latter is formed with $\mathbf{v}_2$ and then subtracted, to yield the stability condition for the threshold energy:

$$d\mathbf{p}_3 \cdot (\mathbf{v}_3 - \mathbf{v}_2) + d\mathbf{p}_4 \cdot (\mathbf{v}_4 - \mathbf{v}_2) + \sum_j \alpha_j \hbar d\mathbf{q}_j \cdot (\mathbf{v}_j - \mathbf{v}_2) = 0. \tag{5.19}$$

A sufficient condition for the validity of (5.19) is $\mathbf{v}_2 = \mathbf{v}_3 = \mathbf{v}_4 = \mathbf{v}_j$. Hence, all particles after the collision must have the same velocity if the process is a minimum incident energy process at threshold. If phonons are involved, this limit is a very restrictive one, since the resulting charge carriers must have the same velocity as the phonons. Since maximum phonon velocities are very small compared to maximum electron velocities, only charge carriers located very near a band extremum result from phonon-coupled threshold ionization processes, and this restricts the phonon processes to those in which the incident electron makes a band transition.

In a direct-gap semiconductor with spherical, parabolic bands, the effective masses can be introduced by m_e and m_h. The threshold criterion requires that $\mathrm{v}_e = \mathrm{v}_h$, in the absence of the phonon processes, which leads to

$$\mathbf{k}_e = \gamma \mathbf{k}_h, \tag{5.20}$$

where $\gamma = m_h/m_e$, since the wave vectors scale as the reciprocal of the effective masses. Thus the energy equation (5.15) becomes

$$E_T = E_c(\mathbf{p}_1) = 2\left(\frac{\hbar^2 k_c^2}{2m_e}\right) + \left(\frac{\hbar^2 k_v^2}{2m_h}\right) + E_G$$

$$= (2 + \gamma)\left(\frac{\hbar^2 k_c^2}{2m_e}\right) + E_G \tag{5.21}$$

and

$$\mathbf{p}_1 = 2\mathbf{p}_c - \mathbf{p}_v = \hbar k_c(2 + \gamma), \tag{5.22}$$

where it has been assumed that both final-state electrons lie in the same band. From the latter equation it may be observed that the energy of the initial state is also given by

$$E_T = E_c(\mathbf{p}_1) = (2 + \gamma)^2\left(\frac{\hbar^2 k_c^2}{2m_e}\right). \tag{5.23}$$

This can now be used to eliminate $\mathbf{k}_c$ from (5.21), with the result that

$$E_{T,e} = E_G \frac{2+\gamma}{1+\gamma}, \tag{5.24}$$

where a second subscript has been added to indicate that this is the electron-initiated threshold energy. The corresponding result for holes is found by interchanging the electron and hole masses, which amounts to replacing γ by $1/\gamma$. Thus the hole result is given by

$$E_{T,h} = E_G \frac{1+2\gamma}{1+\gamma}. \tag{5.25}$$

When $\gamma = 1$, both threshold energies are equal to $1.5E_G$. This is an oft-quoted result, but is restricted to very few semiconductors in actual practice as the masses are quite different in most cases. If γ is very large, the ionization threshold for electrons is just the energy gap, while that for holes is twice the energy gap. For very small γ, the opposite result holds. Thus, in general, it is found that the ionization energy varies for most semiconductors from the energy gap to twice the energy gap for this single band case.

With the foregoing concepts in mind, it is clear that one can easily determine the threshold energies for the ionizing collision by utilization of the equal-velocities criterion if sufficiently accurate energy band structure data is available. The calculation of threshold energies may be performed computationally by a Newton's method when the full $\boldsymbol{k}$-dependent band structure is known. In many cases, the available band structure data are not sufficiently accurate to allow the calculation of resultant carriers lying off the principal axes and a graphical interpolation routine must be used. In Table 5.1, the two lowest threshold energies, as calculated by Anderson and Crowell (1972) and, more recently for Si and Ge by Czajkowski *et al.* (1990), are given for each type of initiating carrier and direction of initial wave vector. All energies are quoted in eV relative to the appropriate band edge and are believed by these authors to be accurate to within ±0.2 eV. It is evident that generally there are a multiplicity of thresholds, so that it is expected that field- and orientation-dependent generation rates are to be seen in the Shockley limit (relatively low electric field near the threshold for ionization). This may have been observed in experiments in GaAs p–n junction devices at relatively low fields near the onset of ionization (Pearsall *et al.*, 1978).

5.1.3 The ionizing collison

The collision rate of the impact ionization process can be calculated in a manner similar to the method used for calculating collision relaxation times. It is assumed that the electrons are described by spherical, parabolic energy bands, although the approach can readily be extended to a more

Table 5.1 Threshold energies for ionization[a] (eV)

Material	*Si*	*Ge*	*GaAs*	*GaP*	*InSb*
Electron initiated					
(100)	1.18 (U)	1.01 (U,D)	2.1 (D*)	2.6 (U*)	0.2 (D)
	1.41 (U,D)	1.06 (U*)	2.3 (U*)	3.0 (U*)	1.6 (D)
(110)	2.1 (U)	1.11 (D*)	1.7 (D)	2.8 (D*)	0.2 (D)
	2.28 (U*)	1.27 (D)	1.9 (D*)	2.9 (D*)	1.6 (D)
(111)	3.1 (U*)	0.76 (U*)	3.2 (D*)	3.0 (D*)	0.2 (D)
	3.39 (U*)	2.5 (D*)	3.6 (D*)	3.4 (D*)	
Hole initiated					
(100)	1.73 (D)	0.97 (D)	1.7 (D*)	2.4 (D)	0.2 (D*)
	2.06 (D*)	1.41 (D*)	1.9 (D)		0.6 (D)
(110)	1.71 (D*)	0.91 (D*)	1.4 (D*)	2.3 (D)	0.2 (D*)
	2.25 (D*)	1.57 (D)	1.6 (D*)	2.6 (D*)	0.4 (D)
(111)	2.73 (D*)	0.88 (D)	1.6 (D*)	2.9 (D*)	0.4 (D*)
	4.4 (D*)	1.0 (D*)	2.3 (D*)	3.6 (D*)	1.5 (D*)
E_G	1.1 (ind.)	0.7 (ind.)	1.4 (D)	2.3 (ind.)	0.2 (D)

Notes:
[a] D direct process – no phonons
U umklapp process involving phonons or a Brillouin zone shift;
* the initiating carrier does not come from the normal conduction or valence band
ind. indirect gap

general band structure. It is also assumed that the dielectric constant is a scalar value and the frequency dependence of this quantity will be ignored (the screening is discussed further in Chapter 7). The approach is then just a semi-classical calculation. The probability per unit time that an inverse Auger process occurs is just

$$\Gamma(\mathbf{k}_e) = \frac{2\pi}{\hbar} \sum_{\mathbf{k}_b,\mathbf{k}_{b'},\mathbf{k}_{e'}} |M(\mathbf{k}_e, \mathbf{k}_b, \mathbf{k}_{b'}, \mathbf{k}_{e'})|^2 \delta_{\mathbf{k}_e+\mathbf{k}_b,\mathbf{k}_{b'}+\mathbf{k}_{e'}} \delta(E_{\mathbf{k}_e} - E_{\mathbf{k}_b} - E_{\mathbf{k}_{b'}} - E_{\mathbf{k}_{e'}}). \quad (5.26)$$

Here $\mathbf{k}_e = \mathbf{k}_1$, $\mathbf{k}_{e'} = \mathbf{k}_3$, $\mathbf{k}_b = \mathbf{k}_2$, $\mathbf{k}_{b'} = \mathbf{k}_4$ in the previous notation. Obviously, the Fermi golden rule is being used to evaluate the scattering process. For the perturbing potential, it is assumed that the incident electron interacts with the bound electron (in the valence band) through a screened Coulomb potential. The incident electron is characterized by $\mathbf{k}_e$, $E(\mathbf{k}_e)$ and the bound electron by $\mathbf{k}_b$, $E(\mathbf{k}_b)$. The screened potential is just

$$V(\mathbf{r}) = \frac{e^2}{4\pi\varepsilon_s|\mathbf{r}_e - \mathbf{r}_b|} \exp(-q_D|\mathbf{r}_e - \mathbf{r}_b|), \quad (5.27)$$

and q_D is the Debye screening wave vector. This potential may also be written in terms of its Fourier transform as

$$V(\mathbf{r}) = \sum_{\mathbf{q}} \frac{e^2}{4\pi\varepsilon_s(q^2 + q_D^2)} e^{i\mathbf{q}\cdot\mathbf{r}}. \tag{5.28}$$

The wave functions are taken to be Bloch functions in the appropriate band. That is, the initiating carrier scatters, but remains in the conduction band, and the second electron is excited from the valence band (leaving a hole) to the conduction band. We can write these wave functions as

$$\begin{aligned}
\psi(\mathbf{k}_e) &= u_c(\mathbf{k}_e)\exp(i\mathbf{k}_e \cdot \mathbf{r}_e),\\
\psi(\mathbf{k}'_e) &= u_c(\mathbf{k}'_e)\exp(i\mathbf{k}'_e \cdot \mathbf{r}_e),\\
\psi(\mathbf{k}_h) &= u_v(\mathbf{k}_h)\exp(i\mathbf{k}_h \cdot \mathbf{r}_h),\\
\psi(\mathbf{k}'_h) &= u_c(\mathbf{k}'_h)\exp(i\mathbf{k}'_h \cdot \mathbf{r}_h).
\end{aligned} \tag{5.29}$$

Thus, the first electron is taken to make a transition from $\mathbf{k}_e$ to $\mathbf{k}'_e$ in the conduction band, while the second electron makes a transition from $\mathbf{k}_h$ in the valence band to $\mathbf{k}'_h$ in the conduction band. The matrix element is then

$$M = \frac{e^2}{\varepsilon_s(q^2 + q_D^2)} I_1 I_2, \tag{5.30}$$

where

$$\begin{aligned}
I_1 &= \int d^3\mathbf{r}_e u_c^*(\mathbf{k}'_e) u_c(\mathbf{k}_e) e^{i(\mathbf{k}_e - \mathbf{k}'_e + \mathbf{q})\cdot\mathbf{r}_e},\\
I_2 &= \int d^3\mathbf{r}_h u_c^*(\mathbf{k}'_h) u_v(\mathbf{k}_h) e^{i(\mathbf{k}_h - \mathbf{k}'_h - \mathbf{q})\cdot\mathbf{r}_h}.
\end{aligned} \tag{5.31}$$

The integration may be split into a summation over unit cells and integration over a single unit cell. By the principle of the closure of a complete set, the summation leads to the conservation of momentum condition, which is also enforced by the summation over $\mathbf{k}'_e$ in (5.26), so that the summation over $\mathbf{q}$ requires

$$\mathbf{q} = \mathbf{k}_h - \mathbf{k}'_h = \mathbf{k}'_e - \mathbf{k}_e. \tag{5.32}$$

With this limitation, the overlap integrals are simpler to evaluate. The first integral in (5.31) is just the type of integral involved in phonon scattering, such as appears in Chapters 3 or 4, and for normalized wave functions (in the parabolic band), the integral is unity. The second integral corresponds to an inter-band transition such as occurs for optical absorption. In this case, the second integral's squared magnitude is the effective oscillator strength, which is given by the f-sum rule (Harrison, 1970) as

$$|I_2|^2 = 1 + \frac{m_h}{m_0}, \tag{5.33}$$

where m_h is the appropriate hole mass, which is assumed to be much larger than the electron mass.

The relaxation time for an ionizing collision can now be calculated by summing over the energies $E(\mathbf{k}_e')$, $E(\mathbf{k}_h)$, and $E(\mathbf{k}_h')$. The first of these is, of course, accomplished with the Kronecker delta function on momentum. The second involves the energy-conserving delta function. To facilitate these integrals, a number of simplifying assumptions are first made. It will be implicitly assumed that the electron in the valence band comes from the heavy-hole band, so that $E(\mathbf{k}_h) \sim -E_T$. Second, it will be assumed that $k_h' \ll k_h$, which follows again from the large size of the hole mass compared with the electron mass. Then, $q^2 = k_h^2 = 2m_h E_{gap}/\hbar^2$. The delta function can then be used to set $E(\mathbf{k}_h')$ in terms of the other energies, and a first integration can be made over $E(\mathbf{k}_e')$ using the joint density-of-states functions as

$$\Gamma_i(E) = \frac{2\pi e^4}{\varepsilon_s^2 \hbar}\left(1 + \frac{m_h}{m_0}\right)\left[\frac{1}{2\pi^2}\left(\frac{2m_e}{\hbar^2}\right)^{\frac{3}{2}}\right]^2$$

$$\times \int_0^{E-E_T} \sqrt{E(\mathbf{k}_e')E(\mathbf{k}_h')}\,\frac{\delta(E_e' - E_T - E_e' - E_h')}{[2m_h E_T/\hbar^2 + q_D^2]^2}\,dE_e'$$

$$= \frac{m_e^3 e^4 (1 + m_h/m_0)}{\hbar^3 \pi^3 \varepsilon_s^2 m_h^2} \int_0^{E-E_T} \frac{\sqrt{E'(E - E' - E_T)}}{(E_T + E_D)^2}\,dE', \tag{5.34}$$

where $E_D = \hbar^2 q_D^2/2m_h$ is an energy corresponding to a hole with the Debye screening wave vector. Equation (5.34) can readily be integrated to give the resulting scattering rate for the ionizing process:

$$\Gamma_i(E) = \frac{m_e^3 e^4}{8\pi^2 \varepsilon_s^2 \hbar^3 m_h^2}\left(1 + \frac{m_h}{m_0}\right)\left(\frac{E - E_T}{E_T + E_D}\right)^2 u_0(E - E_T). \tag{5.35}$$

The use of (5.35) in an ensemble Monte Carlo program is straightforward and relatively easy. In fact, this led to the two data points (which are only an example, the complete curve can easily be calculated) in Figure 5.1.

Experimental results are also illustrated. The comparison with the Baraff theory points out how good a scaling theory the latter really is.

The use of the impact ionizing collision to determine the generation rates corresponds to a soft ionization process, as it introduces only a probability that an ionizing collision will occur if the energy is sufficiently high (Antoncik and Landsberg, 1963, 1967; Kane, 1967; Curby and Ferry, 1973; Ridley, 1987). A hard ionizing collision would always occur as soon as the energy reaches E_T. The "hard" process was used, for example, in the Shockley model of Chapter 4 to introduce optical phonon scattering, but the soft process is found to be important in nearly all breakdown processes in semiconductors. It has been used in GaAs (Shichijo and Hess, 1981), InSb and InAs (Curby and Ferry, 1973), and even in SiO_2 (Ferry, 1988). The problem with including the impact ionization in the Monte Carlo procedure is that it occurs in a very high energy range that the electrons seldom reach. Thus the calculations must be run for a long time to see significant statistical data on impact ionization. An alternative method, introduced by Lebwohl and Price (1971), uses the Monte Carlo process to ascertain with some precision the form of the distribution function up to energies well above the optical phonon energy, so that a relatively well understood falloff of the distribution to higher energies can be extrapolated. Then this analytical formulation of the distribution function is used to evaluate (5.1), for example, to give the ionization coefficient. This hybrid method also works well in predicting the expected generation rates for various semiconductors.

5.2 Intervalley scattering and intervalley transfer

Transfer between different valleys of the full energy band is one of the oldest and most interesting aspects of high-electric-field transport in semiconductors. There are two possible methods by which transfer between nonequivalent sets of valleys can occur. In one, the symmetry-breaking properties of the high electric field are used to break the symmetry between valleys that are normally equivalent. This occurs, for example, in Si at low temperatures when the electric field is not oriented at the same angle with all six mimima of the conduction band. For this case, the carriers in each valley are heated differently and a repopulation appears between the various valleys. If the field is not oriented along one of the high-symmetry crystalline axes, a transverse voltage can even appear. This is due to the fact that the current will not be parallel to the applied field in any valley, and repopulation breaks the cubic symmetry (Shibuya, 1955; Sasaki *et al.*, 1958, 1959), a process termed the Sasaki-Shibuya effect. Because the electric field is not oriented along a principal axis of the valley, the current contributed by each valley is not parallel to the electric field. Normally, the transverse components sum to zero to preserve the cubic symmetry of the crystal, but when the valleys are heated non-equivalently, transfer of electrons between the various valleys upsets this summation, since the populations of the individual valleys are no longer equal. Since the transverse current in a sample is usually required to vanish, a transverse electric field is induced.

Another version of this is when the field is aligned with the principal axis (100), at low temperatures, the two valleys along the field are heated differently than the four valleys normal to the field. Two of the equivalent minima of the conduction band have their longitudinal (heavy) mass in the direction of the field, while the other four minima have their transverse (light) mass in the field direction. In this case the carriers in the four transverse valleys can become much hotter than those in the two longitudinal valleys. Since the mass is much smaller, the velocity can be much larger, and the resultant energy input is much larger. In this case, electrons can transfer from the hot (fourfold) set of valleys to the cold (twofold) set of longitudinal valleys. At low temperatures this effect is sufficiently large to show negative differential conductivity in Si (Gram, 1972). This transfer upsets the normal cubic symmetry of the sixfold valleys of the conduction band.

The second method by which intervalley transfer can play a major role is when only a single conduction band minimum, with a small effective mass, is normally occupied, such as in the case of GaAs (the principal minimum is at the Γ point). With the application of a high electric field, the carriers are heated to relatively high energies. Then, a fraction of the carriers will actually be at energies above the minima of a secondary set of valleys [which lie along the (111) directions at the set of L points of the Brillouin zone in GaAs] of the conduction band. Since the equivalent L minima have relatively high values of the effective mass, their density of states is much larger than that of the central minimum, and intervalley scattering will cause electrons to move to the satellite valleys of the conduction band. Consider, for example, the conduction band of GaAs shown in Figure 2.9. There is a central valley characterized by a small effective mass of $0.067m_0$. In addition, there are subsidiary minima located at both the L point, lying some 0.29 eV higher, and at the X point, lying some 0.5 eV higher. These valleys have a considerably greater effective mass and density of states, as the former is Ge-like while the latter is similar to Si. Under normal circumstances, the central valley is the only one occupied. However, for an applied field of some 3.5 kV/cm at 300 K, electrons begin to transfer to the L valleys (Gunn, 1963), although this mechanism was not immediately recognized as the cause of the observed negative differential conductivity.

5.2.1 The Gunn effect

Early measurements on electron transfer between nonequivalent sets of valleys in the conduction band were made by Gunn (1963) in GaAs, but the effect is also seen in InSb, InAs, PbTe, InGaAs, Ge, and in inversion layers in Si. The resulting negative differential conductance that occurs when the carriers are transferred from low-mass, high-velocity states to high-mass, low-velocity states is referred to as the Gunn effect and is used in transferred electron devices (TEDs, which are used for microwave sources). Transfer occurs as the carriers are heated in an external electric field. Once some of the carriers reach energies near that of the satellite valleys (the L and X valleys), inter-valley scattering can occur. Since the density of states in the satellite valleys is much higher than in the central valley, inter-valley scattering from the Γ valley to the satellite X and L valleys can occur. The transfer in this particular direction is much more pronounced than the reverse process (recall that the scattering rate is essentially a direct measure of the density of final states). Because of the higher mass and density of states in the satellite valleys, the mobility and velocity are much lower and a negative differential conductivity will occur.

To understand why electrons transfer, it is necessary to consider the calculations of electron densities. The density of states per unit energy for any valley is given by

$$\rho(E) = \frac{1}{2\pi^2}\left(\frac{2m_i}{\hbar^2}\right)^{\frac{3}{2}}(E - E_0)^{\frac{1}{2}}, \tag{5.36}$$

where E_0 is the band minimum (which is zero for the Γ valley). Here, m_i is the density of states mass for valley i. For the central Γ valley, the total population (assuming a Maxwellian distribution at temperature T_e) is

$$n_\Gamma = \frac{1}{4}\left(\frac{2m_\Gamma k_B T_{e\Gamma}}{\pi\hbar^2}\right)^{\frac{3}{2}}\exp\left(\frac{E_F}{k_B T_{e\Gamma}}\right). \tag{5.37}$$

This is just the effective density of states for the conduction band and is reduced by the position of the Fermi energy (which, for non-degenerate materials, is below the conduction band edge). For the L valleys, this becomes

$$\begin{aligned} n_L &= 4\cdot\frac{1}{2\pi^2}\left(\frac{2m_L}{\hbar^2}\right)^{\frac{3}{2}}\int_{\Delta_{\Gamma L}}^{\infty} E^{\frac{1}{2}}\exp\left(-\frac{E - E_F}{k_B T_{eL}}\right)dE \\ &= \left(\frac{2m_L k_B T_{eL}}{\pi\hbar^2}\right)^{\frac{3}{2}}\exp\left(-\frac{\Delta_{\Gamma L} - E_F}{k_B T_{eL}}\right), \end{aligned} \tag{5.38}$$

where $\Delta_{\Gamma L}$ is the separation of the upper valleys from the central valley, and it has been assumed that the temperature in the satellite valleys is different than that of the central valley. The ratio of populations in the two sets of valleys is then

$$\frac{n_L}{n_\Gamma} = 4\left(\frac{m_L T_{eL}}{m_\Gamma T_{e\Gamma}}\right)^{\frac{3}{2}} \exp\left(-\frac{\Delta_{\Gamma L}}{k_B T_{eL}}\right) \exp\left(\frac{E_F}{k_B}\left(\frac{1}{T_{eL}} - \frac{1}{T_{e\Gamma}}\right)\right). \tag{5.39}$$

For GaAs, the appropriate values for the parameters are thought to be $4^{\frac{2}{3}} m_L = 1.2 m_0$, $m_\Gamma = 0.067 m_0$, and $\Delta_{\Gamma L} = 0.29$ eV. If the two temperatures are equal, the coefficient of the exponential is about 70. Then $n_L = n_\Gamma$ for an electron temperature of about 950 K, which is not unreasonable for moderate electric fields. For temperatures higher than this, the upper valleys have a larger density of states occupied. Thus when an electron initially in the central valley at an energy $E = \Delta_{\Gamma L}$ is scattered, it is more likely to undergo an inter-valley process and scatter to one of the satellite valleys.

The condition given above is actually more stringent than necessary and should be modified by the respective mobilities if it is actually of interest to calculate the onset of negative differential conductivity. For this, it is only necessary to carry out a simple argument on the net conductivity for the ensemble of electrons. The total conductivity for carriers in the two sets of valleys is given by

$$\sigma = n_\Gamma e \mu_\Gamma + n_L e \mu_L . \tag{5.40}$$

The change in conductivity with electric field is then given by

$$\frac{d\sigma}{dF} = e\mu_\Gamma \frac{dn_\Gamma}{dF} + e\mu_L \frac{dn_L}{dF} = e\frac{dn_L}{dF}(\mu_L - \mu_\Gamma) \tag{5.41}$$

since the total number of carriers is a constant. Here it has been assumed that the mobility is only a very weak function of the electric field, which is the normal case for polar-optical phonon scattering in the central valley of the conduction band. The current in the semiconductor sample is just $J = \sigma F$, so that

$$\frac{dJ}{dF} = \sigma + F\frac{d\sigma}{dF} = n_\Gamma e\mu_\Gamma + n_L e\mu_L + Fe\frac{dn_L}{dF}(\mu_L - \mu_\Gamma). \tag{5.42}$$

If the differential conductivity is negative, equation (5.42) requires that

$$\frac{\mu_\Gamma - \mu_L}{\mu_\Gamma + (n_L/n_\Gamma)\mu_L} \frac{F}{n_\Gamma} \frac{dn_L}{dF} > 1. \tag{5.43}$$

For the case of interest, the density in the central valley is a decreasing function of the field, so that $\mu_\Gamma > \mu_L$ is required. This can be accomplished by having the density of states in the satellite valleys higher than in the central valley and/or the relaxation time shorter in the satellite valleys than in the central valley. Both of these are consistent with each other. Thus the band structure of a material like GaAs is quite favorable for the process. In general, the mobility in the central valley is much larger than in the satellite valleys (about 7000 cm²/Vs in the central valley and about 100 cm²/Vs in the satellite valleys), so that the latter can be ignored, and (5.43) becomes

$$-\frac{dn_\Gamma}{dF} > \frac{n_\Gamma}{F}. \tag{5.44}$$

The latter quantity can be calculated with the help of (5.39) in the following manner:

$$\frac{dn_\Gamma}{dF} = -\frac{dn_L}{dF} = -\left(\frac{\Delta_{\Gamma L}}{k_B T_e}\right)\frac{n_\Gamma}{T_e}\frac{n_L}{n_L + n_\Gamma}\frac{dT_e}{dF}, \tag{5.45}$$

and it has been assumed that the two temperatures are equal. This may be rewritten in the form

$$\left(\frac{\Delta_{\Gamma L}}{k_B T_e}\right)\frac{n_L}{n_L + n_\Gamma}\frac{F}{T_e}\frac{dT_e}{dF} > 1. \tag{5.46}$$

It is not a bad assumption at this point to assume that the electron temperature increases linearly with the electric field (a quadratic variation makes only a small change in the results obtained below). Then the last factor is just unity.

To simplify the discussion, the factor $\Delta_{\Gamma L}/k_B T_e$ will be written as y. Then the resulting (5.46) becomes the simple transcendental equation

$$y > 1 + \frac{n_\Gamma}{n_L} \sim 1 + e^y/70. \tag{5.47}$$

There are two regions of y where this inequality is not satisfied. One is at very small values of y (very high values of electron temperature), which is not of interest. The other region is at large values of y, which correspond to the low electron temperatures of interest. This relation then gives an approximate upper limit on y (lower limit on the electron temperature) which is about 5.8, or $T_e \sim 600$ K. The key assumption here is the form for the dependence of the electron temperature on the field. If a quadratic form is used, essentially the same result is found due to the exponential factor dominating the results. Thus, the details of the dependence of the electron temperature on the field are not really crucial. At 600 K, the factor n_L/n_Γ is only about 0.15, so that negative differential conductivity sets in with as little as 15 percent of the electrons transferred to the satellite valleys.

It is of course preferable to use more exact calculations to determine the effective velocity and the relative populations of the various sets of valleys. In Figure 5.2, the velocity-field curve for GaAs is shown. The theoretical curve is obtained from an ensemble Monte Carlo calculation. The experimental data are the measurements of Bosch and Engelmann (1975). There is some discrepancy near the peak velocity and in the negative differential conductance regime, but it must be remembered that the negative conductance gives rise to field inhomogeneities (Ridley, 1963) which are not reflected in the calculations. Also shown are the relative populations of the Γ and L valleys, and it can be seen that the above estimates are quite good for the required transfer at the onset of negative differential conductivity.

The ideas of the Gunn effect are not so much those of intervalley scattering, even between non-equivalent sets of valleys, as of the instabilities and high-electric field domains that are present in the sample. In a semiconductor in which there is negative differential conductivity, the dielectric relaxation frequency is *negative* if the material is biased in the negative differential conduction regime. In this case, fluctuations in the field and density grow rather than decay. In looking at Figure 5.2, it is apparent that for a dc bias above the field corresponding to the peak velocity, the drift velocity is a decreasing function of the field. Thus, if there is some small region for which a fluctuation makes the local field higher than the remainder of the sample, the velocity is lower in this region. For conservation of the total field (the applied voltage), the field outside the fluctuation region is reduced slightly, which causes the velocity to increase. Thus carriers downstream from the fluctuation (relatively) rapidly move away from the fluctuation, while carriers upstream rapidly approach the fluctuation. The fluctuation will be uniform across the lateral dimension in order to conserve total current, but a *dipole domain* forms. This is an accumulation of charge on the upstream side of the region and a depletion of charge on the downstream side. This supports a further increase of the electric field in the fluctuation region. The overall domain moves with the average velocity of the carriers, since it is a fluctuation in free charge. At some time the fluctuation reaches the anode of the sample, leaves the sample, and the entire process can begin once again. Usually, the most common site for the field fluctuations is at the cathode, and this leads to transit-time domain oscillations with a frequency of $f = v_d/L$, where L is the length of the sample. It is the domain motion and resultant oscillations that are normally associated with the Gunn effect. In this section, it is assumed that the negative differential conductivity is caused by the intervalley transfer. But, it could also have been caused by transfer between nominally equivalent valleys, which are heated differently by the applied electric field. A full treatment of the instabilities and oscillations that arise from the Gunn effect are beyond the treatment here, as they can consume an entire book in their own right (Shaw *et al.*, 1979).

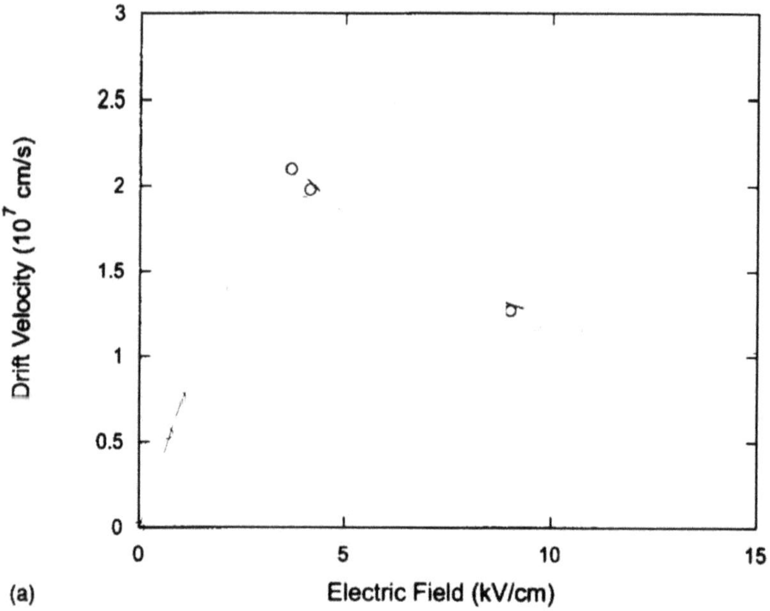

(a)

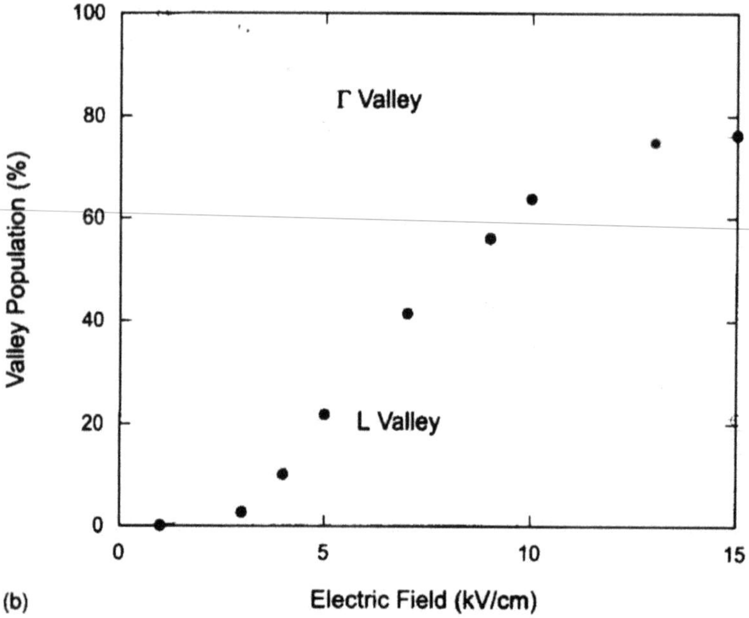

(b)

Figure 5.2 (a) The velocity as a function of electric field in GaAs, at 300 K, shows a peak, and then reduction with field as negative differential conductance begins to occur. (b) The valley populations show that the negative differential conductance arises from inter-valley transfer of hot carriers. (Calculations courtesy of L. Shifren, Arizona State University.)

5.2.2 *Transfer in the "equivalent valleys" of Si*

As mentioned above, it is possible for inter-valley scattering to occur between nominally equivalent valleys in Si, when a high electric field is present. This effect can be greatly enhanced if the six-fold set of valleys is split by some symmetry breaking process. One such process is quantization in the quantum well formed between SiO_2 and the bulk potential between the inversion layer and the *p*-type bulk of a MOSFET (Ando *et al.*, 1982). This was briefly touched upon in the last chapter where we discussed scattering from interface charge and interface roughness. The separation of the various valleys in the quantization occurs because of the ellipsoidal nature of the equivalent energy surfaces of the conduction band. Consider the (100) surface for example. Two valleys of the conduction band, the (100) and ($\bar{1}$00) (the bar implies a minus direction), have their long axes normal to the surface and the quantization is determined by the longitudinal effective mass m_L. On the other hand, the other four valleys all have their "short" axes normal to the surface and the quantization is determined by the transverse effective mass m_T. Since $m_L > m_T$, the set of quantized energy levels arising from the two-fold (m_L) set of valleys (often termed the Δ_2 valleys) will lie below the set arising from the four-fold set of valleys (correspondingly termed the Δ_4 valleys). In fact, the lowest energy level of the four-fold set, denoted E_0', lies well above that of the two-fold set, denoted E_0, and is almost degenerate with the second level of this latter set, E_1.

This symmetry-breaking can also occur when Si is strained. The most common occurrence of this is in heterostructures between Si and an alloy of Si and Ge, noted as $Si_{1-x}Ge_x$. When the latter is grown sufficiently thick, on an Si substrate for example, the strain in the alloy will relax and the lattice constant will adjust to one between that of Si and that of Ge, according to the random alloy theory (Section 2.6). A subsequent thin layer of Si grown on top of the SiGe alloy will be under tensile strain, and this will split the six-fold degenerate valleys precisely as in the MOSFET inversion layer (Abstreiter *et al.*, 1985). The four valleys in the plane of the heterostructure become equivalent, while the two valleys with their longitudinal axes normal to the heterostructure interface are split off, and this splitting is to a lower energy.

The advantage of the symmetry-breaking of the six-fold degenerate conduction band lies in the fact that the lower-lying two-fold set of valleys will exhibit the transverse effective mass for transport in the plane (either the plane of the inversion layer or the plane for transport parallel to the heterostructure interface). This leads to a higher mobility since the transverse mass is smaller than the composite conduction mass discussed in Chapter 3. In addition, negative differential conductance and enhanced velocity overshoot can occur. As the carriers are heated in the Δ_2 valleys, they gain energy and can eventually transfer to the cold Δ_4 valleys in a process quite similar to that of GaAs. Both of these effects can lead to enhanced performance

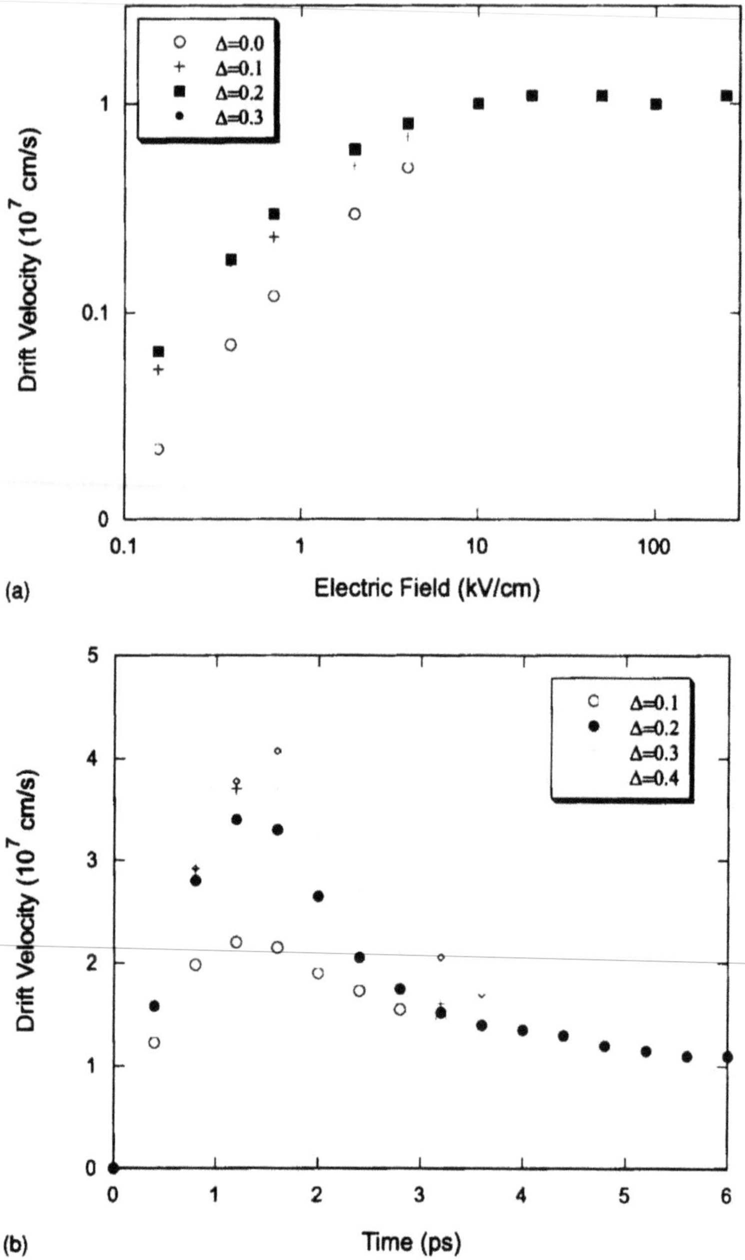

Figure 5.3 (a) The velocity-field curve for strained Si shows negative differential conductance, similar to that of GaAs. In this case, the six-fold degenerate conduction band is split, either by a surface quantization in the MOS structure, or by strain in an $Si/Si_{1-x}Ge_x$ heterostructure (where the SiGe is relaxed). Here, there is little change once $\Delta E > 0.2$ eV. (b) The overshoot velocity is significantly enhanced in the strained Si, due to the enhanced separation between the split set of valleys.

(over unstrained Si) in MOSFETs or in heterojunction transistors (Ismail *et al.*, 1992), a result supported by simulation of such devices (Yamada *et al.*, 1994). In Figure 5.3, we plot the velocity field curve and the velocity overshoot in a 50 kV/cm electric field for various values of the strain. The strain is indicated by the energy separation Δ between the upper Δ_4 valleys and the lower Δ_2 valleys. The energy separation is directly related to the strain and thus to the composition of the SiGe. This is expressed by $\Delta = 0.6x$, where x is the fraction of Ge in the alloy. Hence, an alloy $Si_{0.8}Ge_{0.2}$ produces an energy separation of 0.12 eV.

5.3 Small signal conductivity

It is well known that the microwave conductivity of semiconductors varies as a function of the frequency. However, the functional dependence becomes quite complicated when hot-electron transport is included. This occurs because one must consider not only the momentum relaxation time of the carriers, but also the energy relaxation time, which does not come into the picture at low electric fields and is usually ignored. This fact was developed in Chapter 3, where the velocity auto-correlation function and velocity overshoot were discussed. Even if the saturation velocity is very high in a semiconductor material, for actual microwave devices there is no guarantee that the microwave conductivity will have a desirable property, and this must be checked independently in each case. In this section, we will examine these high frequency properties. First, we examine the free-carrier absorption in a near equilibrium situation, and then progress to the complex high electric field case. Finally, we discuss the ballistic case.

5.3.1 Free-carrier absorption

For incident high-frequency waves of energy insufficient to produce an electron–hole pair, such as microwave radiation, absorption can occur either due to phonons or to free carriers. If the absorption is due to phonons, the lattice energy is changed. By far the more important process is the free-carrier absorption. This absorption can occur only in a partially filled band, where there are free electrons or holes capable of gaining energy from the electromagnetic field. This process is of primary importance in discussing the microwave conductivity of the semiconductor, although it can be of importance even into the infrared regions.

For these properties it is necessary to retain the term containing the explicit time dependence of the distribution function, and the Boltzmann equation, in the relaxation-time approximation, becomes

$$\frac{\partial f}{\partial t} + \frac{1}{\hbar}\mathrm{F} \cdot \nabla_k f = -\frac{f - f_0}{\tau_m}, \tag{5.48}$$

where the general force $\mathbf{F}$, rather than just the electric field, is used. However, for the present, in treating the high-frequency field it will be assumed that only the electric field is of importance, and that this field has the temporal variation $F = F_0 \exp(-i\omega t)$. Moreover, it will be assumed that a steady state exists, so that the velocity of the carriers and the distribution function show this same time variation. Then (5.48) becomes, for $f_1 = f - f_0$,

$$(1 - i\omega\tau_m) f_1 = e\tau_m \mathbf{F}_0 \cdot \mathbf{v} \frac{\partial f_0}{\partial E}. \tag{5.49}$$

The exponential time variation has been suppressed since it is a common factor in all terms. It is now easy to solve for f_1 and then to calculate the current (for $\mathbf{F}_0 \parallel \mathbf{a}_x$) as

$$J_x = e^2 F_0 \int_0^\infty \frac{v_x^2 \tau_m}{1 - i\omega\tau_m} \rho(E) \frac{\partial f_0}{\partial E} dE$$

$$= F_0 \frac{ne^2}{m^*} \left\{ \left\langle \frac{\tau_m}{1 + \omega^2\tau_m^2} \right\rangle + i\omega \left\langle \frac{\tau_m^2}{1 + \omega^2\tau_m^2} \right\rangle \right\}. \tag{5.50}$$

We recognize the high-frequency conductivity as the coefficient of the field in the last line above. Here, only the electron contribution is treated, but (5.50) is readily extended to the case for holes or for both electrons and holes. It may be observed that the effectiveness of the electric field in producing current is reduced at high frequencies. The breakpoint for the high-frequency reduction occurs at $\omega\tau_m = 1$, or when the microwave frequency is the same as the collision frequency. A typical value for the collision time was about 5×10^{-13} s, so that the high frequency effects become important for radian frequencies of the order of 10^{12} Hz. Although the term "collision frequency" has been used, the discussion of Chapter 3 points out that the "time" involved in the relaxation-time approximation is the momentum relaxation time.

The amount by which the conductivity is reduced at high frequencies can be expressed simply by taking the normalized version of the real part of the conductivity, as

$$\mathrm{Re}\left\{ \frac{\sigma(\omega)}{\sigma(0)} \right\} = \frac{1}{\langle \tau_m \rangle} \left\langle \frac{\tau_m}{1 + \omega^2\tau_m^2} \right\rangle. \tag{5.51}$$

For this high-frequency reduction of the conductivity, only the real part has been retained. The reason for this lies in the fact that at high frequency, the current density and the electric field are no longer in phase but are related by

$$\mathbf{J} = \sigma\mathbf{F} - i\omega\varepsilon(\omega)\mathbf{F} = \hat{\sigma}\mathbf{F}, \tag{5.52}$$

where

$$\hat{\sigma} = \sigma - i\omega\varepsilon(\omega). \tag{5.53}$$

In fact, in many cases, the conductivity is a tensor relation, just as in the case of a magnetic field, so the simple scalar form written here should be used with caution. The second term on the right of the latter two equations, which involves the dielectric constant, or the frequency-dependent dielectric function, is the displacement current contribution. The conductivity is actually a complex quantity, but the normal resistive portion is just the real part. The imaginary part of the complex conductivity contributes to the dielectric function. In fact, the latter part is the free-carrier contribution to the dielectric constant, and

$$\varepsilon = \varepsilon(\omega) - \mathrm{Im}\left\{\frac{\sigma}{\omega}\right\} = \varepsilon(\omega) - \frac{\sigma(0)}{\langle\tau_m\rangle}\left\langle\frac{\tau_m^2}{1+\omega^2\tau_m^2}\right\rangle. \tag{5.54}$$

A remarkable result is that the dielectric "constant" can be negative at low frequencies, so that the wave is attenuated and is absorbed by the free carriers. This is the classical skin depth in a conducting medium discussed in electromagnetic field theory.

If it is assumed for the moment that $\omega\tau_m \gg 1$, the large bracket average can be converted to a more suitable form, for which the dielectric function may be written as

$$\varepsilon = \varepsilon_\infty\left(1 - \frac{\omega_p^2}{\omega^2}\right), \tag{5.55}$$

where

$$\omega_p = \sqrt{\frac{\sigma(0)}{\langle\tau_m\rangle\varepsilon_\infty}} = \sqrt{\frac{ne^2}{m^*\varepsilon_\infty}} \tag{5.56}$$

is the *free-carrier* plasma frequency. The semiconductor is reflecting for $\omega < \omega_p$, and the reflection mentioned above is just the classical plasma reflection edge observed in gaseous plasmas.

The absorption coefficient for the microwave field may now be calculated by treating the semiconductor as a lossy dielectric. The assumed electromagnetic wave propagates as

$$\mathbf{F} = \mathbf{F}_0 \exp(i\mathbf{k}_0 \cdot \mathbf{r} - i\omega t), \tag{5.57}$$

where

$$k_0 = \frac{\omega}{c}\sqrt{\frac{\varepsilon}{\varepsilon_0}} \tag{5.58}$$

is the wave propagation vector. If the frequency is below the plasma frequency, the wave is heavily absorbed. The current point of interest is the weak free-carrier absorption at frequencies well above the plasma edge, so that the imaginary part of the conductivity may be ignored. Still, the dielectric function is complex, and

$$\varepsilon = \varepsilon_\infty\left(1 + i\frac{\sigma}{\omega\varepsilon_\infty}\right) = \varepsilon_\infty\left(1 + i\frac{\omega_D}{\omega}\right). \tag{5.59}$$

Here, the dielectric relaxation frequency $\omega_D = \sigma/\varepsilon_\infty$ has been introduced. The absorption is determined by letting $k_0 = k_r + ik_i$. Then, for high frequencies ($\omega\tau_m \gg 1$),

$$k_i = \frac{\omega}{c}\sqrt{\frac{\varepsilon_\infty}{\varepsilon_0}}\frac{\sigma_r}{2\omega\varepsilon_\infty} = \frac{\omega_p^2}{2c\omega^2}\sqrt{\frac{\varepsilon_\infty}{\varepsilon_0}}\left\langle\frac{1}{\tau_m}\right\rangle. \tag{5.60}$$

The absorption coefficient is just $\alpha = 2k_i$. At low frequencies, it is slightly more difficult to calculate the absorption coefficient due to the plasma reflection edge, but the result is easily related to the classical skin depth.

5.3.2 High-field conductivity: single valley conduction

The energy and momentum balance equations for a semiconductor, under the assumption of a displaced Maxwellian distribution function for the carriers, may be written (from Chapter 4) as

$$m^*\frac{\partial \mathbf{v}_d}{\partial t} = e\mathbf{F} - m^*\mathbf{v}_d\Gamma_m(T_e)$$

$$\frac{\partial\langle E\rangle}{\partial t} = e\mathbf{F}\cdot\mathbf{v}_d - \frac{3}{2}\Gamma_e(T_e)k_B(T_e - T). \tag{5.61}$$

These two equations are just (4.92) and (4.93) under the assumption that only a single valley is being considered. The total scattering process in each equation has been defined by an equivalent relaxation rate Γ_i. Here, the mass is the appropriate effective mass for the carrier and the relaxation rates have been written as explicit functions of the electron temperature rather than of the average energy itself. For any of the relaxation mechanisms of interest here, this will always be the case. The relaxation rates in (5.61) are the total relaxation rates for all processes active in the semiconductor of interest and are given by proper sums over all of the individual interactions. These two equations, in general, are valid whenever a moment method can be applied to the Boltzmann equation itself. In the following, the above equations will be linearized about a non-equilibrium steady-state velocity and field, although this approach really does not imply any loss of generality.

In calculating the microwave conductivity, it will be assumed that the density is uniform and that it is the microwave mobility that is of primary interest, but the correction can be made quite simply (we will address this in the next section). To proceed, the field is written in the form

$$\mathbf{F} = \mathbf{F}_0 + \mathbf{F}_1 e^{i\omega t}. \tag{5.62}$$

As a consequence of this field variation, the electronic properties respond with their own sinusoidal variations as

$$\begin{aligned} \mathbf{v}_d &= \mathbf{v}_{d0} + \mathbf{v}_{d1} e^{i\omega t} \\ T_e &= T_{e0} + T_{e1} e^{i\omega t}. \end{aligned} \tag{5.63}$$

In addition, the scattering rates also vary, due to their dependence on the electron temperature. These variations can be expressed as

$$\begin{aligned} \Gamma_m &= \Gamma_{m0} + T_{e1} \left.\frac{\partial \Gamma_m}{\partial T_e}\right|_{T_{e0}} e^{i\omega t} = \Gamma_{m0} + T_{e1}\Gamma'_m(T_{e0}) e^{i\omega t} \\ \Gamma_e &= \Gamma_{e0} + T_{e1} \left.\frac{\partial \Gamma_e}{\partial T_e}\right|_{T_{e0}} e^{i\omega t} = \Gamma_{e0} + T_{e1}\Gamma'_e(T_{e0}) e^{i\omega t}. \end{aligned} \tag{5.64}$$

Here, Γ_{m0} and Γ_{e0} and the various partial derivatives are all evaluated at T_{e0}. In the following, it will also be assumed that the steady state has been reached, so that only the small-signal response is being sought.

Using the expansions above in (5.61), the zero-order (dc) terms can readily be evaluated to be

$$\begin{aligned} e\mathbf{F}_0 &= m^* \mathbf{v}_{d0} \Gamma_{m0} \\ e\mathbf{v}_{d0} \cdot \mathbf{F}_0 &= \tfrac{3}{2} \Gamma_{e0} k_B (T_{e0} - T), \end{aligned} \tag{5.65}$$

from which one may readily obtain $\mathbf{v}_{d0}$, and the *chordal* mobility, as

$$\mu_0 = \frac{e}{m^* \Gamma_{m0}}. \tag{5.66}$$

By "chordal mobility," it is meant that this mobility is strictly defined as the steady-state velocity divided by the electric field, and hence represents a chordal line drawn from the origin of the velocity-field plot to the actual operating point $\mathbf{v}_d(F_0)$. Similarly, the first-order equations yield

$$e\mathbf{F}_1 = m^* \mathbf{v}_{d1} \Gamma_{m0} \left(1 + i\frac{\omega}{\Gamma_{m0}}\right) + m^* \mathbf{v}_{d0} T_{e1} \Gamma'_m,$$

$$e\mathbf{v}_{d1} \cdot \mathbf{F}_0 + e\mathbf{v}_{d0} \cdot \mathbf{F}_1 = \tfrac{3}{2}\Gamma_{e0} k_B T_{e1} \gamma + i\omega m^* \mathbf{v}_{d0} \cdot \mathbf{v}_{d1}, \tag{5.67}$$

where

$$\gamma = 1 + i\frac{\omega}{\Gamma_{e0}} + \frac{\Gamma'_e}{\Gamma_{e0}}(T_{e0} - T). \tag{5.68}$$

The equations above can now be solved, assuming that only the longitudinal conductivity is being sought (the transverse conductivity shows only the chordal mobility as there is no coupling to the energy equation), which is found to be

$$\mu_1 = \frac{v_{d1}}{F_1} = \mu_0 \frac{1 - \eta(\omega)}{1 + i(\omega/\Gamma_{m0}) + \eta(\omega)[1 - i(\omega/\Gamma_{m0})]},$$

$$\eta(\omega) = \frac{(T_{e0} - T)}{\gamma \Gamma_{m0}} \Gamma'_m. \tag{5.69}$$

The steady-state (dc) value of the small-signal mobility is thus related to the chordal mobility by

$$\mu_1 = \mu_0 \frac{1 - \eta(0)}{1 + \eta(0)}. \tag{5.70}$$

As expected, the microwave mobility is a complex quantity, and its frequency dependence is not a simple function, since frequency dependencies arise from both the energy and the momentum relaxation processes. Equation (5.70) also illustrates that the bounds on the microwave conductivity (at low frequency) lie between the $\mu = 0$ value for true velocity saturation and the chordal mobility. The microwave conductivity is given as

$$\sigma_1 = n_0 e \mathrm{Re}\{\mu_1\}, \tag{5.71}$$

where n_0 is the background carrier density. The imaginary part of the mobility contributes to the real part of the dielectric constant, as

$$\varepsilon = \varepsilon_L + \frac{n_0 e}{\omega}\mathrm{Im}\{\mu_1\}. \tag{5.72}$$

It should be remarked that, in general, the imaginary part of the microwave mobility is negative and can reduce the overall conductivity, even making it negative below the electron plasma frequency (discussed in Chapter 7). While it has been popular to assume that the ac conductivity is just the differential mobility of a static velocity-field relationship, this occurs only if ω/Γ_e and ω/Γ_m are both much smaller than unity. It should be remarked that the behavior introduced by the latter two characteristic properties is in the infrared; studies of the effects must be made with far-infrared measurements if the frequency variation is to be studied.

Because there are multiple roots in the characteristic equation for the conductivity (i.e., there are multiple time constants and the Laplace or Fourier transform would yield second-order equations for the resulting time constants), a peak can occur in the microwave conductivity. In fact, the denominator is quadratic in frequency if the full frequency variation (from η) is introduced in (5.69) and the denominator is rationalized. This should be compared with the prediction from the simple theory of Chapter 4, where the transform of (4.124) demonstrates the oscillatory behavior of the velocity correlation function. The peak in the ac mobility at a given frequency (found below) is related intrinsically to the velocity overshoot discussed in Chapter 4. The existence of this peak can be found from (5.69) by taking the derivative of μ_1 with respect to the frequency. Setting this derivative to zero determines the frequencies at which the mobility has maxima (or minima). This procedure, of course, produces a minimum (or maximum if it is the only extremal point) at zero frequency, but in addition, the microwave mobility peaks at the frequency

$$\omega_{\max}^2 \approx \Gamma_{m0}\Gamma_{e0}\left[1 + (T_{e0} - T)\left(\frac{\Gamma_e'}{\Gamma_{e0}} + \frac{\Gamma_m'}{\Gamma_{m0}}\right)\right]. \tag{5.73}$$

As in the discussions of Chapter 4, it is readily apparent that the velocity overshoot, and the peaking of the microwave conductivity at $\omega_{\max}$, are largest in the far-from-equilibrium hot-carrier case. The effects are much smaller when the electron temperature is the same as the lattice temperature.

The behavior described above can be illustrated, but is quite difficult to measure (the effect occurs in a portion of the spectrum that is difficult to measure). In a device, the actual two-terminal current-voltage curve will show a simple saturation rather than negative conductance. The lack of a two-terminal negative conductance at dc is related to Shockley's (1954) "positive conductance" theorem, in which a space–charge domain will form, creating a non-uniform electric field to make the terminal current saturate rather than show the negative differential mobility (Shaw *et al.*, 1979). However, the negative differential conductivity can be observed in ac measurements, which is why the Gunn effect is useful for microwave amplifiers. In saturation, the ac conductance is nearly zero, due to the constant current at low frequency, but rising for the overshoot effects. This is shown in Figure 5.4 for the theoretical transport in cubic SiC (Ferry, 1975). Both the low frequency mobility and the frequency dependent mobility are shown. Where the low frequency mobility differs significantly from the chordal mobility, we may expect velocity overshoot to occur and this is seen in the frequency dependent mobility of Figure 5.4(b), which is for an electric field of 42 kV/cm.

5.3.3 *High-field conductivity: multiple valley conduction*

For a two-valley system (or any set of multiple valleys), the situation is more complex, as it is necessary to account explicitly for the possibility of differential carrier transfer from one set of valleys to the other in response to the ac field. The total complex conductivity is given by

$$\sigma = \sum_i n_i e \mu_i, \tag{5.74}$$

where n_i and μ_i are the carrier density and mobility of the *ith* set of valleys. Because of the repopulation of the levels, one must include an equation for the density balance between the levels, in addition to the above equations for momentum and energy balance. The charge conservation equation gives (for a homogeneous system in which the divergence of the current can be ignored)

$$\sum_i n_i = n, \quad \frac{dn_i}{dt} = \sum_j (\Gamma_{ji} n_j - \Gamma_{ij} n_i), \tag{5.75}$$

where Γ_{ij} is the total scattering rate for population transfer from valley i to valley j.

As above, a small ac field, superimposed on a large dc field, is considered, and the linear variation of the parameters will be determined. The resulting equations for the density variations are

$$\Gamma_{12,0} n_{1,0} = \Gamma_{21,0} n_{2,0} \tag{5.76}$$

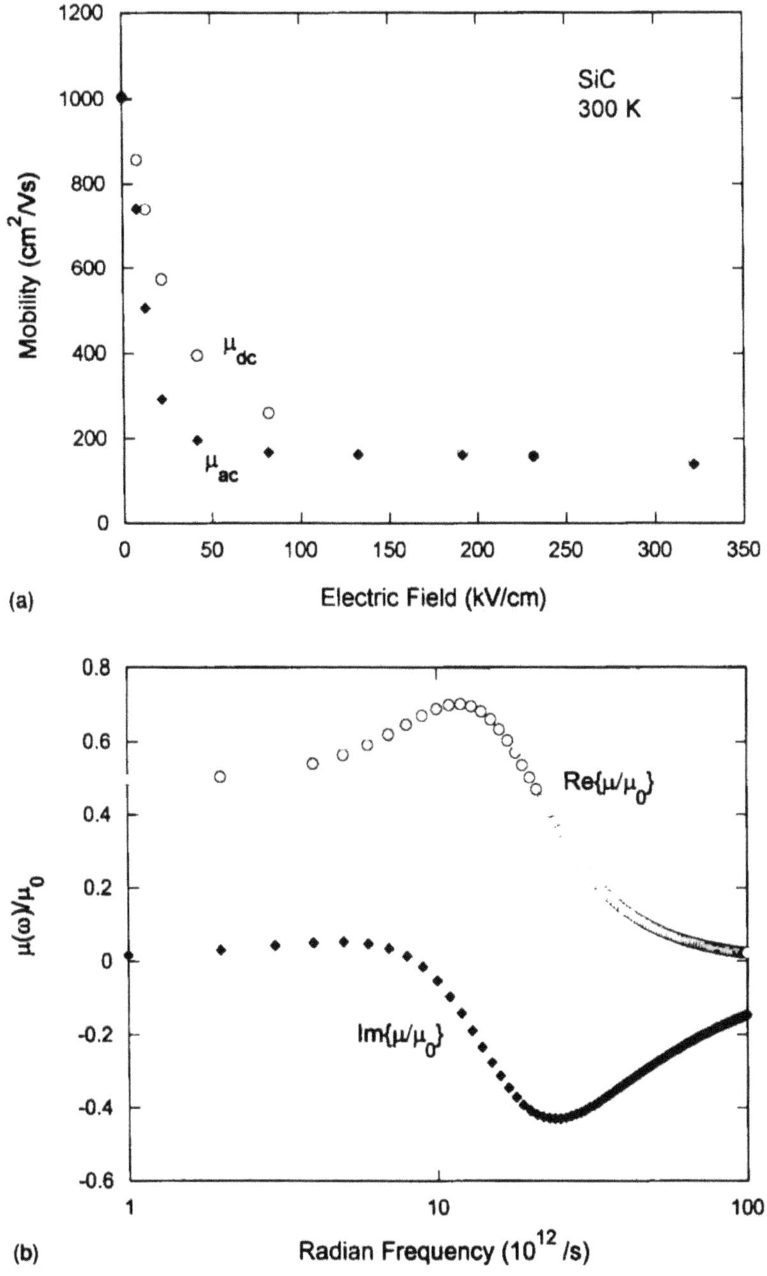

Figure 5.4 (a) The chordal (dc) mobility, defined as the velocity divided by the field, and the low frequency ac mobility for cubic SiC. The transport is calculated using the hydrodynamic equations for this material. Velocity overshoot can be expected where these two mobilities differ significantly. (b) The real and imaginary parts of the ac mobility for a field of 42 kV/cm. Here, these are normalized to the chordal mobility.

and

$$\Gamma_{12,0}n_{1,1}\left(1+i\frac{\omega}{\Gamma_{12,0}+\Gamma_{21,0}}\right)=-n_{1,0}\left(T_{e1,1}\Gamma'_{12}+\frac{\Gamma_{12,0}}{\Gamma_{21,0}}T_{e2,1}\Gamma'_{21}\right), \quad (5.77)$$

$$n_{1,1}=-n_{2,1}. \quad (5.78)$$

The momentum and energy equations must also be modified to account for the influx of the particles, which carry energy and momentum into (and out of) each valley via inter-valley scattering. The momentum balance equation becomes

$$m_i\frac{d\mathbf{v}_{di}}{dt}=e\mathbf{F}-m_i\mathbf{v}_{di}\left(\Gamma_{mi}+\frac{n_j}{n_i}\Gamma_{ji}-\Gamma_{ij}\right), \quad (5.79)$$

and Γ_{mi} refers to the sum of both intra-valley scattering and inter-valley "out-scattering." The energy balance equation becomes

$$\frac{d\langle E\rangle_i}{dt}=e\mathbf{F}\cdot\mathbf{v}_{di}-\frac{3}{2}k_B\Gamma_{ei}(T_{ei}-T)+\langle E\rangle_i\left(\Gamma_{ij}-\frac{n_j}{n_i}\Gamma_{ji}\right). \quad (5.80)$$

The zero-order equations are essentially unchanged, while the first-order equations now become

$$e\mathbf{F}_1=m_i\mathbf{v}_{di,1}\Gamma_{mi,0}\left(1+i\frac{\omega}{\Gamma_{mi,0}}\right)+m_i\mathbf{v}_{di,0}T_{ei,1}\Gamma'_{mi}+im_i\mathbf{v}_{di,0}\omega\frac{n_{i,1}}{n_{i,0}} \quad (5.81)$$

and

$$e\mathbf{F}_1\cdot\mathbf{v}_{di,0}+e\mathbf{F}_0\cdot\mathbf{v}_{di,1}=\frac{3}{2}k_BT_{ei,1}\Gamma_{ei,0}\gamma_i+i\omega m_i\mathbf{v}_{di,0}\cdot\mathbf{v}_{di,1}$$
$$+i\omega\frac{n_{i,1}}{n_{i,0}}\left(\frac{3}{2}k_BT_{ei,0}+i\omega m_i\mathbf{v}_{di,0}\cdot\mathbf{v}_{di,0}\right). \quad (5.82)$$

The parameter γ_i is given by (5.68) for each valley. The above set of equations comprises a coupled pair for the two-valley system. They are not readily separated. However, once $\mathbf{v}_{di,1}$, and $n_{i,1}$, are found, the ac conductivity is readily obtained from

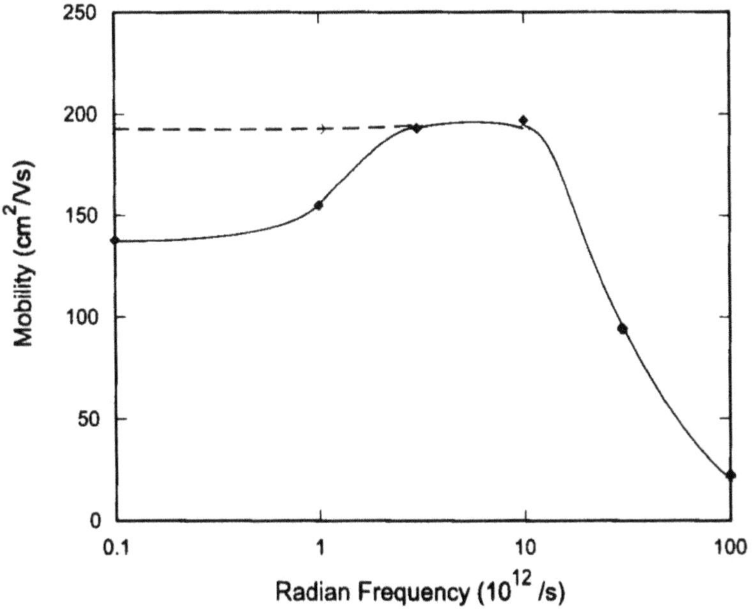

Figure 5.5 The ac conductivity for an Si inversion layer at 300 K. Here, it is assumed that the three lowest sub-bands are occupied (two from the twofold set of valleys, and one from the fourfold set of valleys). An inversion density of 6.0×10^{12} cm^{-2} has been assumed along with a longitudinal field of 20 kV/cm in the channel. The dashed curve is the result if differential repopulation of the sub-bands is not included.

$$\sigma_1(\omega) = \frac{e}{F_1}(n_{1,0}v_{d1,1} + n_{1,1}v_{d1,0} + n_{2,0}v_{d2,1} + n_{2,1}v_{d2,0})$$

$$= \frac{e}{F_1}[n_{1,0}v_{d1,1} + n_{2,0}v_{d2,1} + n_{1,1}(v_{d1,0} - v_{d2,0})]. \qquad (5.83)$$

The evaluation of these equations is usually done numerically on a computer.

An example of the above procedure is shown in Figure 5.5 for electrons in an Si inversion layer. In an *n*-type, quantized inversion layer at the surface of (100)-oriented Si wafers, the six equivalent minima of the bulk silicon conduction band split into two sets of sub-bands, as discussed above. One set consists of the sub-bands arising from the two valleys with their longitudinal (heavy) mass in the direction perpendicular to the surface. The other set of four valleys has the transverse (small) mass perpendicular to the surface. The lowest sub-bands belong to the twofold set of valleys. Under a high electric field, the population shifts to the fourfold set of valleys since the twofold set gets hotter and the fourfold set has a higher density of states.

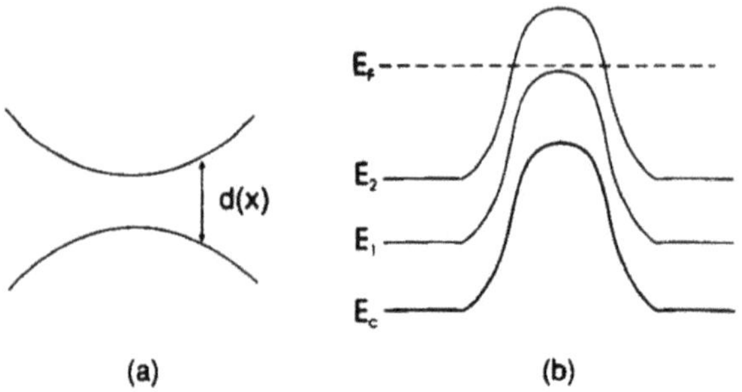

Figure 5.6 (a) The variation in width of a quantum point contact. This is considered to be an adiabatic transition if $d(x)$ varies slowly on the scale of the Fermi wavelength. (b) Schematic variation of the energy levels, and the band edge, for a saddle potential within the quantum point contact.

Both of these effects induce inter-valley scattering. At an electric field for which both sets of valleys are significantly populated, the microwave conductivity will show a significant peaking due to differential repopulation and an effect similar to velocity overshoot (in the temporal behavior).

5.3.4 *Microwave effects in a quantum point contact*

In semiconductor microstructures, such as quantum point contacts, the energy levels are quantized. There exists the opportunity for free-carrier absorption to cause dramatic changes in the conductance. In essence, this is a more sophisticated application of the effect of ac field-induced repopulation that was discussed just above. We consider this effect in a quantum point contact (QPC). If the QPC geometry is sufficiently smooth (on the scale of the Fermi wavelength), the boundaries do not induce any scattering between different modes of the structure. In this case, it is then possible to separate the longitudinal and transverse parts of the Schrödinger equation and solve each component independently. This is phrased the "adiabatic" approximation (Glazman *et al.*, 1988). This is demonstrated in Figure 5.6. In panel (a), a smoothly varying potential is shown, in which the width of the QPC varies as $d(x)$. For a sufficiently smooth variation, the transversely quantized energy levels just increase as the width decreases. This is shown in panel (b), where it is assumed that the narrowing arises from a saddle potential which causes the bottom of the conduction band to increase inversely with $d(x)$. The energy levels also increase and propagation is cut off as they pass through the Fermi energy. In a saddle potential, the transverse potential is nearly like a harmonic oscillator, so that the energy level spacing remains constant as they are pushed upward by the confinement. *It is the upper energy levels that are reflected first as the constriction narrows.*

5.3.4.1 Optical absorption

The absorption of a high frequency field occurs through transitions between the quantized energy levels. For example, in a QPC with hard wall boundary conditions (*different than that illustrated in Figure 5.6*), transitions between the n and $n + 1$ levels occur at the photon energy

$$\hbar\omega = \frac{\hbar^2}{2m^*d^2(x)}[(n+1)^2 - n^2] = \frac{\hbar^2(2n+1)}{2m^*d^2(x)}. \tag{5.84}$$

Thus, the energy separation between the levels increases as one moves into the QPC, and $d(x)$ becomes smaller. Since the transition frequency is x-dependent, it is not at all clear that an absorption spectrum will show any distinct features. However, (5.84) gives the answer to this (Grincwajg *et al.*, 1994). In the region away from the QPC, the energy level separation is small, and the absorption will take place at relatively low frequencies. On the other hand, the constriction causes the energy levels to move apart, so that absorption in the QPC occurs at relatively high frequencies. Therefore, spectroscopy of the QPC is possible and highly effective, especially as only propagating modes contribute to the high frequency peaks, since they are the only ones able to reach the center of the constriction.

One interesting application of this is the characterization of the lateral confinement potential in the QPC. For hard wall potentials, (5.84) shows us a series of peaks equally spaced by

$$\Delta\omega = \frac{\hbar}{m^*d^2(x)}. \tag{5.85}$$

The number of observed peaks corresponds to the number of propagating modes (Grincwajg *et al.*, 1994). On the other hand, if the confining potential is of harmonic oscillator form, with a parabolic energy, then there is only a single peak as

$$\hbar\omega = E_{n+1} - E_n = [(n + 1 + \tfrac{1}{2}) - (n + \tfrac{1}{2})]\hbar\omega_0 = \hbar\omega_0. \tag{5.86}$$

(This is the case shown in Figure 5.6.) The differences between these two signals leads to the possibility of studying the shape of QPC potentials through variations in the optical absorption spectrum as the size of the potential is varied by, e.g., an applied potential.

5.3.4.2 Photoconductance oscillations

It is clear from Figure 5.6 that, if one applies a bias across the QPC, one can measure the photoconductance of the structure. Normally, the bias is considered to raise the Fermi level (and the energies of the quantum states) on the left by $eV/2$, and to lower this level on the right by $-eV/2$. Thus, the conductance is set by those states within this bias range from the Fermi level. Absorption of a photon on the left-hand side of the barrier in Figure 5.6(b) can raise an electron from a level, which passes through the QPC, to a level which is reflected by the QPC. Hence, this absorption leads to *negative* photo-conductance, as it reduces the number of particles passing through the QPC (Grincwajg *et al.*, 1995). On the other hand, absorption of a photon on the right-hand side of the barrier puts this electron into a conducting state, which causes a *positive* photo-conductance. As the applied bias eV is varied, the relative probabilities of these two events varies as the bias changes the shapes of the detailed potential in the region of the QPC. Consequently, one might expect to find an oscillatory photo-conductance as a function of applied bias, or as a function of constriction width as this latter is varied by a gate potential.

A final interesting prediction is that in a system with two QPCs, the photon interaction may well lead to coherent interference effects (Gorelik *et al.*, 1994). The transition from a conducting level to a reflecting level is similar to impurity scattering in which back-scattering is induced by the impurity interaction in this quasi-one-dimensional situation. There is an important difference, however, between impurity scattering and photon absorption. Normally, impurity scattering is elastic. This optical absorption, however, is quite inelastic, as the electron absorbs the photon with its consequent energy. In general, this would destroy interference effects. However, the electromagnetic field of the photon corresponds to a coherent state if monochromatic radiation is used, and there is a single phase of the photon. Therefore, it is thought that the electron phase memory will be preserved in the absorption process and interference effects are possible. Gorelik *et al.* (1994) have calculated this interference effect in the wide region between two QPCs. In this region, absorption pumps electrons from propagating states to non-propagating states, just as before. Now, however, within this confined region, these bound states begin to be sufficiently occupied that they decay back to the propagating states. A balance is created between these two populations and a steady-state occurs. Interference in this regard refers to the resonant behavior of this process as the photon energy is varied, and length-dependent oscillations will occur in the measured photo-conductance. This effect is closely related to the time-dependent "ringing" oscillations that occur in tunneling into quantum dots and resonant-tunneling diodes (Wingreen *et al.*, 1993; Jauho *et al.*, 1994; Bruder and Schoeller, 1994), even when the barriers are partially transmitting. Initial experimental work on the ac conductance of single-electron quantum dots has been reported by Kouwenhoven *et al.* (1994) and in semiconductor superlattices by Keay *et al.* (1995).

PROBLEMS

1. The curves of the generation rate in InSb (Curby and Ferry, 1973) show both a low-field and a high-field region. Using the known parameters for transport in this material, determine the best fit to the ionization energy and phonon mean free path for these data.
2. Using the ensemble Monte Carlo program developed to treat transport in InSb and InAs in Chapter 4, introduce the ionizing collision and compute the ionization rates for these two materials as a function of the electric field.
3. From the parameters found for the transport in InP at the end of Chapter 4 (using the drifted hydrodynamic equation approach), determine the frequency response of the low-field mobility at an applied electric field of 4 kV/cm.
4. From the mobility data for GaAs, compute an average momentum relaxation time as a function of the impurity density. Calculate and plot the free-carrier absorption at 77 K as a function of the impurity density.

REFERENCES

Abstreiter, G., Brugger, H., Wolf, T., Jorke, H., and Herog, H. J., 1985, *Phys. Rev. Lett.*, 54, 2441.

Anderson, C. L., and Crowell, C. R., 1972, *Phys. Rev.*, B5, 2267.

Ando, T., Fowler, A., and Stern, F., 1982, *Rev. Mod. Phys.*, 54, 437.

Antoncik, E., and Landsberg, P. T., 1963, *Proc. Phys. Soc.*, 82, 337.

Antoncik, E., and Landsberg, P. T., 1967, *Czech. J. Phys.*, B17, 735.

Baraff, G. A., 1962, *Phys. Rev.*, 128, 2507.

Bosch, B. G., and Engelmann, R. W. H., 1975, *Gunn-Effect Electronics* (New York: John Wiley).

Bruder, C. and Schoeller, H., 1994, *Phys. Rev. Lett.*, 72, 1076.

Curby, R. C., and Ferry, D. K., 1973, *Phys. Stat. Sol. (a)*, 15, 319.

Czajkowski, I. K., Allam, J., Silver, M., Adams, A. R., and Gell, M. A., 1990, *IEE Proc. F: J. Optoelectron.*, 137, 79.

Ferry, D. K., 1975, *Phys. Rev.*, B12, 2361.

Ferry, D. K., 1988, in *The Physics and Technology of Amorphous SiO_2*, Ed. by Devine, R. (New York: Plenum Press) 365–373.

Glazman, L. I., Lesovik, G. B., Khmel'nitskii, D. E., and Shekhter, R. I., 1988, *Pis'ma Z. Eksp. Teor. Fiz.*, 48, 218 [translation in *JETP Lett.*, 48, 238].

Gorelik, L. Y., Grincwajg, A., and Jonson, M., 1994, *Phys. Rev. Lett.*, 73, 2260.

Gram, N. O., 1972, *Phys. Lett.*, A38, 235.

Grincwajg, A., Jonson, M., and Shekhter, R. I., 1994, *Phys. Rev. B*, 49, 7557.

Grincwajg, A., Gorelik, L. Y., Kleiner, V. Z., and Shekhter, R. I., 1995, *Phys. Rev. B*, 52, 12168.

Gunn, J. B., 1963, *Sol. State Commun.*, 1, 88.
Harrison, W. A., 1970, *Solid State Theory* (New York: McGraw-Hill).
Ismail, K., Meyerson, B. S., Rishton, S., Chu, J., Nelson, S., and Nocera, J., 1992, *IEEE Electron Dev. Lett.*, 13, 229.
Jauho, A.-P., Wingreen, N. S., and Meir, Y., 1994, *Phys. Rev. B*, 50, 5528.
Kane, E. O., 1967, *Phys. Rev.*, 159, 624.
Keay, B. J., Allen, S. J., Jr., Galan, J., Kaminski, J. P., Campman, K. L., Gossard, A. C., Bhattacharya, U., and Rodwell, M. J. W., 1995, *Phys. Rev. Lett.*, 75, 4098.
Kouwenhoven, L. P., Jauhar, S., Orenstein, J., McEuen, P. L., Nagamune, Y., Motohisa, J., and Sakaki, H., 1994, *Phys. Rev. Lett.*, 73, 3443.
Lebwohl, P. A., and Price, P., 1971, *Sol. State Commun.*, 9, 1221.
Lee, C. A., Logan, R. A., Batdorf, R. L., Kleimack, J. J., and Wiegmann, W., 1964, *Phys. Rev. A*, 134, 761.
McKay, K. B., and McAfee, K. B., 1953, *Phys. Rev.*, 91, 1079.
Okuto, Y., and Crowell, C. R., 1972, *Phys. Rev.*, B6, 3076.
Pearsall, T. P., Capasso, F., Nahory, R. E., Pollak, M. A., and Chelikowsky, J. R., 1978, *Sol.-State Electron.*, 21, 297.
Ridley, B. K., 1963, *Proc. Phys. Soc. London*, 82, 954.
Ridley, B. K., 1987, *Semicond. Sci. Technol.*, 2, 116.
Ryder, E. J., 1953, *Phys. Rev.*, 90, 766.
Sasaki, W., Shibuya, M., and Mizuguchi, K., 1958, *J. Phys. Soc. Jpn.*, 13, 456.
Sasaki, W., Shibuya, M., Mizuguchi, K., and Hatoyama, G. M., 1959, *J. Phys. Chem. Sol.*, 8, 250.
Shaw, M. P., Grubin, H. L., and Solomon, P., 1979, *The Gunn-Hilsum Effect* (New York: Academic Press).
Shibuya, M., 1955, *Phys. Rev.*, 99, 1189.
Shichijo, H., and Hess, K., 1981, *Phys. Rev.*, B33, 4197.
Shockley, W., 1951, *Bell Sys. Tech. J.*, 21, 990.
Shockley, W., 1954, *Bell Sys. Tech. J.*, 33, 799.
Shockley, W., 1961, *Sol.-State Electron.*, 2, 36.
Wingreen, N. S., Jauho, A.-P., and Meir, Y., 1993, *Phys. Rev. B*, 48, 8487.
Wolff, P. A., 1954, *Phys. Rev.*, 95, 1415.
Yamada, T., Zhou, J.-R., Miyata, H., and Ferry, D. K., 1994, *Semicond. Sci. Technol.*, 9, 775.

Chapter 6

Optical properties

The properties of excess carriers in semiconductors are some of the more interesting aspects of microelectronics. The role these excess carriers play is important from the simplest transistor to the most elegant photo-voltaic device (and conversely, from the simplest photo-conductive device to the most elegant bipolar transistor). In this chapter, the role of these optically created, non-equilibrium carriers is discussed. While the approach depends nominally upon the Boltzmann equation, it is more usual to work with the continuity equation, incorporating current primarily through the relaxation-time approximation. The aspects of diffusion of minority carriers will be treated first, as it is of primary importance to later discussions of optically generated carriers and their transport. The discussion then turns to optical absorption and recombination. Finally, a discussion of ultra-fast laser excitation of semiconductors is presented, with simulations using the ensemble Monte Carlo technique.

Excess carriers are the result of the creation of new non-equilibrium carriers, under the condition $\hbar\omega > E_{gap}$, for which one must account for the creation of new electron–hole pairs by the high-frequency radiation. The quantity E_{gap} here is a generic gap. Both direct transitions, in which the carrier momentum is conserved directly, and indirect processes, in which the emission or absorption of an optical phonon is required to conserve momentum, are discussed. In this process, the electrons may be excited from the valence band to form electron–hole pairs. However, it could as easily refer to the gap between a dopant atom's energy level and the appropriate band for free carrier motion. In the latter case, one free carrier is formed with the counter-balancing charge localized on the dopant atom. The treatments are quite similar and differ only in obvious details.

6.1 Diffusion of excess carriers

In the case of either optical or electrical injection of minority carriers, one must still maintain some semblance of charge neutrality within the semiconductor. If excess carriers are injected into the semiconductor by some means, quite large deviations from the thermal equilibrium densities can occur. This is well known in, for example, bipolar transistors and in optical absorption in semiconductors. Some of these carriers can be bound to trap or impurity levels, but they will still contribute to the overall charge neutrality. The total carrier densities may be written in terms of the excess carriers, as

$$n = n_0 + \Delta n, \quad p = p_0 + \Delta p, \tag{6.1}$$

and the net charge density is just (n_0 and p_0 balance the impurity concentrations)

$$\rho = e(\Delta p - \Delta n). \tag{6.2}$$

From Poisson's equation, the induced electric fields may be computed. These fields will be quite large and are the source of the built-in potentials in *p–n* junctions. In most cases for bulk semiconductors, however, the fields will decay in a manner which forces charge neutrality upon the semiconductor, at least locally. This may be seen from the continuity equation, as

$$\frac{\partial \rho}{\partial t} = -\nabla \cdot \mathrm{J} = -\sigma \nabla \cdot \mathrm{F} = -\frac{\sigma \rho}{\varepsilon_s}. \tag{6.3}$$

From this equation, it may be seen that the charge non-neutrality decays as

$$\rho = \rho(0) e^{-t/\tau_D}, \quad \tau_D = \frac{\varepsilon_s}{\sigma}. \tag{6.4}$$

The last relation defines the reciprocal of the dielectric relaxation frequency. The charge deviation from neutrality decays in a time of the order of a few picoseconds, provided that the differential conductivity is positive. The latter proviso is the normal situation, except for a few special cases. In the last chapter, negative differential conductivity was found. The Gunn effect arises from intervalley transfer and leads to negative differential conductivity. Such an effect actually leads to *growth* of the charge perturbation, and to domains which transit the device. Normally, however, in near-equilibrium situations, this is unusual and one is led to the conclusion that the number of excess electrons and the number of excess holes must balance one another to maintain space–charge neutrality.

The current flow in a semiconductor must be maintained according to the continuity equation introduced in (6.3). Here, the interest is in finding a specific equation that provides for the continuity of each carrier independently. The coupling of the two equations for these different charges will ultimately be through the assumption of charge neutrality. This is a rather nefarious assumption. It is generally true that the total device is space–charge neutral, so that the only electric fields that exist are within the semiconductor device. On the other hand, there is no real reason to assume that this neutrality exists at each point within the semiconductor, other than the result (6.4). Thus, in a steady-state case, which is the primary one of interest here, (6.4) leads us to assume a balance between excess electrons and holes at each point of the semiconductor. It is clear that charge neutrality does not exist at every point of a *p–n* junction, for example, but this charge non-neutrality, and the built-in fields, are for the majority carriers, not the injected minority carriers. There are other cases, such as domains in negative differential mobility materials, in which non-neutrality exists in the majority carriers. Our major interest, however, in this section is the case of injected minority carriers for which charge neutrality (actually, $\Delta n = \Delta p$) may be assumed to hold in steady state. Thus the processes of generation and recombination of carriers in the semiconductor must be added to (6.3) to account for the injection and recombination of the excess carriers. Thus, for electrons,

$$\frac{\partial n}{\partial t} = \frac{1}{e}\nabla \cdot \mathbf{J}_e + G_e - \frac{n}{\tau_n}$$

$$= G_e - \frac{n}{\tau_n} + \mu_e \mathbf{F} \cdot \nabla n + n\mu_e \nabla \cdot \mathbf{F} + D_e \nabla^2 n, \tag{6.5}$$

and for holes

$$\frac{\partial p}{\partial t} = -\frac{1}{e}\nabla \cdot \mathbf{J}_h + G_h - \frac{p}{\tau_p}$$

$$= G_h - \frac{p}{\tau_p} - \mu_h \mathbf{F} \cdot \nabla p - p\mu_h \nabla \cdot \mathbf{F} + D_h \nabla^2 p. \tag{6.6}$$

The quantities G_e and G_h are the generation rates for electrons and holes, respectively, and τ_n and τ_p are the recombination times for electrons and holes, respectively. If electrical neutrality is to be maintained within the semiconductor, the terms in the divergence of the electric field may be ignored, as they are identically zero from Gauss's law. The resulting equations depend only on the excess carrier densities, as the generation and recombination rates balance each other in equilibrium. It is further assumed that the recombination lifetimes are their equilibrium values, and (6.5)–(6.6) may be rewritten as ($g = G - G_0$)

$$\frac{\partial \Delta n}{\partial t} = g_e - \frac{\Delta n}{\tau_n} + \mu_e \mathbf{F} \cdot \nabla(\Delta n) + D_e \nabla^2(\Delta n), \tag{6.7}$$

and

$$\frac{\partial \Delta p}{\partial t} = g_h - \frac{\Delta p}{\tau_p} - \mu_h \mathbf{F} \cdot \nabla(\Delta p) + D_h \nabla^2(\Delta p). \tag{6.8}$$

6.1.1 Extrinsic material

In strongly extrinsic material, it is important to treat only the equation for the minority carriers when considering minority carrier diffusion. This assertion will be justified in greater detail in the next section, but note here that it is the discrepancy between τ_n and τ_p that requires the use of the minority carrier properties. The excess majority carriers are relatively small in comparison with the background doping, but the excess minority carriers can be much larger than the background minority carrier concentration when the overall excess densities are small. The requirement of space–charge neutrality equates the two excess densities, but it is the sensitivity of the overall response to the minority carriers that drives the need to use the latter's properties. Here the case for which $n \gg p$ will be treated, and attention will be concentrated on the behavior of the minority holes. In addition, the excess generation in the bulk will be ignored, taking instead the excess hole density to be defined by a boundary condition at $x = 0$, as in a bipolar transistor. Thus, in one dimension, (6.8) becomes

$$\frac{\partial \Delta p}{\partial t} = -\frac{\Delta p}{\tau_p} - \mu_h F_x \frac{\partial(\Delta p)}{\partial x} + D_h \frac{\partial^2(\Delta p)}{\partial x^2}. \tag{6.9}$$

For small electric fields, the second term on the right-hand side may be ignored (we return to this term later). The current flow is then due entirely to diffusion of the minority carriers from a region of large concentration to a region of low concentration. In the steady-state situation, Δp is a function of position alone and is determined by the solutions to

$$D_h \frac{\partial^2(\Delta p)}{\partial x^2} - \frac{\Delta p}{\tau_p} = 0. \tag{6.10}$$

A semi-infinite semiconductor for $x > 0$ will be assumed, and the aforementioned boundary condition will be set to $\Delta p = \Delta p_0$ at $x = 0$. The solution to (6.10) is then given by

$$\Delta p = \Delta p_0 e^{-x/L_p}, \tag{6.11}$$

where

$$L_p = \sqrt{D_h \tau_p}\,. \tag{6.12}$$

This latter expression is the diffusion length for holes. A second solution occurs with a positive exponential, but this term must have its coefficient set to zero in order to satisfy a condition that the number of excess carriers is finite as x approaches large positive values. The role of these two solutions is of course reversed for $x < 0$, as will be seen below.

The general solution, in which the electric field term is retained in (6.9), can now be discussed. In general, one would expect this field to have some spatial variation due to the injected charge, which then leads to a spatially varying conductivity. However, this spatial variation in the field must lead to a condition of non-neutral space charge, which has been asserted not to occur. Thus we are left with a conundrum that requires more exact treatments to account for the spatially varying field and its consequent space–charge variation. Here, the approach will stick strictly to the assumption that the material is space–charge neutral and that the electric field is constant as a consequence. In the steady state, (6.9) may be written as

$$D_h \frac{\partial^2(\Delta p)}{\partial x^2} - \mu_h F_x \frac{\partial(\Delta p)}{\partial x} - \frac{\Delta p}{\tau_p} = 0. \tag{6.13}$$

Again, it will be assumed that the excess holes are provided at $x = 0$ in the amount Δp_0. If solutions are sought in the form $\exp(-\lambda x)$, then the values for λ are given by

$$\lambda = -\frac{1}{2L_p}\left(\frac{F}{F_c}\right) \pm \frac{1}{2L_p}\sqrt{\left(\frac{F}{F_c}\right)^2 + 4}. \tag{6.14}$$

In this equation, the electric field quantity $F_c = L_p/\mu_h\tau_p$ has been introduced. This critical field is the value for which the minority hole will drift under the influence of the field a distance equal to the distance it diffuses in the concentration gradient. This field is therefore the critical field for which drift and diffusion effects are equivalent.

It is now possible to examine these two solutions in a manner that brings in diffusion both along and against the electric field. These two cases are referred to as downstream and upstream diffusion, respectively. The electric field is assumed to be directed in the positive x direction, so that the solution for $x > 0$ is the downstream diffusion effect, and the solution for $x < 0$ is the upstream diffusion effect. For $F \ll F_c$, we can expand the radical in (6.14) and

$$\lambda = \pm\frac{1}{L_p}, \tag{6.15}$$

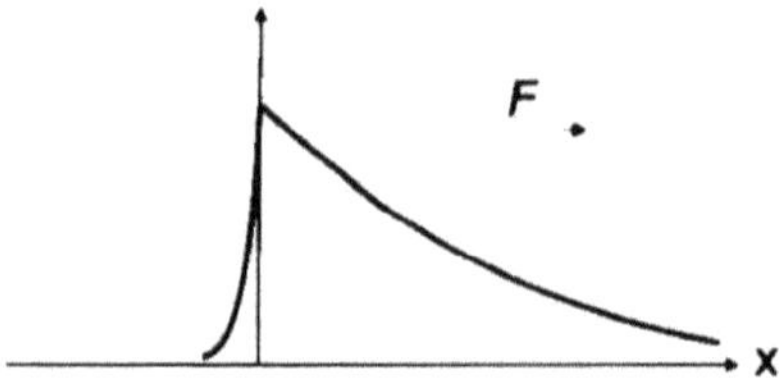

Figure 6.1 An excess density of holes ΔP_0 is injected at $x = 0$. The downstream diffusion ($x > 0$) is greatly assisted by drift in the field, while the upstream diffusion ($x < 0$) is greatly retarded by the field.

so that the carriers move by simple diffusion in both directions without being affected by the field. On the other hand, if $F \gg F_c$, we find that

$$\lambda_d = \frac{1}{L_p}\left(\frac{F_c}{F}\right) \equiv \frac{1}{L_d}, \quad \lambda_u = -\frac{1}{L_p}\left(\frac{F}{F_c}\right) = -\frac{eF}{k_B T} \equiv -\frac{1}{L_u}. \tag{6.16}$$

Here, L_u is the upstream "diffusion" length for diffusion opposed by the electric field. The upstream excess carrier density dies off quite rapidly over a distance determined solely by the electric field and not by the diffusion parameters. On the other hand, the downstream diffusion length L_d can be quite long, as it is primarily the drift length defined by the distance that the excess holes will drift in the field prior to recombining. These are illustrated in Figure 6.1.

6.1.2 Ambipolar diffusion

We now turn our attention to material that is more general and may not be strongly extrinsic. This may be initially extrinsic material with a large excess carrier density, or nearly intrinsic material. For completeness, the term in the spatial variation of the electric field will be retained, since it may be large under some circumstances. Equations (6.5) and (6.6) are rewritten in terms of the excess densities, and for one dimension this gives

$$\frac{\partial(\Delta n)}{\partial t} = -\frac{\Delta n}{\tau_n} + D_e\frac{\partial^2(\Delta n)}{\partial x^2} + \mu_e F\frac{\partial(\Delta n)}{\partial x} + n\mu_e\frac{\partial F}{\partial x},$$

$$\frac{\partial(\Delta p)}{\partial t} = -\frac{\Delta p}{\tau_p} + D_h\frac{\partial^2(\Delta p)}{\partial x^2} - \mu_h F\frac{\partial(\Delta p)}{\partial x} - p\mu_h\frac{\partial F}{\partial x}. \tag{6.17}$$

The first of these equations is multiplied by $\sigma_e = ne\mu_e$ and the second by $\sigma_h = pe\mu_h$ and then the two are added together. This yields a single equation, which is simplified by assuming that $\Delta n = \Delta p$ and $\tau_n = \tau_p$ only in the time derivative and recombination terms. The term involving the spatial variation of the electric field completely drops out of the resulting equation, and we are therefore free to make the latter assumption in the remaining terms, so that

$$\frac{\partial(\Delta p)}{\partial t} = -\frac{\Delta p}{\tau_p} + \left(\frac{\sigma_e D_h + \sigma_h D_e}{\sigma_e + \sigma_h}\right)\frac{\partial^2(\Delta p)}{\partial x^2} - \left(\frac{\sigma_e \mu_h - \sigma_h \mu_e}{\sigma_e + \sigma_h}\right) F \frac{\partial(\Delta p)}{\partial x}. \tag{6.18}$$

The *ambipolar* diffusion constant may now be defined as

$$D_a = \frac{\sigma_e D_h + \sigma_h D_e}{\sigma_e + \sigma_h} = \frac{D_e D_h (n + p)}{n D_e + p D_h}, \tag{6.19}$$

and the *ambipolar* mobility is similarly defined as

$$\mu_a = \frac{\sigma_e \mu_h - \sigma_h \mu_e}{\sigma_e + \sigma_h} = \frac{\mu_e \mu_h (n - p)}{n \mu_e + p \mu_h}. \tag{6.20}$$

It is clear from the latter results that, for strongly extrinsic material, it is correct to use the diffusion constant and mobility of the minority carriers in describing the diffusion behavior of the minority carriers. On the other hand, in intrinsic material, where $n = p$, the mobility and field effects drop out completely, and one needs only a simple diffusion model in which the effective diffusion constant is an average over those of each type of carrier.

6.2 Interband optical absorption

At sufficiently high frequencies, such as in the optical region, it is possible for the electromagnetic field to induce electrons from the valence band (or from an impurity level lying in the gap) to move into the conduction band, creating a new electron–hole pair. This is the optical absorption usually studied in semiconductors. There are two primary cases that must be studied, which are quite different in nature. One is the direct transition that can occur in materials such as GaAs, in which the lowest band gap is a direct gap. This direct transition can also occur at the L and X points, and at the zone center in indirect materials as well. However, in indirect materials, such as Si and Ge, the lowest band gap between the conduction and valence bands is indirect since the minimum of the conduction band is at a point in the Brillouin zone different from that at which the maximum of the valence band

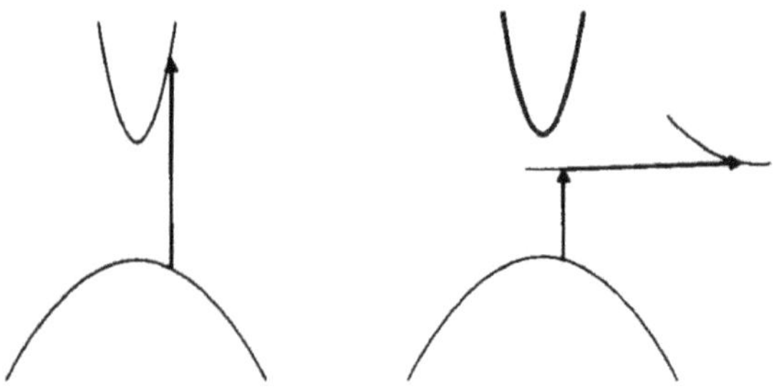

Figure 6.2 (a) Direct optical absorption in which an electron moves from the valence to the conduction band, creating an electron–hole pair. (b) Indirect optical absorption, in which the photon excites the electron to a *virtual* state (non-propagating state in the band gap), where it can be excited to the conduction band by a phonon.

occurs. These are shown in Figure 6.2. For optical absorption to occur across the indirect gap, another process must become involved in order to provide the large wave vector change in the electron in moving from the maximum of the valence band to the minimum of the conduction band. As we shall see below, the photon that is absorbed has a quite small wave vector, on the scale of the wave vectors included in the Brillouin zone, so that a lattice phonon must be involved to generate the change in wave vector. This is a higher-order quantum process and is therefore much less likely, but still is the dominant absorption mechanism in the indirect gap materials. In this section, the optical absorption processes that lead to the generation of electron–hole pairs are considered.

6.2.1 Direct transitions

The quantity that we want to calculate is the absorption coefficient, which describes how much of the optical electromagnetic energy is turned into electron–hole pairs. To do this, it is first necessary to compute the transition probability for the inter-band transition. This quantity was first encountered in Chapter 2, in dealing with the effective mass theorem. It was assumed there that one could use an effective mass defined by the curvature of a single conduction band if there were no inter-band processes. For the optical properties of interest here, it is now necessary to actually evaluate these matrix elements. A slightly different course, which shows how these matrix elements arise in a natural manner, will be followed to obtain the desired relationships and then discuss their evaluation in a semiconductor. Once the transition probabilities have been determined, the number of electrons that are induced to make the cross-gap transition per unit time can be calculated, and this, in turn, may be related to the incident power of the wave that is absorbed in the semiconductor. This will then give the absorption coefficient.

In semiconductors, the direct gap absorption process is the more likely to occur since it involves only a single transition by an electron to create the electron–hole pair, and hence involves only a single interaction process. This type of absorption is the mechanism for the creation of electron–hole pairs across the fundamental E_0 gap (we have previously called this E_{gap}, here we adopt the common optical notation) in semiconductors such as GaAs, InAs, and so on, and for higher-energy processes in the indirect semiconductors. Because of the width of the energy gaps in most semiconductors, optical sources in the near infrared and visible regions of the spectrum are normally required. In Figure 6.2(a), the basic process by which this transition occurs is illustrated.

The process by which the transition rate is calculated, and by which we shall be able to identify the matrix elements that are needed, is time-dependent perturbation theory (Schiff, 1955; Ferry, 1995). For a time-varying, sinusoidal forcing function at frequency ω_0, the transition coefficient is sought for transitions from the state $E(\mathbf{k}_1)$ to the state $E(\mathbf{k}_2)$, in which the states are defined by their energies and their wave vectors. The energies will lie in different bands in this approach, but the values to be used will continually remain within the first Brillouin zone. It is of primary interest to take electrons from the essentially full valence band to the empty conduction band, and the transition coefficient is given by

$$a_{12} = -\frac{H_{12}}{\hbar}\left[\frac{e^{i(\omega_{21}+\omega_0)t}-1}{\omega_{21}+\omega_0}+\frac{e^{i(\omega_{21}-\omega_0)t}-1}{\omega_{21}-\omega_0}\right], \tag{6.21}$$

where $\hbar\omega_{21} = E(\mathbf{k}_2) - E(\mathbf{k}_1)$, and

$$H_{12} = \langle 2|H_{opt}|1\rangle = \int d^3r\,\psi_{k_1}^{*} H_{opt}\psi_{k_2} \tag{6.22}$$

is the matrix element for the transition. Equation (6.21) is another form of the Fermi golden rule, and if the long-time limit is taken, the term in brackets reduces to a function that is approximately $2\pi\delta(\hbar\omega_0 \pm \hbar\omega_{21})$. In this limit, it is of the same form as the Fermi golden rule invoked in Chapters 3 and 4, and (6.22) defines just the matrix element. In general, each of the wave functions in (6.22) is the appropriate Bloch function for the band under consideration. In general, the transition coefficient is small unless one of the denominators goes to zero, which means that it is a "resonant" process. For this case, standard perturbation theory approaches suggest that (6.21) be rewritten in terms of a function such as $(\sin x)/x$ and take the limiting case of resonance (the long-time limit). This procedure leads to the transition rate

$$c_{12} = \frac{|a_{12}|^2}{t} = \frac{4|H_{12}|^2 t}{\hbar^2} \frac{\sin^2[(\omega_{21} - \omega_0)t/2]}{(\omega_{21} - \omega_0)^2 t^2}. \tag{6.23}$$

It is now necessary to calculate the matrix element.

In the presence of an electromagnetic wave, the Hamiltonian for the electrons is given in a quite simple form by

$$H = \frac{1}{2m^*}(-i\hbar\nabla + e\mathbf{A})^2 + V(\mathbf{r}) + \varphi(\mathbf{r}), \tag{6.24}$$

where $\mathbf{A}$ and ϕ are the vector potential and the scalar potential, respectively, of the wave, and the other two terms have their normal meaning. The individual components of the electromagnetic field are related to the potentials through (in the Lorentz gauge)

$$\mathbf{F} = -\nabla\varphi - \frac{\partial \mathbf{A}}{\partial t}, \quad \mathbf{B} = \nabla \times \mathbf{A}. \tag{6.25}$$

Other gauges are certainly possible, and have important usage for special applications. These other gauges are related to the choice (6.25) through a gauge transformation. In general, the scalar potential is relatively slowly varying over a unit cell of the lattice (it generally gives rise to certain electrostatic-like behaviors). The leading term in (6.24), and the only term usually treated in optical absorption, arises from the cross product of the two factors in the momentum relation [the leading term in (6.24) is the generalized momentum]. Thus (6.24) may be rewritten in the approximate form

$$H \cong \frac{\hbar^2}{2m^*}\nabla^2 + V(\mathbf{r}) + \frac{i\hbar e}{m^*}\mathbf{A} \cdot \nabla. \tag{6.26}$$

It may now be recognized that the last term is the perturbing potential of the electromagnetic wave, and this term varies with time at a frequency ω. For the wave functions, it is sufficient to use the normalized Bloch functions for the respective bands (valence and conduction). The transition of interest is that of an electron from near the top of the valence band to near the bottom of the conduction band. For this case, the wave function denoted by the symbol "1" in (6.22) refers to the valence band, and the energy is measured from the top of the valence band. The wave function denoted by the symbol "2" is that of the conduction band, and its respective energy is measured from the bottom of the conduction band. We take the vector potential to vary as (retaining only the single term for absorption)

$$\mathbf{A} = \frac{A_0}{2}\mathbf{a}e^{i(\mathbf{k}_0\cdot\mathbf{r}-\omega_0 t)} + \textit{complex conjugate}, \tag{6.27}$$

and the matrix element is now given by

$$\begin{aligned} H_{12} &= \frac{ie\hbar A_0}{2m^*N}\int d^3\mathbf{r}u_c^*(\mathbf{r})e^{-i\mathbf{k}_c\cdot\mathbf{r}+i\mathbf{k}_0\cdot\mathbf{r}}(\mathbf{a}\cdot\nabla)u_v(r)e^{i\mathbf{k}_v\cdot\mathbf{r}} \\ &= \frac{ie\hbar A_0}{2m^*N}\int d^3\mathbf{r}u_c^*(\mathbf{r})e^{-i\mathbf{k}_c\cdot\mathbf{r}+i\mathbf{k}_0\cdot\mathbf{r}+i\mathbf{k}_v\cdot\mathbf{r}}[\mathbf{a}\cdot\nabla u_v(r) + i\mathbf{a}\cdot\mathbf{k}_v u_v(r)]. \end{aligned} \tag{6.28}$$

Since the u are periodic in the unit cell, the integration can be reduced to that only over a single unit cell plus a summation over all the unit cells. The summation over the cell positions in the exponent reduces just to the closure relationship for the generalized Fourier series that the Bloch functions represent. Thus, to within a reciprocal lattice vector G, we have

$$\mathbf{k}_0 + \mathbf{k}_v - \mathbf{k}_c = 0. \tag{6.29}$$

It is easily shown that the magnitude of the wave vector for the electromagnetic wave is orders of magnitude smaller than that of the electron and hole states in (6.29). One can therefore essentially ignore $\mathbf{k}_0$ and say that the optically induced transition must be "vertical" with $\mathbf{k}_c = \mathbf{k}_v$. This is the vertical transition that is drawn in Figure 6.2(a).

Using the selection rule above, and summing over the appropriate number of cells within the lattice, the matrix element (6.28) now becomes

$$H_{12} = \frac{ie\hbar A_0}{2m^*}\int_{cell} d^3\mathbf{r}u_c^*(\mathbf{r})[\mathbf{a}\cdot\nabla u_v(r) + i\mathbf{a}\cdot\mathbf{k}_v u_v(r)]. \tag{6.30}$$

Normally, the first term in brackets dominates the matrix element (the second integral does not vanish, as it is the total Bloch functions that are orthogonal for different bands, not just the cell periodic parts). There are cases, however, where the first term will vanish due to the particular symmetry of the wave functions involved, and the second term provides a weak contribution to the optical absorption. These cases, where the first term vanishes by symmetry, are termed *forbidden* transitions. This terminology arises from the form of the leading term in (6.30), which appears to be a dipole-like interaction, and thus the terminology for dipole interactions is adopted from that for atomic spectroscopy. From this, we expect that the transition must involve a change in angular momentum, such as p to s symmetry. Fortunately, in materials like GaAs, the top of the valence band is p-like and the bottom of the conduction band is s-like, so this symmetry rule is satisfied. Here it may be said that the transition is symmetry allowed. Where the dipole transition is allowed, the dominant part of the matrix element is

$$H_{12} = \frac{ie\hbar A_0}{2m^*}\int_{cell} d^3\mathbf{r}u_c^*(\mathbf{r})\mathbf{a}\cdot\nabla u_v(r) = \frac{eA_0}{2m^*}\mathbf{a}\cdot\mathbf{P}_{12}, \tag{6.31}$$

and the dipole matrix element has been defined by the last term. Once the form of the wave functions is known, one can easily evaluate the matrix element. The wave functions themselves are given by the band structure calculations discussed in Chapter 2. Once the energy bands are found at any point in Brillouin zone, the wave functions for each of the bands can be found as a linear combination of the appropriate s and p orbitals on the atoms making up the crystal. As mentioned, the top of the valence band at the zone center is predominantly p symmetry, while the bottom of the conduction band, say in GaAs, at the zone center is predominantly s symmetry. Normally, the cell periodic part of the wave function would then yield zero in the integral in (6.31), but the dipole operation modifies the p symmetry of the valence orbitals to couple to the s symmetry of the conduction orbitals. However, we must know the exact normalization and contribution of each of the atomic orbitals in order to evaluate (6.31). Usually, the momentum matrix element is determined experimentally in terms of an *oscillator strength* for the transition, and this notation will be followed below.

To proceed, the matrix element (6.31) is introduced into the transition rate (6.23), which yields

$$c_{cv} = \frac{e^2A_0^2t}{m^{*2}\hbar^2}|\mathbf{a}\cdot\mathbf{P}_{cv}|^2\frac{\sin^2[(\omega_{cv}-\omega_0)t/2]}{(\omega_{cv}-\omega_0)^2t^2}, \tag{6.32}$$

where

$$\hbar\omega_{cv} = E_c(\mathbf{k}) - E_v(\mathbf{k}) = E_{gap} + \frac{\hbar^2k^2}{2}\left(\frac{1}{m_e}+\frac{1}{m_h}\right) \equiv E_{gap} + \frac{\hbar^2k^2}{2m_r}. \tag{6.33}$$

For monochromatic radiation, a sum must still be made over all of the available states $\mathbf{k}$ that are allowed in the process, which will serve to introduce the density of final states for the transition. For this, the angular averaged dipole element is introduced by the definition

$$\int d\Omega\,|\mathbf{a}\cdot\mathbf{P}_{cv}|^2 = 4\pi\langle P_{cv}^2\rangle, \tag{6.34}$$

and the connection between the $(\sin x)/x$ function as a representation of the delta function, with $x = (\hbar t/m_r)k^2$, may then be adopted. Thus

$$\begin{aligned} P(\omega) &= \int d^3\mathbf{k}c_{cv} = \frac{e^2A_0^2t}{m^{*2}\hbar^2}\int\frac{d^3\mathbf{k}}{4\pi^3}|\mathbf{a}\cdot\mathbf{P}_{cv}|^2\frac{\sin^2[(\omega_{cv}-\omega_0)t/2]}{(\omega_{cv}-\omega_0)^2t^2} \\ &= \frac{2e^2A_0^2k\langle P_{cv}^2\rangle m_r}{\pi^2m^{*2}\hbar^3}\int\frac{\sin^2x}{x^2}dx = \frac{2e^2A_0^2k\langle P_{cv}^2\rangle m_r}{\pi m^{*2}\hbar^3}. \end{aligned} \tag{6.35}$$

At this point, the oscillator strength is introduced as the ratio of the kinetic energy in the transition to the photon energy:

$$f_{osc} = \frac{2\langle P_{cv}^2\rangle}{m^*\hbar\omega_0}, \tag{6.36}$$

and, reintroducing the energy from the wave vector, the absorption strength is finally found to be

$$P(\omega) = \frac{e^2A_0^2(2m^*)^{\frac{3}{2}}f_{osc}\omega_0}{2\pi m^*\hbar^3}(\hbar\omega_0 - E_{gap})^{\frac{1}{2}}. \tag{6.37}$$

The oscillator strength is normally of order of unity for the allowed transitions and varies with the mass as $1 + (m_0/m_h)$. Finally, it is necessary to normalize (6.37) to the number of incident photons arrival rate in order to calculate the absorption coefficient. The number of photons absorbed within the semiconductor sample, for a thickness of d, is just $P(\omega)d$, which is related to the absorption coefficient by $\alpha P_0 d$, where P_0 is the flux of incoming photons. The latter quantity is readily calculated from the power incident on the sample, as (here S is the Poynting vector of the plane wave)

$$P_0 = \frac{\mathbf{S}\cdot\mathbf{a}_k}{\hbar\omega_0} = \frac{\omega_0A_0^2}{2\hbar}\sqrt{\frac{\varepsilon_\infty}{\mu_0}} = \frac{\omega_0A_0^2}{2\hbar}G_\omega, \tag{6.38}$$

where the optical admittance G_ω of the semiconductor sample has been introduced. From the relations above, the absorption coefficient may finally be found to be

$$\alpha = \frac{e^2(2m^*)^{\frac{3}{2}}f_{osc}}{\pi m\hbar^2G_\omega}(\hbar\omega_0 - E_{gap})^{\frac{1}{2}}. \tag{6.39}$$

The square root variation with energy above the fundamental gap is that expected simply from counting the number of final states available for the transition, as the density of states in the conduction band varies as the square root of the energy. The mass in the numerator is the reduced mass, which is a weighted contribution of both the electron and hole effective masses. The mass in the denominator, however, arose from the Hamiltonian (6.26). This is prior to the introduction of the effective mass approximation, as it still contains the full crystal potential. Thus the mass in the denominator of (6.39) is really the free electron mass m_0. The actual values that appear will vary with doping, temperature, and other parameters. In particular, if the material is heavily doped, the absorption edge is shifted to higher energies, the so-called Burstein shift. This is because an electron must make the transition from a full valence-band state to an empty conduction-band state. If the material is degenerate, the Fermi level lies in the band, and the absorption edge will be shifted by the amount of the Fermi level into the band (Burstein, 1954).

If the dipole transition is forbidden by the symmetry of the wave functions involved in the conduction- and valence-band states, the absorption can still occur through the second term of (6.28). In fact, while the dipole term will normally vanish when the two wave functions have the same symmetry, this is just the requirement needed to have the overlap integral in the second term contribute. The latter term can be written as

$$H_{12} = -\frac{e\hbar A_0}{2m}\mathbf{a}\cdot\mathbf{k}\int_{cell} u_c^* u_v d^3\mathbf{r}. \tag{6.40}$$

Taking the oscillator strength as

$$f' = \left|\int_{cell} u_c^*(\mathbf{r}) u_v(\mathbf{r}) d^3\mathbf{r}\right|^2, \tag{6.41}$$

and averaging over the angle in the scalar product, the power absorbed is found to be

$$\begin{aligned} P(\omega) &= \frac{4\pi e^2 A_0^2 f'}{3m^2}\int \frac{k^4 dk}{4\pi^3}\frac{\sin^2[(\omega_{cv}-\omega_0)t/2]}{(\omega_{cv}-\omega_0)^2 t^2} \\ &= \frac{e^2 A_0^2 k^3 f' m_r}{6\pi^2 m^2 \hbar}, \end{aligned} \tag{6.42}$$

under the same approximations as those used in obtaining (6.37). Reinserting the energy dependence from the momentum terms, one finally arrives at

$$P(\omega) = \frac{e^2 A_0^2 (2m_r)^{\frac{5}{2}} f'}{12\pi^2 m^2 \hbar}(\hbar\omega_0 - E_{gap})^{\frac{3}{2}}, \tag{6.43}$$

and

$$\alpha = \frac{e^2 (2m_r)^{\frac{5}{2}} f'}{6\pi^2 m^2 \hbar^2 G_\omega} \frac{(\hbar\omega_0 - E_{gap})^{\frac{3}{2}}}{\hbar\omega_0} \tag{6.44}$$

is the absorption coefficient for the *forbidden* transition in the direct optical gap. The net dependence on the frequency is still as the square root of the frequency, but the actual variation is somewhat more complicated. In general, the absorption for the forbidden transition is less than that of the allowed transition. However, it increases more rapidly above the absorption edge.

6.2.2 *Indirect transitions*

In the above section, it was observed that when an electron makes an inter-band transition by absorbing a photon, conservation of crystal momentum enforces a strict selection rule onto the momentum of the electron, which in essence requires a vertical transition to occur. However, in indirect gap semiconductors, this would exclude optical absorption at the band-gap energy. Such transitions are, in fact, observed to occur, so that some additional process is involved. This process is the absorption or emission of a short-wavelength phonon whose momentum lies near the Brillouin zone boundary (in essence an inter-valley phonon). The new selection rule between the initial and final states then becomes

$$\mathbf{k}' = \mathbf{k} \pm \mathbf{q}, \tag{6.45}$$

where $\mathbf{q}$ is the phonon wave vector. The energy conservation condition is then given by the requirement

$$E_c(\mathbf{k}') - E_v(\mathbf{k}) = \hbar\omega_0 \pm \hbar\omega_{\mathbf{q}}, \tag{6.46}$$

where $\omega_{\mathbf{q}}$ is the frequency of the zone-edge phonon. Here, the momentum of the photon has been neglected on the assumption that it is considerably smaller than that of either the electron or the phonon. The phonon transitions of this type also contribute to free carrier absorption, but this may be handled by essentially classical approaches, as shown in the last chapter. In Figure 6.2(b), the indirect transition is illustrated.

The process shown in Figure 6.2(b) involves the absorption of a photon and the absorption of a phonon. The final state of the electron is in the minimum of the conduction band that is lowest for the fundamental gap absorption. This minimum is at L for Ge and near X for Si. The photon absorption ends in a state that lies in the forbidden gap. Normally, this state is not allowed, but it actually has a short (usually very short) lifetime associated with it, and it is termed a localized, or *virtual*, state. It is generally assumed that the electron can live in this state long enough to emit (or absorb) a phonon necessary to complete the indirect process. One can ask, however, just where these localized states come from. Certainly, impurities (donors or acceptors) will introduce some states within the band gap, and defects also introduce some states. But, the energy bands themselves also introduce localized states. We can illustrate this simply by taking the two-band model discussed in Section 2.3.1. Then, (2.66) can be rewritten as (the spin–orbit interaction is ignored in the two-band model)

$$E'(E' - E_{gap}) - k^2P^2 = 0. \tag{2.66}$$

Certainly, for $k^2 > 0$, the mirror-image energy bands given by (2.68) are found, for which

$$E' = \frac{E_{gap}}{2}\left[1 \pm \sqrt{1 + \frac{4k^2P^2}{E_{gap}^2}}\right]. \tag{2.68}$$

Here, the top of the valence band is taken as the zero of energy, just as in Section 2.3.1. These bands are shown in Figure 6.3 as the solid lines. These

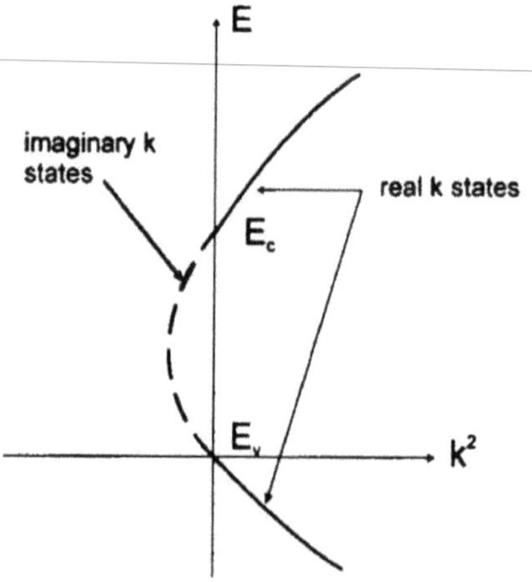

Figure 6.3 The complex band structure is plotted as a function of the square of the momentum. The solid lines for $k^2 > 0$ are the normal bands for propagating waves. The dashed curve for $k^2 < 0$ corresponds to the complex band, where the states are localized states and are not propagating wave states.

lines are not linear in k^2 because of the non-parabolicity. However, these are not the only possible solutions of (2.66). In general, we solve for the band structure by assuming that the electrons are represented by propagating waves for which $\psi \sim e^{ikx}$. But, this is not true for *localized* states. In fact, there is another solution to (2.66) which arises for $k^2 < 0$; that is, for when k is imaginary and the wave function dies off in space as $\psi \sim e^{-|kx|}$. In this case, the solution is given by

$$k = \pm iP\sqrt{E'(E_{gap} - E')}\,. \tag{6.47}$$

This solution is shown as the dashed line in Figure 6.3. It can be seen that the curve for imaginary k joins smoothly into the two curves for real k. Those states where k is imaginary lie in the band gap, where propagating waves are forbidden, but localized states can exist quite easily. The overall band structure is known as the *complex* band structure. The application here, however, is that of indirect absorption, and it is clear that a transition into a localized state is quite easy, and the electron can exist in this state sufficiently long to scatter to real states at the zone edge by a phonon interaction. Thus, the indirect absorption process is quite complicated, and involves two quantum interactions, which means a second-order process.

The perturbation treatment must be expanded to account for the two quite different processes that are required to complete the indirect transition. The perturbing field incorporates the electromagnetic vector potential, but also a potential interaction between the electron and the phonon, as discussed in Chapter 4. As before, the perturbing total vector potential may be written in the form

$$\mathbf{A} = \frac{\mathbf{A}_0}{2} e^{i(\mathbf{k}_0 \cdot \mathbf{r} - \omega_0 t)} + \frac{\mathbf{A}_q}{2} e^{i(\mathbf{q} \cdot \mathbf{r} - \omega_q t)}, \tag{6.48}$$

where the first term is the optical field and the second term is the lattice field. The theory of second-order perturbation theory is found in most quantum mechanics texts. The problem lies in computing the matrix elements, since we have generally talked just about Bloch functions for the allowed energy bands. A wave-function representation is needed for the localized state, and although this can be calculated from most band-structure calculations, it is not usually done. As above, the matrix element will be absorbed into an oscillator strength anyway, so we shall not worry here about the details of the matrix elements.

In general, there will be four terms that contribute to the indirect optical absorption process. First, the electron can either emit or absorb the phonon, which of course gives two distinct terms. Then the phonon process can either follow the photon absorption or can occur prior to the photon absorption. In the first of these two possibilities, the virtual state is characterized by wave functions near the center of the zone, so that the photon oscillator strength is that appropriate for the Γ point, and the phonon processes involve conduction-band wave functions. In the second set of processes, the virtual state is one corresponding to a predominantly valence state near the zone edge, so that the oscillator strength is that corresponding to the E_1 (L point) or E_2 (X point) transition, and the phonon process involves valence-band wave functions. This leads to different strengths for the four terms that can contribute to the indirect absorption. However, the energies of both states enter into the calculation for the following reason. The localized states have a relatively constant level that serves as a reference level. For a given optical energy, the transition at the zone center has a "defect" hole energy, which is the kinetic energy of the hole state measured from the valence-band maximum. On the other hand, the transition at the zone edge involves the "defect" electron energy, which is the kinetic energy of the electron in the final state at the zone edge. It is necessary to sum over these two defect energies to fully account for the joint density of states, since the total defect energy ($\hbar\omega_0 - E_{gap}$) can be distributed between the electron and hole states in an infinity of ways. Thus the initial- and final-state energy differences must be introduced as

$$\hbar\omega_i = E_c(\mathbf{k}_i) - E_v(\mathbf{k}_i), \quad \hbar\omega_f = E_c(\mathbf{k}_f) - E_v(\mathbf{k}_f), \tag{6.49}$$

and one of the two energies lies in the virtual state in each case. Each of the four terms has the general form

$$c_{cv} = \frac{4|H_0|^2|H_q|^2}{\hbar^2 t}\frac{\sin^2[(\omega_f - \omega_0 \pm \omega_q)t/2]}{(\omega_f - \omega_0)^2(\omega_f - \omega_0 \pm \omega_q)^2}, \tag{6.50}$$

and the matrix elements are appropriate to the position of the transition. This must now be integrated over the initial and final energies of the hole and electron and have the properties of the phonons inserted. In general, all unknown sines are incorporated into the definitions of an arbitrary oscillator strength. Here, only the final results will be given in terms of the energy variation for the absorption coefficient (a more complete derivation is available in Smith, 1969)

$$\alpha \sim \frac{x}{e^x - 1}(\hbar\omega_0 - E_{gap} + \hbar\omega_q)^2, \quad x = \frac{\hbar\omega_q}{k_B T} \tag{6.51}$$

for the absorption of a phonon, and

$$\alpha \sim x(\hbar\omega_0 - E_{gap} - \hbar\omega_q)^2 \tag{6.52}$$

for the emission of a phonon. By comparison with (6.39), it may be observed that there is a considerable temperature dependence for the indirect transition, which is not found in the direct transitions. In addition, the dependence upon photon frequency is different for the indirect transition, and has a much faster variation, although the overall strength is much weaker. The major part of the direct transition variation is in the actual turn-on edge, with little variation above this edge due to the variation as the square root of the energy. On the other hand, the energy variation above threshold for silicon varies quite significantly, even though the overall strength of the absorption is very weak.

6.3 Recombination

If the carriers excited by the inter-band optical absorption process are left to themselves, they will eventually return to an equilibrium state; that is, the electrons and holes will recombine as the excited electron drops across the energy gap to the empty valence state. In this way the numbers of electrons and holes present in the semiconductor will return to their equilibrium levels. As discussed earlier, the rate of creation of electron–hole pairs is introduced through a generation rate G_0. This will now allow the continuity equation for the deviation of the electrons from equilibrium to be written as

$$\frac{d(\Delta n)}{dt} = G_0 - \frac{n}{\tau_n} = -\frac{\Delta n}{\tau_n}, \tag{6.53}$$

where τ_n is the recombination time and may be a function of the hole (or electron) density. If a steady state is reached, the time derivative term vanishes, and $G_0 = n_0/\tau_n$. This leads us to interpret G_0 as the equilibrium generation rate. The recombination time has been assumed to be independent of the level of the excess carrier generation, and this is not always true, as will be seen below. Further, an equivalent equation can be written for the holes that are produced by the optical absorption process, and this is what leads to ambipolar diffusion discussed in an earlier section. The recombination rate τ_n, and its counterpart for the holes, represent a rate of decay and are the lifetime of excess electron–hole pairs.

There are a number of processes by which recombination can occur. A direct transition can occur in which the energy of the electron–hole pair is either radiated as electromagnetic energy or else goes into the lattice vibrations as a shower of phonons are emitted. These processes are the inverse of direct absorption transitions. The phonon process is relatively rare in most semiconductors because of the number of phonons required, which makes it a high-order perturbative process and therefore unlikely. These two processes can also occur through intermediate impurity levels, called traps, which are located in the forbidden gap region. Thus the electron hops downward through a series of levels, while the hole climbs upward through a series of levels, the so-called Shockley–Read–Hall process (Shockley and Read, 1952; Hall, 1951, 1952). This process is in fact far more likely in the indirect gap materials, since the localized (in real space) states from impurities or defects have an energy level that samples nearly all of the Brillouin zone and therefore breaks the *k*-conservation requirement of the transition.

6.3.1 Radiative recombination

If α is the absorption coefficient at a particular frequency, then the mean free path of the photons in the semiconductor is just α^{-1}, and the mean photon lifetime is just $(\alpha c/\varepsilon_\infty^{\frac{1}{2}})^{-1}$ in the semiconductor. This lifetime can be used to determine the rate of carrier generation in the semiconductor, and this fact can be used to balance generation and recombination in the steady state. This will then determine the dependence of the electron and hole recombination times on the individual densities. To begin, the photon lifetime must be multiplied by the number of photons in the frequency range $d\nu$ centered about the frequency ν $(= \omega/2\pi)$. This is given by the Planck radiation formula for a thermal source, as

$$N_{ph}(\nu)d\nu = \frac{8\pi\nu^2\varepsilon_\infty^{\frac{3}{2}}d\nu}{c^3(e^{h\nu/k_BT}-1)}. \tag{6.54}$$

Then we can now write the net generation rate as

$$G(\nu) = \frac{N_{ph}}{\tau_{ph}} = \frac{8\pi\alpha\nu^2\varepsilon_\infty}{c^2(e^{h\nu/k_BT}-1)}. \tag{6.55}$$

In thermal equilibrium, the recombination rate is equal to the generation rate, and this will hold true in any steady-state situation. This fact now gives the equilibrium recombination rate by assuming that the semiconductor is subject to thermal background photons according to (6.55). The recombination rate is found by then equating it to (6.55) and integrating over all frequencies, which gives ($x = h\nu/k_BT$)

$$R_0 = \frac{8\pi k_B^3T^3}{c^2\hbar^3}\int_0^\infty \frac{x^2\varepsilon_\infty\alpha}{e^x-1}dx. \tag{6.56}$$

The dielectric function has been left inside the integral, as it may have some frequency dependence of its own. The major contribution to the integral in (6.56) arises from states right near the band edge, since α is zero below the band edge and the exponential in the denominator falls off rapidly for higher frequencies. Thus the recombination rate benefits from only a narrow band of thermal excitation energies quite near the band-gap energy. However, (6.56) now gives the equilibrium recombination rate for electron–hole pairs once the absorption coefficient is known for the material. In fact, if $\alpha = \alpha_0(h\nu - E_{gap})^r$, the energy factor can be expanded to give the recombination rate as

$$R_0 \approx \frac{8\pi\alpha_0\varepsilon_\infty(k_BT)^{2+r}}{c^2\hbar^3}\left(\frac{E_{gap}}{k_BT}\right)^2 \exp\left(-\frac{E_{gap}}{k_BT}\right)\left[\Gamma(r+1) + \frac{2k_BT}{E_{gap}}\Gamma(r+2)\right]. \qquad (6.57)$$

Once the appropriate constants are either calculated or measured, the recombination rate is then easily calculated. It is clear, however, that low values of absorption coefficient equate to small recombination rates and therefore long recombination times.

In non-equilibrium conditions, the recombination rate may be written as $R = Apn$, where A is a constant, for direct recombination since one electron and one hole recombine in eliminating an electron–hole pair. Thus, the recombination rate is dependent on the numbers of electrons and holes. In equilibrium, however, this same form leads to $R_0 = An_0p_0 = An_i^2$, and

$$R = R_0\frac{pn}{n_i^2}. \qquad (6.58)$$

In the following, it will be assumed that $n_0 < p_0$, so that concern is with the minority electron concentration. When excess carriers are generated, the recombination rate (after the excess generation has been turned off) may be written as

$$\frac{d(\Delta n)}{dt} = -\frac{\Delta n}{\tau_n} = R_0 - R_0\frac{pn}{n_i^2} = -R_0\frac{pn - n_i^2}{n_i^2}$$
$$= -R_0\frac{(\Delta n)^2 + (n_0 + p_0)\Delta n}{n_i^2}. \qquad (6.59)$$

For low densities of excess carriers and a strongly extrinsic material (Δn, $n_0 \ll p_0$), the lifetime is relatively constant and is given by

$$\frac{1}{\tau_n} = R_0\frac{p_0}{n_i^2} = \frac{R_0}{n_0}. \qquad (6.60)$$

However, for large excess densities, which exceed the background concentration of holes, the recombination rate depends on the excess density, and the decay is non-exponential in nature. In this case, the decay may be written as

$$\Delta n(t) = \frac{\Delta n(0)}{1 + R_0 t \dfrac{\Delta n(0)}{n_i^2}}. \tag{6.61}$$

The form in (6.61) is only valid, however, for the situation in which the excess density remains larger than the background doping concentration, so that as the excess density decays to the latter level, the temporal behavior changes over to the normal exponential decay.

6.3.2 *Trap recombination*

As mentioned above, it is also possible for electrons and holes to recombine through a series of intermediate levels, lying in the forbidden band gap. These levels are provided by impurities, defects (or ***traps***), or just the complex band structure. Still, the recombination through these levels can occur only if there is some mechanism for the carrier to give up energy, which is essentially the band-gap energy. This can occur through radiation of the energy either to photons or to phonons. There exists still a third mechanism, known as the Auger process, in which the energy liberated by the electron–hole pair is taken up by a second electron being excited to a state high in the conduction band (this is the inverse of the impact ionization process). In general, the recombination time, or the trap capture time, is a suitable combination of these processes. The trap recombination time is also the appropriate mechanism by which electrons recombine with ionized donors (or holes with the ionized acceptors). In this section, we concentrate primarily on the latter process, and the band-to-band process is easily generalized from it. For radiative processes we can write the continuity equation just as earlier. Here, the generation rate will depend on the number of neutral donors ($N_d > N_a$ is assumed here), while the recombination rate will depend on the number of electrons as well as the number of ionized donors. Thus these two quantities may be written as

$$R_d = R_0 \frac{nN_{di}}{n_0 N_{di0}} = R_0 \frac{n(n + N_a)}{n_0(n_0 + N_a)},$$

$$G_d = R_0 \frac{N_{dn}}{N_{dn0}} = R_0 \frac{N_d - N_a - n}{N_d - N_a - n_0}. \tag{6.62}$$

The recombination time can then be written as

$$\frac{1}{\tau_n} = \frac{R_0}{\Delta n}\left[\frac{(n_0 + \Delta n)(n_0 + \Delta n + N_a)}{n_0(n_0 + N_a)} - \frac{N_d - N_a - n_0 - \Delta n}{N_d - N_a - n_0}\right]$$

$$= \frac{R_0}{n_0}\left[\frac{N_d - N_a}{N_d - N_a - n_0} + \frac{n_0 + \Delta n}{n_0 + N_a}\right]. \quad (6.63)$$

It is clear that only in the most unusual situations will the lifetime deviate from a nearly constant value. One such situation is excess generation across the band gap in addition to excitation from the trap level. In general, though, the lifetime is that of the exponential decay. Impurity level absorption is often used for far-infrared detectors, but radiative recombination rates are typically very low and not the primary cause of the recombination.

For phonon-aided recombination, the energy lost by the carriers goes into the lattice vibrations, which is a more likely process for the small energy by which the impurity is usually separated from the conduction band. The lifetime has the same general density dependence as that for the radiative rate, except that the equilibrium constant R is not that for radiative absorption. Rather, the constant R must be evaluated for the phonon-assisted generation process.

For an n-type semiconductor, the Auger process can also occur with capture of the free carriers by the donor levels. Here the excess energy is taken up by a second electron, which is excited to a higher-energy state in the conduction band. The recombination rate will be proportional to the square of the free-electron concentration, since there are two electrons involved in the overall process. The generation rate, on the other hand, is merely proportional to the number of neutral donors, as above. The inverse process of Auger recombination is impact ionization of the donor levels, and this also requires free electrons. Thus, the forward and reverse processes are balanced to calculate proper rate constants. As before,

$$R_a = R_0\frac{n^2 N_{di}}{n_0^2 N_{di0}} = R_0\frac{n^2(n + N_a)}{n_0^2(n_0 + N_a)},$$

$$G_a = R_0\frac{n(N_d - N_a - n)}{n_0(N_d - N_a - n_0)}. \quad (6.64)$$

These may now be combined to give the Auger recombination lifetime as

$$\frac{1}{\tau_{na}} = \frac{R_0}{\Delta n}\left[\frac{(n_0 + \Delta n)^2(n_0 + N_a + \Delta n)}{n_0^2(n_0 + N_a)} - \frac{(n_0 + \Delta n)(N_d - N_a - n_0 - \Delta n)}{n_0(N_d - N_a - n_0)}\right]$$

$$= \frac{R_0}{n_0}\left(1 + \frac{\Delta n}{n_0}\right)\left[\frac{N_d - N_a}{N_d - N_a - n_0} + \frac{n_0 + \Delta n}{n_0 + N_a}\right]. \quad (6.65)$$

If n_0 is relatively small, such as occurs at low temperatures, Auger recombination can readily occur in a non-exponential fashion. Auger recombination is thought to occur at low temperatures in highly-doped material, such as in the base and emitter of bipolar transistors. In some cases this is actually felt to be cross-band-gap recombination (in which the number of ionized donors is replaced with the number of holes). The latter case is particularly important at low temperatures in laser-excited semiconductors where the density of electron–hole pairs is quite large.

6.4 Photoconductivity

In semiconductors at low temperatures, or semiconductors with only a small free carrier density, the number of free carriers can be increased significantly by illuminating the material with optical radiation of photon energy greater than the band gap. At higher temperatures, the conductivity change in this process is relatively small unless the illuminating light is quite intense, which will be discussed in the next section. Also at low temperatures, carriers localized on the impurity atoms can be excited into the conduction (or valence) band optically. In each of the various processes, the excess carriers generated by the incident radiation lead to an increase in conductivity, a process known as photoconductivity. If I_p is the incident radiation intensity (watts per square meter), and I_p' is the radiation intensity within the semiconductor, then the power absorbed per unit length in the material is just $\alpha I_p'$. The number of electron–hole pairs produced is $\eta\alpha I_p'/\hbar\omega_0$, where η is the "quantum efficiency" defined in terms of the fraction of the absorbed photons that actually produce electron–hole pairs. I_p' depends on I_p through the dielectric constant ε_∞ of the semiconductor, the thickness d of the material, and the absorption coefficient α. Only if the absorption is small and the sample infinitely thick do the last two parameters drop out of the relation between the radiation intensity just inside and outside the sample.

Consider a thin semiconductor slab, subject to an applied electric field in the plane of the slab that produces a small current. With a thin slab, the intensity of the radiation will be essentially uniform throughout the material (a simplification that will be removed shortly). The material is assumed to be n-type, so that the continuity equation for the minority carriers generated by the incident radiation is just

$$\frac{d(\Delta p)}{dt} = G_{ex} - \frac{\Delta p}{\tau_p}. \tag{6.66}$$

A steady-state leads to the density

$$\Delta p = G_{ex}\tau_p = \frac{\eta\alpha I_p'\tau_p}{\hbar\omega_0}. \tag{6.67}$$

This excess hole density (and electron density for band-gap illumination) leads to a net conductivity

$$\Delta\sigma = \sigma - \sigma_0 = (n_0 + \Delta p)e\mu_e + (p_0 + \Delta p)e\mu_h - (n_0\mu_e + p_0\mu_h)e$$
$$= e\Delta p\mu_h(1 + b), \tag{6.68}$$

where $b = \mu_e/\mu_h$ is the mobility ratio. This change in conductivity leads to an increase in the current through the sample, and hence to a photocurrent, which for a sample of thickness d and width w is given by

$$i_{ph} = wd\Delta\sigma F = ewd\left(\frac{\eta\alpha I_p'}{\hbar\omega_0}\right)\tau_p\mu_h(1 + b)F. \tag{6.69}$$

The result in (6.69) is adequate for relatively low levels of illumination where the excess free-carrier density remains small compared to the background free-electron concentration. If the intensity is quite large, however, the recombination time begins to be dependent on the actual total density of carriers. For this, one must modify (6.67) by the inclusion of a more exact limit from (6.59) as

$$\Delta p = \sqrt{\frac{\eta\alpha I_p' n_i^2}{R_0\hbar\omega_0}}. \tag{6.70}$$

In this case, the photocurrent varies as the square root of the incident intensity rather than linearly in the intensity. This is because the rising intensity causes a corresponding increase in the recombination rate, which is a competing process.

In non-steady-state situations, (6.66) leads to a single exponential time constant in the photo-current rise and fall, and the time constant is the recombination time. If the excitation level is high, the behavior evident in (6.70) can also lead to a non-exponential temporal evolution of the photo-current, as discussed above.

6.4.1 *Transverse photovoltage*

In the case where the sample is not thin, the intensity can vary throughout the thickness of the material. This leads to a spatially varying excess density, as the number of excess carriers depends on the illumination at a particular point. The spatial variation in the excess holes throughout the thickness will lead to the generation of diffusion potentials within the sample, and the resulting voltage differential between the front and back surfaces is termed the *transverse photovoltage*. The illumination is taken to occur at the top surface, and arrives normal to this surface, the $x = 0$ plane, with the sample located in the space $x > 0$. The slab is taken to be large in the y, z directions, but has a thickness d in the x direction such that $\alpha d > 1$. The primary photo-current will be taken to be in the y direction, but the concern here is the photovoltage itself. In the x direction the variation in density leads to the drift-diffusion equation

$$J_x = e\mu_h(bn + p)F + e(b - 1)D_h\frac{\partial(\Delta p)}{\partial x}. \tag{6.71}$$

Since no terminal current flows in this direction, this current density must be set to zero, which causes the drift field to counterbalance the diffusion forces. This leads us to the local field

$$F_x = -\frac{(b-1)D_h}{\mu_h(bn + p)}\frac{\partial(\Delta p)}{\partial x} = -\frac{k_BT}{e}\left(\frac{b-1}{bn+p}\right)\frac{\partial(\Delta p)}{\partial x}. \tag{6.72}$$

If the number of excess carriers is small compared to the background n-type doping, the transverse photovoltage may be found from (6.72) as

$$V_{tpv} = -\int_0^d F_x dx = \frac{k_BT(b-1)}{ebn_0}[\Delta p(d) - \Delta p(0)]. \tag{6.73}$$

Note that this is the voltage *at the back surface* relative to the front surface, which is the opposite polarity of that normally measured. Since $\Delta p(0) > \Delta p(d)$, the voltage at the back is negative with respect to the front surface (or if the back surface is grounded, the front surface develops a positive voltage). This can be seen from the electric field in (6.72); since the excess hole density decays away for $x > 0$, the x component of the field is positive, which requires the front surface voltage to be positive compared with the back surface. When the sample is optically thick ($\alpha d \gg 1$, $d \gg L_p$, where the latter is the appropriate diffusion length), the density at the back surface is essentially negligible compared to that at the illuminated front surface. The photovoltage then becomes linearly dependent on the excited excess electron–hole concentration.

For large excess densities, one cannot ignore the contribution of the optically created electron–hole pairs to the denominator term in (6.72). For the case of intense illumination, where the number of excess carriers is larger than the background free-carrier contribution, the transverse photovoltage in now given by

$$V_{tpv} = \frac{k_B T(b-1)}{e(b+1)} \ln\left(\frac{\Delta p(d)}{\Delta p(0)}\right). \tag{6.74}$$

In both cases the photovoltage has the same sign, but the dependence on the excess concentration at the front surface varies. If the sample is not optically thick, and particularly in the high-intensity case, it is necessary to find the hole concentration at the back surface, and this requires us to solve a diffusion equation for the carrier density throughout the sample. An example of this is provided in the next section.

6.4.2 Photoelectromagnetic effects

If a magnetic field is now applied *parallel* to the surface, new effects will occur that will modify the induced voltages and currents. We apply this field in the z-direction and take the sample to be of width w in this direction. The photovoltage will induce currents in the y direction as well, since the diffusion is across the magnetic field and we have to consider the Hall effect. In the following, primarily the low-intensity limit will be discussed, so that attention can be concentrated solely on the minority holes. It will also be assumed that the magnetic field is low as well, $\omega_c\tau \ll 1$. The hole current density is given by

$$J_{hy} = \frac{pe^2}{m_h}(\tau F_y - \omega_{ch}\tau^2 F_x) = \mu_h(peF_y - BJ_{hx}), \tag{6.75}$$

where $J_{hx} = \sigma F_x$. Similarly,

$$J_{ey} = \mu_e(neF_y + BJ_{ex}). \tag{6.76}$$

The total current in the x direction must vanish, which leads to $J_{hx} = -J_{ex}$. The hole current density in the x direction consists of drift and diffusion terms as

$$J_{hx} = ep\mu_h F_x - eD_h\frac{\partial(\Delta p)}{\partial x} = -e\sigma_h\frac{D_e - D_h}{\sigma_e + \sigma_h}\frac{\partial(\Delta p)}{\partial x} - eD_h\frac{\partial(\Delta p)}{\partial x}$$

$$= -eD_a\frac{\partial(\Delta p)}{\partial x}, \tag{6.77}$$

where (6.72) has been used to replace the electric field and D_a is the ambipolar diffusion coefficient.

In the transverse photovoltage detector outlined above, the y-directed current is set to zero in order to measure the transverse photovoltage.

However, one cannot simply set the current density J_y to zero, since this would leave F_y a function of x through (6.75) and (6.76). This is an unphysical situation, since there is no time-varying magnetic field, and therefore $\nabla \times \mathrm{F} = 0$. Since F_x *is* not a function of the y-direction, we cannot have F_y a function of the x-direction. Therefore, it is necessary to have the *total* current vanish as

$$w\int_0^d (J_{ey} + J_{hy})dx = 0. \tag{6.78}$$

Inserting (6.75) and (6.76) into this latter equation leads to

$$F_y = \frac{B}{d(\sigma_e + \sigma_h)}\int_0^d (\mu_e + \mu_h) J_{hx} dx = \frac{BD_a}{dn_0}\left(\frac{b-1}{b}\right)[\Delta p(0) - \Delta p(d)]. \tag{6.79}$$

This field differs from the x-component by the Hall angle as expected.

It is often more useful to measure the y component of current rather than either of the voltages. In this configuration one actually measures the short-circuit current, which is obtained by setting $F_y = 0$. The current is then

$$\begin{aligned} i_y &= w\int_0^d (J_{ey} + J_{hy})dx = -w\mu_h(b+1)B\int_0^d J_{hy} dx \\ &= weD_a\mu_h(b+1)B[\Delta p(0) - \Delta p(d)]. \end{aligned} \tag{6.80}$$

This current is normally termed the photoelectromagnetic current and is the quantity primarily measured.

To complete this treatment, the diffusion equation must finally be solved to ascertain the exact variation of the excess hole density across the thickness of the semiconductor slab. The continuity equation for the excess carriers leads to the ambipolar diffusion equation

$$D_a\frac{d^2(\Delta p)}{dx^2} - \frac{\Delta p}{\tau_p} = -Re^{-\alpha x}, \tag{6.81}$$

where

$$R = \frac{\eta\alpha I'_p}{\hbar\omega_0} = \frac{\eta\alpha I_p(1 - R_s)}{\hbar\omega_0} \tag{6.82}$$

and R_s is the surface reflectance. Equation (6.81) is easily solved for the case for which the density at the back surface is negligible to give ($L_a = D_a\tau_p$)

$$\Delta p = Ae^{-x/L_a} - \frac{R\tau_p}{\alpha^2 L_a^2 - 1}e^{-\alpha x}, \tag{6.83}$$

where A is a constant to be evaluated at the front surface $x = 0$. The flow into the surface, which counters the excess density generation at that point, must disappear by surface recombination, a process similar to bulk recombination and characterized by a surface recombination velocity s. Thus,

$$-eD_a\frac{d(\Delta p)}{dx}\bigg|_{x=0} = -es\Delta p(0), \tag{6.84}$$

which leads to

$$A = \frac{R\tau_p}{\alpha^2 L_a^2 - 1}\frac{\alpha L_a^2 + s\tau_p}{L_a + s\tau_p}. \tag{6.85}$$

These results now give

$$\Delta p(0) = \frac{\eta I_p(1 - R_s)\tau_p}{\hbar\omega_0(L_a + s\tau_p)}\frac{\alpha L_a}{\alpha L_a + 1} \tag{6.86}$$

and

$$i_s = \frac{e\mu_h(b + 1)B\eta I_p(1 - R_s)L_a^2}{\hbar(L_a + s\tau_p)}\frac{\alpha L_a}{\alpha L_a + 1}. \tag{6.87}$$

This result is for the case of the optically thick sample, where the excess carrier concentration vanishes before reaching the back surface. Hence, the photoelectromagnetic current is linear in the magnetic field as well as in the intensity of illumination. This is a quite sensitive method of detecting optical signals, as the magnetic field increases the sensitivity over that of normal photo-detectors.

6.5 Sub-picosecond optical excitation

Since the mid-1970s, researchers have had the use of ultra-short laser pulses, first in the picosecond regime and later in the few femtosecond regime. These have been an important tool for studying the dynamics of non-equilibrium carriers. The short laser pulse can be used to initiate a

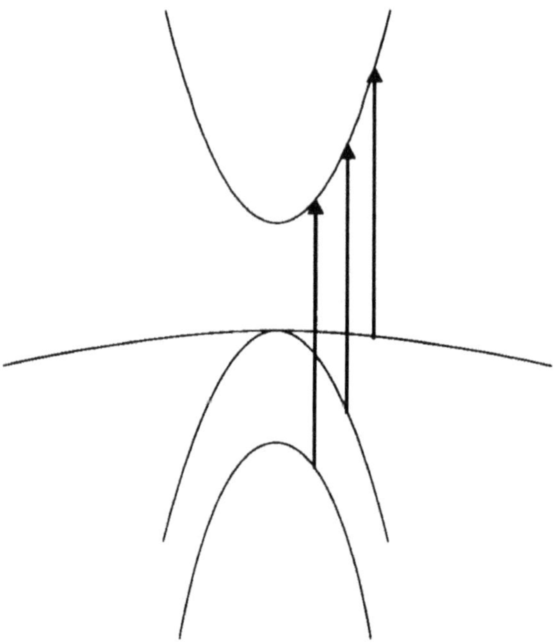

Figure 6.4 Excitation of electrons from the valence band to the conduction band can, in general, flow from all three upper valence bands.

far-from-equilibrium electron–hole plasma, which then relaxes, and ultimately recombines. Light emission and light scattering can be used to study the population dynamics, the phonon properties, and a variety of other aspects. By the same token, experimental research has benefited from careful theoretical analysis, usually by the ensemble Monte Carlo technique. Several models have been explored to explain the role of processes such as screening of the LO phonon–electron interaction, hot phonons, and electron–hole interactions. Space limitations prohibit us from exploring this in exhaustive detail, as books have been written solely on this topic (see, e.g., Shah, 1992). In this section, we review some of the properties of the far-from-equilibrium dynamic carrier system, and the various properties that can be studied.

6.5.1 Excitation and intervalley transfer

In general, the laser pulse excites carriers with a vertical transition. However, this transition can begin in any of three valence band states for a sufficiently energetic photon. We show this in Figure 6.4. There are three dominant paths indicated: (i) the transition is from the heavy-hole band high into the conduction band, (ii) the transition is from the light-hole band, but still relatively high into the conduction band, and (iii) the transition is from the split-off band low into the conduction band. The excess energy $(\hbar\omega_0 - E'_{gap})$ is split between the electron and the hole that are produced, where E'_{gap} is the actual gap for the bands in question. In general,

$$\hbar\omega_0 = E'_{gap} + \frac{\hbar^2 k^2}{2}\left(\frac{1}{m_e} + \frac{1}{m_h}\right). \tag{6.88}$$

The fraction of energy that goes into the electron is given by

$$f_e = \frac{m_h}{m_e + m_h}. \tag{6.89}$$

On the other hand, the fraction of photons that go into each transition depends upon the absorption coefficient for each transition, and therefore upon the joint density of states. Thus, the dominant transition is that originating in the heavy-hole band.

Although carriers are generated mono-energetically, and the time scales in picosecond and femtosecond processes are very short for a standard Fermi–Dirac, or Maxwellian, distribution to be defined, most early models assumed that the distribution was quickly assumed by the carriers. Moreover, many researchers assumed that the electron and hole temperatures were equal. The rationale is to explain the observed luminescence from the sample after the laser pulse. In fact, one expects the carriers to get into a thermal distribution and then the luminescence depends upon the joint number of electrons and holes available for recombination. By measuring this luminescence spectrum, it was felt that one was measuring the key dynamics of the electron–hole plasma. In the real case, a number of ensemble Monte Carlo simulations have been carried out by many groups to explore the actual equilibration of the hot plasma. Here, we discuss just a few of these results for their elucidation of the physics of the process.

In Figure 6.5, we plot the excited electron distribution following a 0.2 picosecond, 1.64 eV laser pulse (electrons enter the conduction band below the threshold for inter-valley transfer) at low temperature. Here, we take the masses as $m_e = 0.063m_0$, $m_{hh} = 0.54m_0$, and $m_{lh} = 0.07m_0$. This leads to carriers excited from the heavy-hole band to appear at about 118 meV (in the conduction band, which is shown in the figure). Carriers from the light-hole band appear at 70 meV. However, more than 95 percent of the carriers are excited from the heavy-hole band due to its much larger effective mass and higher density of states (we note that there is no excitation from the split-off band as $E_{gap} + \Delta > \hbar\omega_0$). Consequently, it is quite difficult to see the excitation from the light-hole band in the figure. Instead, the peaks lying below the main excitation peak are phonon replicas, in which carriers have relaxed downward in energy by the emission of one and two

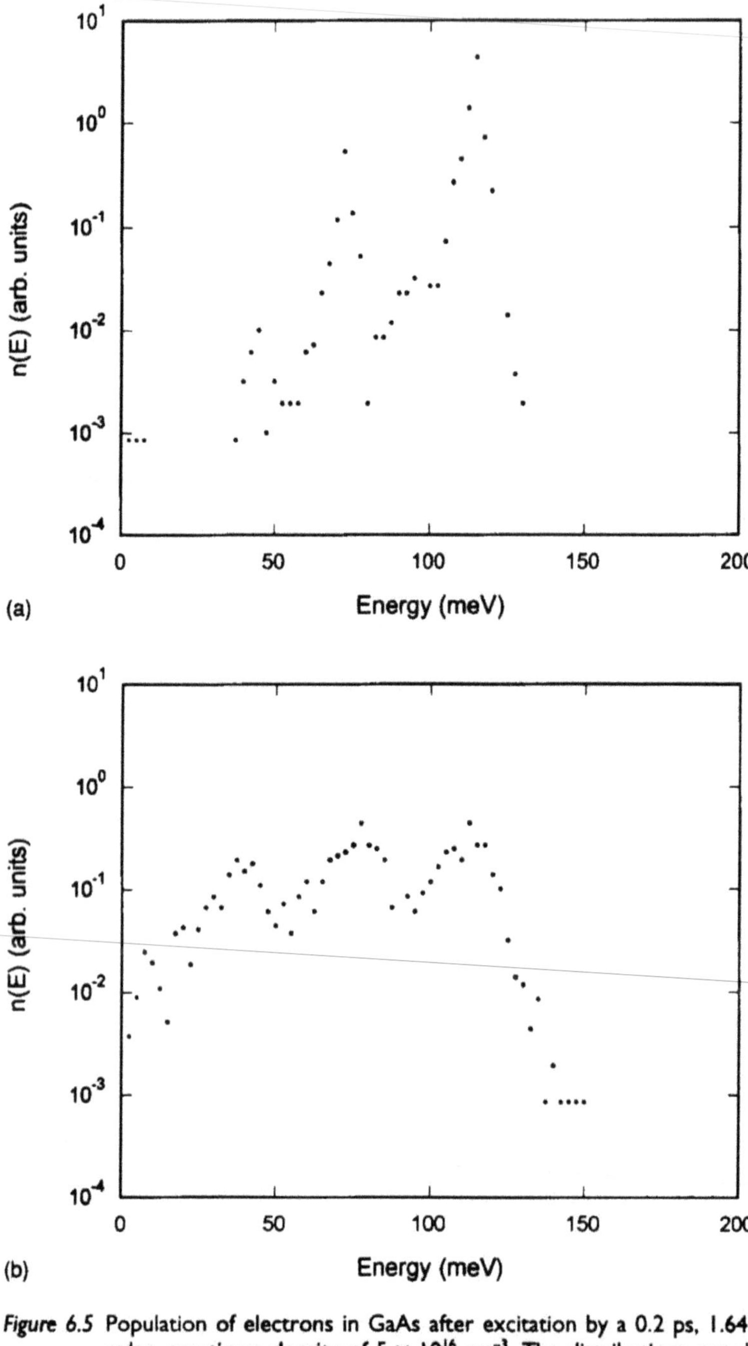

Figure 6.5 Population of electrons in GaAs after excitation by a 0.2 ps, 1.64 eV laser pulse, creating a density of 5×10^{16} cm^{-3}. The distributions are shown after (a) 100 fs and (b) 500 fs, illustrating that, at this low density, the thermal distribution is not yet created even at 500 fs. (From Osman and Ferry, 1987.)

phonons. Two different times, after the pulse, are shown. At the low densities assumed (5×10^{16} cm^{-3}), these phonon peaks are still observable after 0.5 ps, as seen in Figure 6.5(b). On the other hand, if the density is increased by a factor of 20, the results in Figure 6.6 show that these phonon replicas are washed out quickly, with only a barely visible remnant at 50 fs. From this, we conclude that the electron–electron (and the electron–hole interaction) are quite important in establishing a thermal distribution (Osman and Ferry, 1987). We will deal with the carrier–carrier scattering in the next chapter.

For higher photon energies, the excited carriers will be be found much higher in the conduction band. If the electron excess energy is sufficiently large, they can scatter to the satellite X and L valleys, as discussed in the last chapter. In GaAs, the L valleys lie 0.29 eV above the Γ minimum, and the X valleys lies an additional 0.2 eV higher. The scattering rate to the satellite valleys is much higher than the scattering rate within the Γ valley, so that the electrons will transfer to these valleys, and be stored here for a period which basically depends upon the difference in the Γ–L and L–Γ scattering rates, for example. Carriers high enough to transfer to the X valleys usually relax to the L valleys as they cool, and then subsequently make it back to the Γ valley after further cooling. The observed luminescence of Γ electrons recombining with holes will show the effect of storage times in the satellite valleys. This gives us some experimental studies of the populations, and in principle will aid in determining the inter-valley coupling constants upon which the scattering rates depend.

6.5.2 *Non-equilibrium phonons*

One of the most interesting techniques of studying non-equilibrium systems is through light-scattering from the excitations in the system. In the above section we talked about time-resolved optical absorption and luminescence as a probe for studying the hot carrier distribution created by the photons. In a sense, these techniques measure the actual carrier density and its distribution. Another method is to probe the non-equilibrium phonon distribution. As the carriers cool from their high energies at which they are created, they emit a cloud of phonons. For example, an electron created 360 meV into the conduction band will emit an average of 9–10 phonons as it cools. This causes a buildup of the phonon distribution which can decay only through three-phonon interactions (e.g., the zone center LO mode decays via two zone-edge phonons, which carry the energy to the heat bath). A method of probing this phonon distribution is by *Raman scattering*. In an optically active medium such as GaAs, light couples readily to the lattice through the charge polarization fields. These are the same fields which lead to the polar modes of the phonons. As we will see in the next chapter, this coupling provides the difference between the high-frequency and the low-frequency

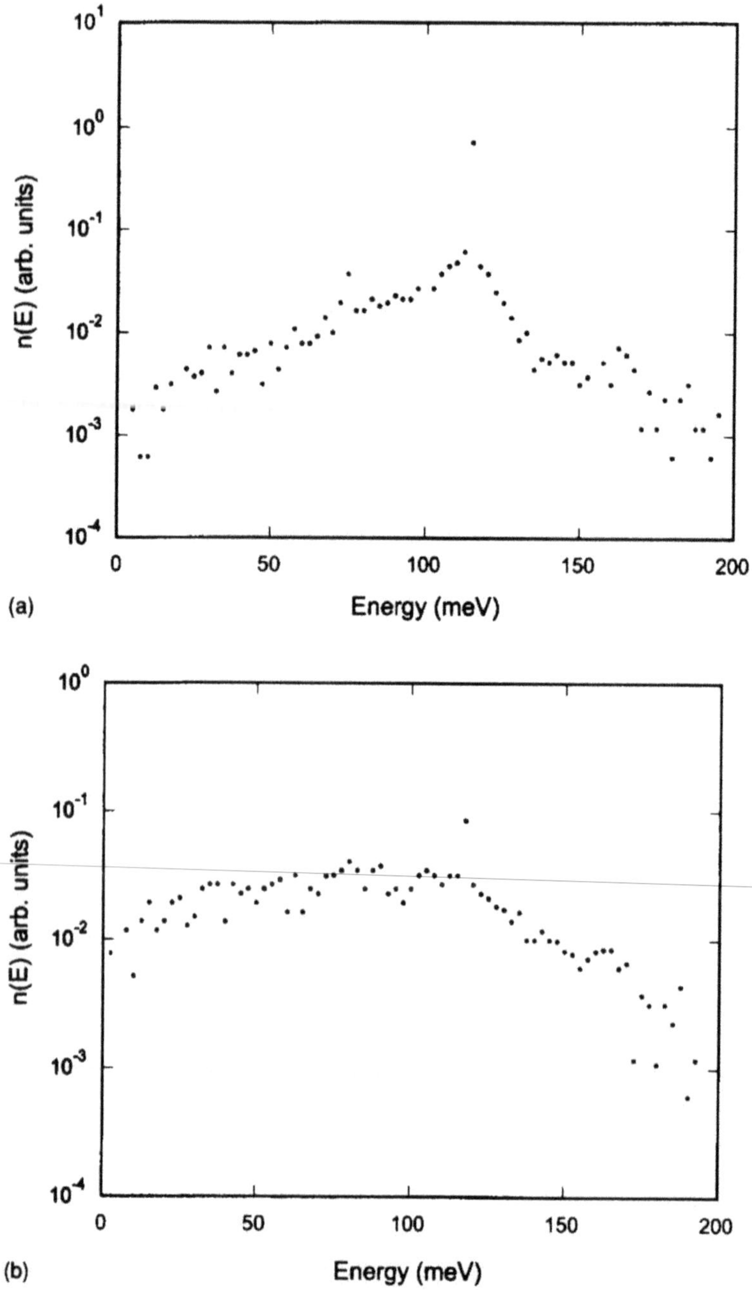

Figure 6.6 Population of electrons in GaAs after excitation by a 0.2 ps, 1.64 eV laser pulse, creating a density of 1×10^{18} cm^{-3}. The distributions are shown after (a) 50 fs and (b) 250 fs, illustrating that, at this high density, the thermal distribution is rapidly created. (From Osman and Ferry, 1987.)

dielectric "constants" of the polar material. Hence, light can be scattered by these polarizations. In first order, scattering by the acoustic modes is referred to as Brillouin scattering, while scattering by the polar optical modes is called Raman scattering. In this process, the incoming photon is back-scattered by the phonons (that is, by the lattice polarization). The photon can emit a phonon or absorb a phonon. The outgoing photon which results from the emission of a phonon, is termed the Stokes shifted signal. That arising from the absorption of a phonon is the anti-Stokes signal. In general, the energy and momentum must be conserved through

$$\hbar\omega_0 = \hbar\omega' \pm \hbar\omega_q, \quad \mathbf{k}_0 - \mathbf{k}_0' = \mathbf{q}. \tag{6.90}$$

The most common situation is the back-scattering, so that $\mathbf{k}_0 = -\mathbf{k}_0'$, and $q = 2(\omega_0/c)\,\varepsilon_\infty^{\frac{1}{2}}$, where c is the speed of light in vacuum. By measuring the Stokes signal I_S, which is proportional to N_q, and the anti-Stoke's signal I_{AS}, which is proportional to $1 + N_q$, we can get an estimate of the phonon distribution as

$$N_q \sim \left(\frac{I_{AS}}{I_S} - 1\right)^{-1}. \tag{6.91}$$

While the above is the basic premise, the phonon distribution that is driven out of equilibrium depends upon the electron dynamics, so that the Raman probe only samples a fraction of this distribution. For example, suppose we excite GaAs with a 2 eV photon at 300 K. Then the Raman scattering wave vector is $q = 6.74 \times 10^5$ cm^{-1}. On the other hand, the phonon wave vectors that can be emitted by the electron at the initial injection energy (of 0.52 eV, assuming a heavy hole mass of $0.54m_0$ and a conduction band effective mass of $0.067m_0$) can emit phonons in the range

$$\sqrt{k^2 + \frac{2m_e\omega_{LO}}{\hbar}} - k < q < \sqrt{k^2 + \frac{2m_e\omega_{LO}}{\hbar}} + k \tag{6.92}$$

or

$$3.24 \times 10^5 < q < 1.94 \times 10^6 \text{ cm}^{-1}. \tag{6.93}$$

Hence, the Raman signal senses a phonon population on the lower edge of that actually created by the energetic carriers. However, as the carriers cool, the entire generated spectrum shifts to lower wave vectors as the average energy (and the average k) decreases. In Figure 6.7, we illustrate a pictorial representation of the phonon distribution versus both q and the time

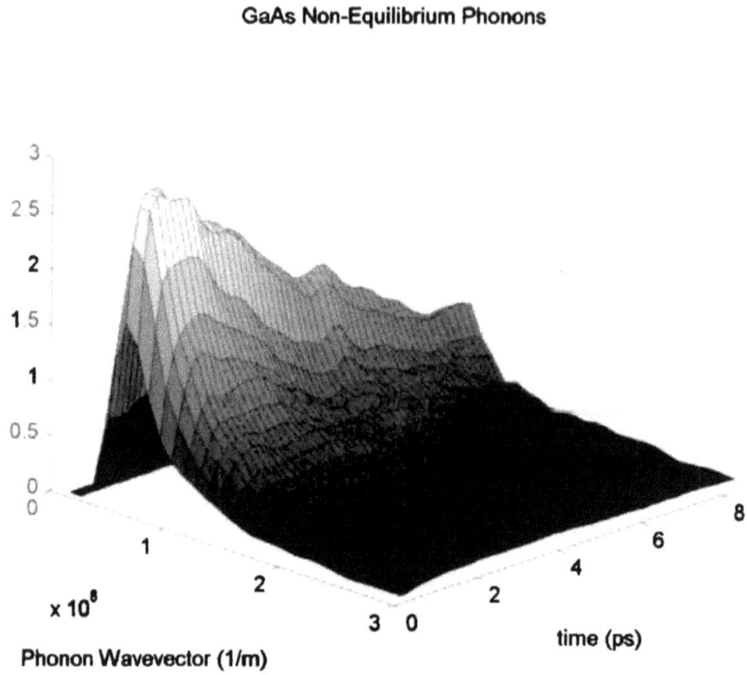

Figure 6.7 The non-equilibrium phonon distribution in GaAs subjected to an intense laser pulse. The details are discussed in the text, but it can be seen that the distribution rises and then decays. The curve is normalized to the equilibrium phonon population for the LO modes. (Calculations courtesy of L. Shifren, Arizona State University.)

after the laser pulse for a 1.9 eV photon (0.1 ps duration) incident on GaAs at 300 K. It is assumed that the carrier density produced from the laser is 5×10^{17} cm^{-3}, and a phonon lifetime of 3 ps is used. From this, one can see how the distribution first builds up and then decays as the phonons recombine. The peak of the phonon distribution occurs at 6.5×10^5 cm^{-1}, while the Raman signal probes a wave vector of 6.4×10^5 cm^{-1}, which in this case fairly well lines up with the non-equilibrium distribution produced by the electrons.

The lifetime of the excess polar optical, or even the non-polar optical, phonons is related to the rate at which the phonon can decay into two acoustic modes as well as the rate at which it is generated by the inverse process. Thus, a continuity equation for the optical phonons can be written

$$\frac{dN(q)}{dt} = G - \Gamma, \tag{6.94}$$

where Γ is a relaxation rate (calculated below) and G is the generation rate. This latter quantity includes the excitation by three phonon processes as well as the excitation by the relaxing electrons. For the moment, we ignore this latter process, and consider only the intrinsic phonon processes. The matrix elements for G are the same as those for the relaxation process, except that the relaxation process varies as $N(q)[N_{LA} + 1][N_{LO-LA} + 1]$, assuming that the zone edge phonons involved are from the LA modes, and the generation process varies as the product $[N(q) + 1]N_{LA}N_{LO-LA}$. Hence, we can write (6.94) as

$$\frac{dN(q)}{dt} = \hat{\Gamma}[(N(q)+1)N_{LA}N_{LO-LA} - N(q)(N_{LA}+1)(N_{LO-LA}+1)]$$

$$= -\hat{\Gamma}N(q)\left[1 + N_{LA} + N_{LO-LA} - \frac{N_{LA}(N_{LA}+1)}{N_{LA} - N(q)}\right]. \tag{6.95}$$

Here, the subscript refers to the particular phonon energy that must appear in the Bose–Einstein distribution. In equilibrium, the term in the square brackets vanishes. Thus, expanding $N(q)$ as $N_{LO} + \delta N$, and assuming that the other two distributions are in equilibrium, this last expression becomes

$$\frac{d\delta N}{dt} = -\hat{\Gamma}\delta N[1 + N_{LA} + N_{LO-LA}] \equiv -\frac{\delta N}{\tau_{LO}}. \tag{6.96}$$

We now turn to a calculation of this quantity.

6.5.2.1 Phonon lifetime

Optical phonons do not readily migrate to the surface, as their non-dispersive nature near the zone center gives them a very low group velocity. Rather, their energy is dissipated through the anharmonic term of the lattice potential. Normally, the phonons are found from the quadratic terms in the potential (Ferry, 1991), but the anharmonic terms are the cubic terms. In the Fourier representation, the cubic terms may be written as

$$H_3 = \frac{1}{6\sqrt{N_a}} \sum_{qq'q''} \mathbf{u}_q \cdot (\mathbf{u}_{q'} \cdot \hat{C}_{qq'q''} \cdot \mathbf{u}_{q''})\delta(\mathbf{q} + \mathbf{q}' + \mathbf{q}''), \tag{6.97}$$

where $\hat{C}$ is a third rank tensor elastic "constant." In general, the phonon of mode q decays into two phonons of modes $\mathbf{q}'$ and $\mathbf{q}''$. Thus, the wave amplitudes u involve the creation (or annihilation) operator for each mode, and the general matrix element for this interaction can be written, in analogy with the scattering processes developed in Chapters 3 and 4, as

$$|M(H_3)|^2 = \frac{\hbar^3}{8\rho_a V\bar{M}^2} \sum_{q'q''} \frac{C^2_{qq'q''}}{\omega_q \omega_{q'} \omega_{q''}} N_q(N_{q'}+1)$$

$$\times (N_{q''}+1)\delta(\mathbf{q}+\mathbf{q}'+\mathbf{q}'')\delta(-\omega_q+\omega_{q'}+\omega_{q''}). \qquad (6.98)$$

The energy conserving δ-function has been indicated here, although it is not strictly part of the matrix element, but is part of the overall Fermi golden rule. For most of the expressions of interest, the wave vectors belong to different modes of the crystal lattice vibrations. For our purposes, **q** usually belongs to the LO modes at the zone center, and **q**′ and **q**″ belong to the LA branch, or LA + LO branches near the zone edge. In most cases, however, the LO zone center mode cannot decay into two TA modes, as the energies of the latter are often too low, but a LA + TA combination is possible provided the propagation directions are not parallel so that the momentum conservation condition can be met.

The transition rate is then given by the full form of the Fermi golden rule, which we may write as

$$\Gamma_{LO} = \frac{2\pi}{\hbar} \sum_{q} |M(H_3)|^2$$

$$= \frac{\pi\hbar^2}{4\rho_a V\bar{M}^2} \sum_{qq'q''} \frac{C^2_{qq'q''}}{\omega_q \omega_{q'} \omega_{q''}} N_q(N_{q'}+1)$$

$$\times (N_{q''}+1)\delta(\mathbf{q}+\mathbf{q}'+\mathbf{q}'')\delta(-\omega_q+\omega_{q'}+\omega_{q''})$$

$$= \frac{\pi\hbar^2}{4\rho_a V\bar{M}^2} \frac{C^2_{qq'q''}}{\omega_{LO}\omega_{LA}(\omega_{LO}-\omega_{LA})}$$

$$\times N_{LO}(N_{LA}+1)(N_{LO-LA}+1) \sum_{q'} \delta(-\omega_q+\omega_{q'}+\omega_{q+q'}). \qquad (6.99)$$

The last summation, of course, is a three-dimensional summation over momentum space, which can be written as

$$\frac{Vq^2_{LA}}{\pi\hbar\omega_{LA}},$$

so that the final form of the relaxation rate is given as

$$\Gamma_{LO} = \frac{\pi\hbar}{4\rho_a \bar{M}^2} \frac{C^2_{qq'q''}q^2_{LA}}{\omega_{LO}\omega^2_{LA}(\omega_{LO}-\omega_{LA})} N_{LO}(N_{LA}+1)(N_{LO-LA}+1). \qquad (6.100)$$

Here, the quantity M is an average mass of the atoms in the crystal, and it has been assumed that the strain coefficient C is an average value as well.

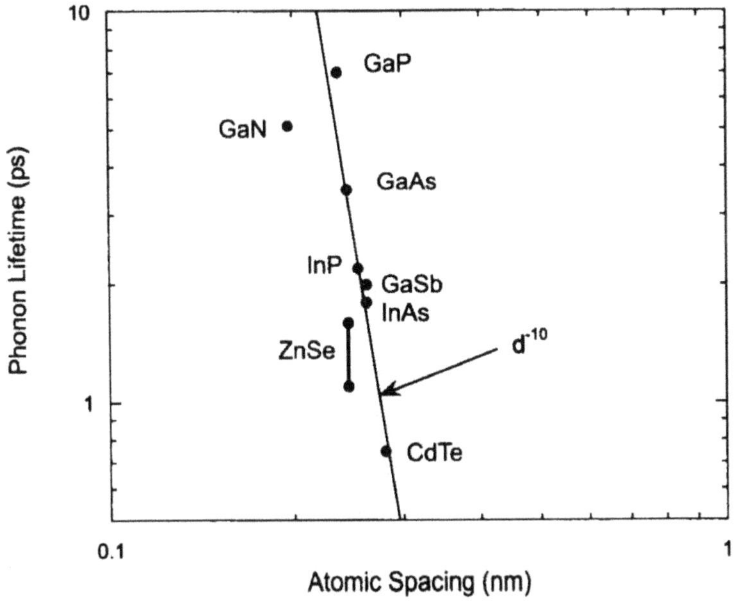

Figure 6.8 The experimentally measured lifetime for the LO phonons in a variety of tetrahedrally coordinated semiconductors. The d^{-10} scaling is expected on theoretical grounds.

Typical values of this latter quantity are 5×10^{11} N/m^2, but vary by an order of magnitude around this value for different interactions and materials. Comparing to (6.96), we can write the LO phonon lifetime as

$$\frac{1}{\tau_{LO}} = \frac{\pi\hbar}{4\rho_a \bar{M}^2} \frac{\tilde{C}^2_{qq'q''} q^2_{LA}}{\omega_{LO}\omega^2_{LA}(\omega_{LO} - \omega_{LA})}[1 + N_{LA} + N_{LO-LA}]. \tag{6.101}$$

In general, the strain constant C is thought to scale as the stiffness constant, or with the bulk modulus (Weinrich, 1965). The latter is asserted to scale as d^{-5} (Harrison, 1980), where d is the inter-atomic spacing of the lattice. In Figure 6.8, we plot the measured phonon lifetimes for a number of semiconductor crystals at 300 K, and it can be seen that the scaling as $C^2 \sim d^{-10}$ is fit rather well for these materials.

6.5.2.2 Treating non-equilibrium phonons in Monte Carlo

Under the conditions in which the non-equilibrium phonons can build up, one needs to take this effect into account in simulations of the transport and dynamics of these photo-generated carriers. For this, one needs to incorporate the time-dependent phonon population within the scattering processes that are encountered by the hot carriers, in particular the polar LO mode scattering. Generally, this can be done relatively easily if one ignores multiple phonon scattering and treats the phonon Boltzmann equation (6.96) within the phonon relaxation time approximation (e.g., assumes a relatively constant phonon lifetime). There are three parts to this process. First, one has to account for a temporal change in the scattering rate of electrons by the polar modes, due to evolution of the phonon population. This can be achieved using a secondary self-scattering process, as outlined in Section 4.8.4. The initial scattering rate $S_{LO}(\mathbf{k})$, defined by (4.59) is evaluated with a *fictitious*, large value for N_q. This value is arbitrarily chosen to be larger than the expected maximum value of the actual time evolving phonon distribution (Joshi *et al.*, 1990). Then, the actual $N(q,t)$ is evaluated during the polar mode scattering event and compared with the assumed maximum value. Using an additional random number r, the scattering process is accepted if

$$r < N(q,t)/N_q. \tag{6.102}$$

If this relation is not satisfied, then the collision is discarded and the particle is treated as if it had undergone another self-scattering process.

The need to have the time evolving $N(q,t)$ brings us to the second requirement. We must have a method for following the time evolution of the polar LO phonon distribution during the Monte Carlo process. To achieve this, we create a distribution function grid in momentum space, and follow the evolution of the phonon distribution on this grid (Lugli *et al.*, 1987). That is, we maintain a histogram defined on a grid in q-space. When an LO phonon is emitted by an electron, the phonon momentum is determined and the value in the appropriate cell in q-space is incremented by 1. When a phonon is absorbed by the electron, the phonon momentum is determined and the value in the appropriate cell in q-space is decreased by 1. This requires knowing just how many phonons should be in each cell in the equilibrium situation. The density of states in q-space is simply

$$\frac{V}{8\pi^3}. \tag{6.103}$$

There is no factor of "2" here, as the phonons are not limited by the Pauli exclusion principle and the number in each cell is given by N_q. The volume V appropriate for this equation is determined from the simulated carrier density n and the number of electrons in the Monte Carlo simulation N. These determine the simulated volume as $V = N/n$. The cell in q-space is determined by the grid size $\Omega_q = \Delta q_x \Delta q_y \Delta q_z$. With the equilibrium phonon distribution N_{q0}, we can determine the expected number of phonons in each cell from

$$n_{\Omega} = \frac{N}{8\pi^3 n} N_{q0}\Omega_q. \tag{6.104}$$

With a fixed number of electrons in the ensemble of the Monte Carlo, there is a tradeoff to be made in selecting Ω_q and the other parameters. If the excess phonon population is expected to be only a few percent of the background, then a sufficiently large Ω_q needs to be used that a change of ± 1 does not constitute a larger variation than the few percent being investigated. This has to do mainly with achieving good statistics in the Monte Carlo simulation.

The third important point to be addressed is that of allowing the phonon distribution to relax back to the equilibrium value N_{q0}. This is done using (6.96). The ensemble Monte Carlo technique has a natural second time scale whereby the simulation is stopped and all ensemble averages are updated (Section 4.8.3). At these times, the phonon population in each cell of the grid is relaxed following the time discretization of (6.96). During the time step Δt on this second time scale, all emission and absorption processes (by the electrons) are monitored for each cell and this gives a generation rate $g(q) = \Gamma_e(q) - \Gamma_a(q)$ for each cell. This is then used with the LO phonon lifetime τ_{LO} to update the phonon distribution as

$$N(q,t + \Delta t) = N(q,t) + g(q)\Delta t - \frac{\Delta t}{\tau_{LO}}[N(q,t) - N_{q0}]. \tag{6.105}$$

The second term on the right is the net increase in population in the particular cell, while the last term is the relaxation to equilibrium.

6.5.3 *Single-particle Raman scattering*

The laser can also be scattered by the free carriers that are actually produced by the laser itself. This process is termed single-particle scattering, which has been long known as an effective probe of the velocity distribution of electrons moving under the influence of an electric field (Mooradian and McWhorter, 1970). Generally, the luminescence must be separated from the total spectrum to achieve the Raman scattering part. By using a polarization of the incident and scattered beams that scatters only from the single-particle excitations associated with spin density fluctuations (Klein, 1983; Abstreiter *et al.*, 1983), one can identify and measure the distribution function. The single-particle scattering (SPS) is inversely proportional to the effective mass, so that the experiment predominantly samples carriers in the Γ valley, even though holes are produced by the laser excitation and carriers can scatter to satellite valleys. The problem is the maintenance of a stable dc field on the sample during the experiment, and this is achieved by placing the active region in a heterostructure *p–i–n* diode. Here, for example, the p- and n-regions are AlGaAs, which pass the incident photons without absorption, and the intrinsic active region is GaAs. The p–i–n builds in an electric field, which can be further modified by applied bias.

The measured SPS spectrum can be converted to an actual carrier distribution function along the direction of light propagation. For light scattering from electrons in semiconductors, momentum and energy conservation require that

$$\hbar\omega = \hbar\mathbf{q}\cdot\mathbf{v} + \frac{\hbar^2 q^2}{2m^*}, \tag{6.106}$$

where $\hbar\omega$ is the energy shift of the scattered photons, and $\mathbf{v}$ is the velocity of the carriers along the propagation direction. A simple expression for the SPS cross-section is given by (Grann *et al.*, 1996)

$$\frac{d^2\sigma}{d\omega d\Omega} \sim \frac{1}{(m^*)^2}\int d^3\mathbf{k} f(\mathbf{k})\delta\left(\omega - \frac{E_{k+q} - E_k}{\hbar}\right). \tag{6.107}$$

This says that the Raman signal at a given frequency shift ω, and therefore at a given velocity $\mathbf{v}$, is proportional to the number of electrons at this velocity. Thus, by measuring the spectrum of the single-particle scattering, one can estimate quite effectively the distribution function.

The distribution function is one of the things that can be calculated quite effectively with the ensemble Monte Carlo technique. In fact, while computing the transport for the p–i–n structure, one can buildup the Raman scattering signal as a function of the time. Carriers which are scattered from the intrinsic GaAs quantum well into the GaAlAs do not give an effective Raman signal. Similarly, particles excited into higher satellite valleys of the conduction band do not contribute. In Figure 6.9, we plot the simulated distribution function, along the field direction, and the SPS Raman scattering for light propagation in this same direction. It may be seen that there is a reasonably good agreement in the two sets of data, which provides good support for this as a viable tool for studying carrier distribution functions.

6.5.4 Quantum well and quantum wire properties

In quantum confined systems such as quantum wells, quantum wires and/or quantum dots, the relaxation of carriers from high energy states to the ground state must proceed through inter-subband processes as well as the intra-subband processes. In particular, the inter-subband processes are important in many devices, including the quantum well lasers and detectors. Carriers in quantum well lasers are injected into the active (recombination) region well above the band edge from the confinement layers. The inter-subband

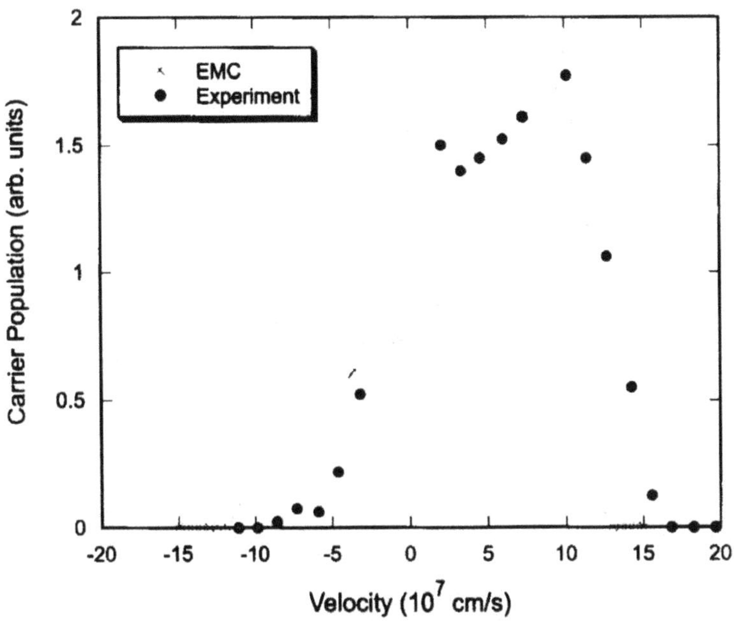

Figure 6.9 Comparison of the ensemble Monte Carlo carrier distribution and the Raman single-particle scattering for a 1.0 μm intrinsic GaAs layer between *p*- and *n*-layers of AlGaAs at 300 K. The Raman signal measures the carrier distribution along the photon propagation direction. The net field in the structure is 25 kV/cm. (After Grann *et al.*, 1995, with permission.)

carrier relaxation describes just how rapidly the carriers get to the bottom of the quantum well and laser action can subsequently proceed. In these quantum-confined systems, there is a band-gap discontinuity between the quantum well material, such as GaAs, and the confining region, such as AlGaAs. The motion of carriers is constrained to lie in the plane of the quantum well, or along the axis of the quantum wire (see, for example, Section 3.3.4 for the changes in impurity scattering in a confined system). In addition, when the two materials (well and confining material) have different elastic properties and/or dielectric functions, we expect the phonon spectra to become much more complicated with the presence of interface modes. These latter arise from the elastic discontinuities, or from the discontinuity in the dielectric function (Menendez, 1989; Kim and Stroscio, 1990; Rücker *et al.*, 1992). For well widths of 10 nm (in GaAs) and below, effects due to surface modes become quite pronounced, whereas for well widths of 20 nm and above, the polar optical interaction approaches that due to bulk optical phonons.

In the previous section, the importance of non-equilibrium phonons was discussed. In the quantum-confined systems, this effect becomes much more

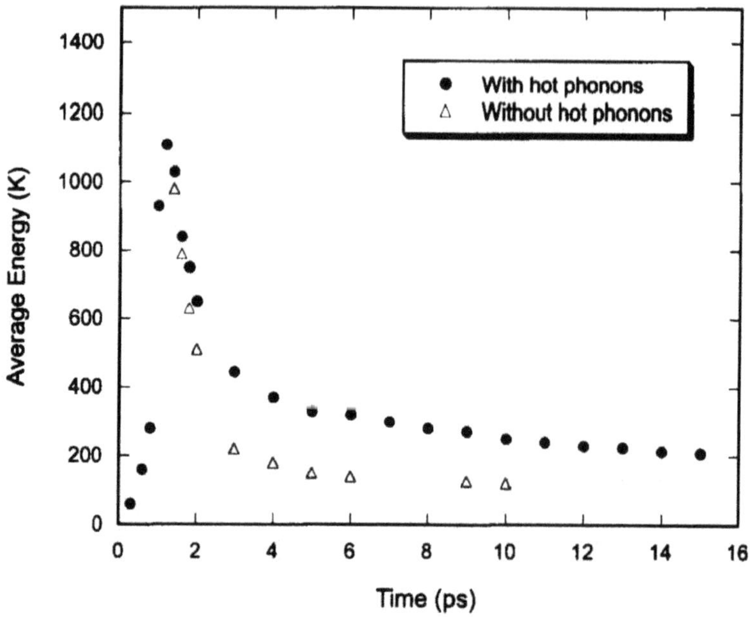

Figure 6.10 Average energy, in temperature K, for photo-excited carriers in a 15 nm GaAs quantum well. The injected density, and the background density, are both 2.5×10^{11} cm^{-2}. The carriers are injected with an excess energy of 0.25 eV. The non-equilibrium phonons are important in creating a "bottleneck" in the relaxation, and the long cooling time is a reflection of the lifetime of the phonons. (After Lugli and Goodnick (1987), with permission.)

important, as the local carrier density is higher, and the phonons are not as effective in decaying into the bulk modes. In these systems, the buildup of the non-equilibrium phonons creates a "bottleneck" in which the phonon population increases to a level such that the rates of emission and absorption of the phonons are comparable, leading to a decrease in the overall cooling process for the carriers. From (4.59), the rates for emission and absorption become comparable when

$$N_q \sim N_q + 1, \quad N_q \gg 1. \tag{6.108}$$

For shorter times, the hot carriers relax by the emission of a cloud of phonons. But, when the phonon population builds to a level satisfying (6.108), re-absorption is just as likely as emission. The carrier density can only cool further as the non-equilibrium phonon distribution relaxes to thermal equilibrium (Kash *et al.*, 1985). An example of the simulated effect of non-equilibrium phonons on the photo-excited carrier relaxation in a GaAs/AlGaAs quantum well at 100 K is shown in Figure 6.10. Here, the Monte

Carlo simulation is for a 15 nm GaAs quantum well with a background doping of 2.5×10^{11} cm^{-2}, and an equal number of carriers are photo-excited by a 0.6 ps laser pulse, centered at 1 ps after the beginning of the simulation (Lugli and Goodnick, 1987). The photo-excited carriers are injected into the quantum well at an excess energy of 0.25 eV. The average temperature of the distribution is plotted as a function of time, both with and without the consideration of the hot phonons. The bottleneck sets in already at 2 ps and the remaining decay of the temperature is that of the phonon lifetime of 7 ps (at 100 K). If the hot phonons are ignored, the carriers rapidly relax to their thermal equilibrium at 100 K. This result compares well with the data of Ryan *et al.* (1986).

There are additional effects that can come into play in quantum-confined systems. For example, the excess energy of the photo-excited carrier will depend upon the subband to which it is excited. Hence, its decay will be separate for each subband at first. But, inter-subband scattering will bring the various distributions into equilibrium with each other, so that for longer times, a single hot distribution exists for all subbands (Dür *et al.*, 1996). This can also lead to a bottleneck-like effect. For narrow wells, the subbands are widely separated in energy, and the carrier relaxation can be quite short, on the order of a picosecond or less for upper subbands (Julien *et al.*, 1988; Tatham *et al.*, 1989). In wide wells, the cooling is slowed by the onset of a single thermal hot carrier distribution, which must relax on its own, and the lifetime of carriers in the upper subbands can be 40 picoseconds and more (Oberli *et al.*, 1987; Levenson *et al.*, 1990; Murdin *et al.*, 1994). A thorough discussion of the range of these experiments is beyond the scope of the current work, but it is mentioned for its interesting effect.

PROBLEMS

1. Using the temperature-dependent mobility found for silicon, plot both the mobility and the diffusion constant as a function of temperature in the range 50 to 300 K for both electrons and holes.
2. Calculate and plot the direct optical absorption process in GaAs, including both the allowed and forbidden transitions for photons in the energy range 1.3 to 1.8 eV at 77 K. You may have to include band non-degeneracy to excite both of these transitions properly.
3. For the optical absorption curve for silicon, fit the energy variation with the assumed variations for phonon-assisted indirect absorption. Is it possible to determine the ratio of emission to absorption processes from this fit?
4. In a particular semiconductor, it is found that the recombination time varies as $\exp(-E_i/k_BT)$. Discuss how such an exponential decay of the

recombination time with temperature can be obtained and how it is related to the trap energy.

5. A particular semiconductor material has a background doping of $N_a = 10^{13}$ cm^{-3} and an electron mobility of 3800 cm^2/Vs, with $b = 2.5$. The semiconductor is illuminated uniformly over its surface with a 1 mW/cm^2 laser beam. If the semiconductor material is 0.1 cm thick, $\alpha = 10^4$ cm^{-1}, and $\tau_p = 1$ μs, calculate the ratio of photocurrent density to dark-current density for an applied field of 5 V/cm. Assume the dielectric constant of the semiconductor is 16 and the quantum efficiency is 10 percent.
6. For the same parameters of Problem 5, compute the transverse photovoltage.
7. For the same parameters given in Problem 5, compute the short circuit current of the photoelectromagnetic effect.

REFERENCES

Abstreiter, G., Cardona, M., and Pinczuk, A., 1983, in *Light Scattering in Solids IV*, Cardona, M., and Güntherodt, H. J., Eds. (New York: Springer) 5.

Burstein, E., 1954, *Phys. Rev.*, 93, 632.

Dür, M., Goodnick, S. M., and Lugli, P., 1996, *Phys. Rev. B*, 54, 17794.

Ferry, D. K., 1991, *Semiconductors* (New York: Macmillan).

Ferry, D. K., 1995, *Quantum Mechanics* (Bristol: Institute of Physics Publishing).

Grann, E. D., Tsen, K. T., Sankey, O. F., Ferry, D. K., Salvador, A., Botcharev, A., and Morkoç, H., 1995, *Appl. Phys. Lett.*, 67, 1760.

Grann, E. D., Tsen, K. T., Ferry, D. K., Salvador, A., Botcharev, A., and Morkoç, H., 1996, *Phys. Rev. B*, 53, 9838.

Hall, R. N., 1951, *Phys. Rev.*, 83, 228.

Hall, R. N., 1952, *Phys. Rev.*, 87, 387.

Harrison, W. A., 1980, *Electronic Structure and the Properties of Solids* (San Francisco: Freeman).

Joshi, R. P., Grondin, R. O., and Ferry, D. K., 1990, *Phys. Rev. B*, 42, 5685.

Julien, F.H., Lourtioz, J.-M., Herschkorn, N., Delacourt, D., Pocholle, J. P., Papuchon, M., Planel, R., and Le Roux, G., 1988, *Appl. Phys. Lett.*, 53, 116.

Kash, J. A., Tsang, J. C., and Hvam, J. M., 1985, *Phys. Rev. Lett.*, 54, 2151.

Kim, K. W., and Stroscio, M. A., 1990, *J. Appl. Phys.*, 68, 6289.

Klein, M. V., 1983, in *Light Scattering in Solids I*, Cardona, M., Ed. (New York: Springer) 151.

Lugli, P. and Goodnick, S. M., 1987, *Phys. Rev. Lett.*, 59, 716.

Lugli, P., Jacoboni, C., Reggiani, L., and Kocevar, P., 1987, *Appl. Phys. Lett.*, 50, 1251.

Levenson, J. A., Dolique, G., Oudar, J. L., and Abram, I., 1990, *Phys. Rev. B*, 41, 3688.

Menendez, J., 1989, *J. Lumin.*, 44, 285.

Mooradian, A., and McWhorter, A. L., 1970, *in Proc. 10th Intern. Conf. On Physics of Semiconductors*, Keller, S. P., Hansel, J. C., and Stern, F., Eds. (Oak Ridge: U.S. Atomic Energy Agency) 380.

Murdin, B. N., Knippels, G. M. H., van der Meer, A. F. G., Pidgeon, C. R., Langerak, C. J. G. M., Helm, M., Heiss, W., Unterrainer, K., Gornik, E., Geerinck, K. K., Hovenier, N. J., and Wenckebach, W. T., 1994, *Semicond. Sci. Technol.*, 9, 1554.

Oberli, D. Y., Wake, D. R., Klein, M. V., Klem, J., Henderson, T., and Morkoç, H., 1987, *Phys. Rev. Lett.*, 59, 696.

Osman, M. A., and Ferry, D. K., 1987, *Phys. Rev. B*, 36, 6018.

Rücker, H., Molinari, E., and Lugli, P., 1992, *Phys. Rev. B*, 45, 6746.

Ryan, J. F., Taylor, R. A., Turberfield, A. J., and Worlock, J. M., 1986, *Surf. Sci.*, 170, 511.

Schiff, L. I., 1955, *Quantum Mechanics*, 2nd Ed. (New York: McGraw-Hill).

Shah, J., Ed., 1992, *Hot Carriers in Semiconductor Nanostructures* (San Diego: Academic Press).

Shockley, W. and Read, W. T., 1952, *Phys. Rev.*, 87, 835.

Smith, R. A., 1969, *Wave Mechanics of Crystalline Solids* (London: Chapman and Hall).

Tatham, M. C., Ryan, J. F., and Foxon, C. T., 1989, *Phys. Rev. Lett.*, 63, 1637.

Weinrich, G., 1965, *Solids: Elementary Theory for Advanced Students* (New York: Wiley).

Chapter 7

The electron–electron interaction

In recent years, as device sizes have become smaller and carrier densities have become larger, the role of the electron–electron interaction has become very important in transport within semiconductor devices. It is not that this effect has automatically become dominant. In fact, the role of the electron–electron interaction was introduced in the treatment of the drifted Maxwellian approximation in Chapter 4 and the screening of the polar optical phonon was also discussed. However, for the time and space scales that exist in the sub-micron and ultra-submicron devices, it is no longer adequate simply to adopt a screening length (or time). The details of the electron–electron (or hole–hole, as the case may be) interaction become relevant to an overall understanding of the device behavior. The main difficulty in dealing with the electron–electron interaction lies in the nonlinear behavior of the interaction potential and in the long range of this Coulomb potential. In the past, several approaches to the study of electron–electron scattering have been presented. Almost all of these were based upon the assumption that the interaction potential is screened at some characteristic length, usually the Debye length, and that the distribution function is a Maxwellian (inherent in the use of the Debye length). Yet with the high densities which occur in modern devices, the Debye length is often smaller than the inter-electronic distance. The effects are complicated by the fact that in polar materials, the rather high free-carrier plasma frequency that accompanies the high densities couples strongly to the polar-optical phonon vibrations to produce coupled modes of electron–plasmon–phonon interactions. To treat all of these effects properly, it is necessary to divorce ourselves from the simple screening approach and treat the full frequency- and wavelength-dependent dielectric function.

The momentum- and frequency-dependent dielectric function has been investigated quite extensively for a variety of physical systems. The main results are that screening is a dynamic process and the full $(\mathbf{q},\omega)$ variation of the dielectric function must be included. In this chapter, an approach will be presented that is based on one form of the dielectric function, the Lindhard function, and the role of electron–electron and electron–plasmon scattering will be discussed. Several calculations of the effect of this scattering will be presented to illustrate the importance of treating the interaction properly. Before proceeding though, as the development of the Lindhard screening function is complex, it is worthwhile first to review how the simple screening approach is developed.

If a charge is placed in a system, it induces a local potential that causes other charges in the system to move. These charges move in such a manner as to counteract this potential, and the net result of the movement of the entire ensemble of charges is a potential that does not behave as the simple Coulomb potential of the initial charge. In essence, this is a self-consistent process, in which the charge produces a potential, which modifies the total charge, which modifies the total potential, and so on. In the end, everything must be put together to provide a single, self-consistent description that describes both the total charge distribution and the total potential. The modification of the Coulomb potential of the single charge is said to be the *screening aspect.* In a semi-classical semiconductor system, the variation of the charge density is directly related to the local potential. The density is related to the Fermi energy, in a non-degenerate semiconductor, as

$$n = N_c \exp\left(\frac{E_F - E_c}{k_B T}\right), \tag{7.1}$$

where E_c is the conduction band edge and

$$N_c = \frac{1}{4}\left(\frac{2m^* k_B T}{\pi\hbar^2}\right)^{\frac{3}{2}} \tag{7.2}$$

is the *effective density of states* in the conduction band. Any local potential can be written in terms of the edge of the conduction band as $-e\phi = E_c - E_{c0}$. The local density may be written as

$$n = n_0 \exp\left(\frac{e\phi}{k_B T}\right). \tag{7.3}$$

Here, n_0 is associated with the unperturbed conduction band edge E_{c0}. The total charge in the system is then the initial charge, located at $r = 0$ for convenience, and the local redistributed charge that moves in response to the initial charge. The approach that will be followed (and which is standard) is to recognize that the initial charge provides a normal Coulomb potential with $\phi = -e/4\pi\varepsilon r$. The screening charge modifies this to have $1/r \sim g(r)/r$, and then it is necessary to find the equation for $g(r)$. If an attempt to use Gauss's law were to be made, an integral equation for $\phi(r)$, or $g(r)$, results. The foregoing procedure is necessary since the initial point charge is normally introduced to Poisson's equation through boundary conditions. Poisson's equation for the charge variation about the initial charge (for $r > 0$) is given by

$$\nabla^2\phi = \frac{e(n - n_0)}{\varepsilon} = \frac{en_0}{\varepsilon}(e^{e\phi/k_BT} - 1) \approx \frac{n_0e^2}{\varepsilon k_BT}\phi \tag{7.4}$$

in the linearized approximation. If the assumed form for $\phi(r)$ *is* now substituted into (7.4), the resulting equation for $g(r)$ *is* found to be, in spherical coordinates,

$$\frac{\partial^2 g(r)}{\partial r^2} - q_D^2 g(r) = 0, \quad q_D^2 = \frac{n_0e^2}{\varepsilon k_BT}. \tag{7.5}$$

Hence, the screening function $g(r) \sim \exp(-q_Dr)$, and the total potential is just

$$\phi(r) = -\frac{e}{4\pi\varepsilon r}e^{-q_Dr}. \tag{7.6}$$

In the following, this approach will be used to compute the full frequency- and wave-vector-dependent screening function, but the linear response approach still will be used. It will be shown how the more exact treatment can still be reduced to the simple form obtained in (7.6), but also how more complicated forms may also arise. We will then examine how the coupled modes of electrons and phonons, as well as the plasmon modes arise in the system. Finally, an extensive treatment of the scattering by these coupled mode and single-particle interactions is given.

7.1 The dielectric function

The dielectric function describes how the various charged species move in response to external (or internal) potentials and provides for the screening of those potentials. In compound semiconductors such as GaAs and AlGaAs, these charged species are the electrons and holes and the atoms themselves. Since there is an ionic contribution to the bonding of these atoms, this also contributes to the total polarization and dielectric properties. Thus the dielectric function may generally be written as

$$\varepsilon(\mathbf{q},\omega) = \varepsilon_0 + \delta\varepsilon_L + \delta\varepsilon_e, \tag{7.7}$$

where ε_0 is the permittivity of free space. The second term is the lattice contribution and the last term is the electronic contribution to the dielectric function, which here includes both free carriers and the entire set of valence electrons. In this equation, the dielectric function is dependent on the wave vector q, as its variation provides major effects in the process of screening. Equation (7.7) must show the limiting behavior that, at sufficiently high frequency, neither the electrons nor the lattice can follow any time-varying perturbation, and only high-frequency permittivity is important. The latter quantity differs from the free-space value by the contribution of the valence electrons (which will be determined below), and is usually termed the "optical dielectric constant" (times the free-space permittivity ε_0). Thus, it is to be expected that the electronic term will be decomposed into two parts – contributing the high-frequency permittivity and the free-carrier screening function.

In the absence of free carriers, the permittivity must approach its static value at low frequency, which includes the entire lattice and valence electronic contributions, while for high frequencies the lattice contribution vanishes, and at still higher frequencies the electronic contribution also vanishes. In purely covalent materials such as Si and Ge, there is no polar contribution to the optical phonon vibrations, and therefore no static lattice contribution to the dielectric function. The difference between the optical dielectric constant and the static value is just the contribution of the polar modes of the lattice vibrations, which may be written as

$$\delta\varepsilon_L = [\varepsilon(0) - \varepsilon(\infty)]\frac{\omega_{TO}^2}{\omega_{TO}^2 - \omega^2}, \tag{7.8}$$

where

$$\frac{\omega_{LO}^2}{\omega_{TO}^2} = \frac{\varepsilon(0)}{\varepsilon(\infty)} \tag{7.9}$$

is the Lyddane–Sachs–Teller relation connecting the longitudinal and transverse optical phonon frequencies in polar material. Thus, at low frequencies $\varepsilon(\omega) \to \varepsilon(0)$, the static value; for high frequencies the lattice contribution vanishes as desired.

Generally, the electronic contribution is calculated in the absence of the lattice contribution. For this, (7.8) is set to zero, but the dielectric permittivity that enters into (7.7) is just the free-space permittivity ε_0. To incorporate all of the electronic contributions, a form must now be derived to describe the dielectric contribution of both the free carriers and the valence band electrons.

7.1.1 The Lindhard potential

The electronic contribution within the linear response approximation essentially means calculating only to the lowest order the screening effects of the electrons upon an external perturbing potential. To do this, a form for the perturbing potential δU will be assumed, which here is a sinusoidally varying potential that has been slowly turned on from $t = -\infty$. The fluctuation in density caused by this perturbing potential is calculated, and the resulting self-consistent potential fluctuation Φ caused by the small fluctuation in density is determined. Finally, the total perturbing potential is related to the applied potential and the self-consistent potential Φ, so that the screening effect can be identified, and the dielectric function determined (Ziman, 1964). This in done in Fourier transform space, so that the response is that for a particular frequency and wave vector.

The perturbing potential at frequency ω and momentum q may be expressed in terms of a time-varying function as

$$\delta U(\mathbf{q},\omega) = U_0 e^{i(\mathbf{q}\cdot\mathbf{r}-\omega t)+\alpha t}, \tag{7.10}$$

where α is a damping constant to account for the dissipative part of the carrier–carrier interaction. In more usual parlance, α is just a small parameter that assures that the potential (7.10) vanishes as $t \to -\infty$.

The approach that will be followed is based on time-dependent perturbation theory. The perturbation couples an initial (plane wave) state of wave vector **k** to one of wave vector **k** ± **q**. In performing the Fourier transform, the *spatial Fourier component* at wave vector q picks out essentially the single element from the summation that has the desired wave vector. All other components have a rapidly varying phase that averages to zero. The wave function for an electron state may then be expressed as

$$\Psi_k = |\mathbf{k}\rangle + b_{\mathbf{k}+\mathbf{q}}|\mathbf{k}+\mathbf{q}\rangle + b_{\mathbf{k}-\mathbf{q}}|\mathbf{k}-\mathbf{q}\rangle. \tag{7.11}$$

The second and third terms represent the deviation from the equilibrium state, and Dirac notation is used ($|\mathbf{k}\rangle = e^{i\mathbf{k}\cdot\mathbf{r}}$). First-order perturbation theory leads us to express the coefficients as (we deal only with the second term for the moment)

$$\begin{aligned} b_{\mathbf{k}+\mathbf{q}} &= \frac{\langle \mathbf{k}+\mathbf{q}\,|\,e\delta U\,|\,\mathbf{k}\rangle}{E(\mathbf{k}+\mathbf{q}) - E(\mathbf{k}) + \hbar\omega - i\hbar\alpha} \\ &= \frac{eU_0 e^{-i\omega t+\alpha t}}{E(\mathbf{k}+\mathbf{q}) - E(\mathbf{k}) + \hbar\omega - i\hbar\alpha}. \end{aligned} \tag{7.12}$$

It will be noted that only a single state contributes to the particular term in the time-dependent perturbation of interest. The change in electron density that is produced by this perturbation is just

$$\delta\rho = -e\sum_{\mathbf{k}} f(\mathbf{k})(|\Psi_{\mathbf{k}}|^2 - 1), \tag{7.13}$$

where the squared magnitude of the wave function produces the probability that a state of wave vector k exists at point r. Weighting this with the summation over the occupation probability $f(\mathbf{k})$ produces $n(\mathbf{r})$, while the factor of unity produces n_0, and these two factors correspond to the terms in (7.4). After some simple algebraic manipulation, the charge density may be found to be (in the linear approximation)

$$\delta\rho = -e\sum_{\mathbf{k}} f(\mathbf{k})(b^*_{\mathbf{k}+\mathbf{q}}e^{-i\mathbf{q}\cdot\mathbf{r}} + b_{\mathbf{k}+\mathbf{q}}e^{i\mathbf{q}\cdot\mathbf{r}} + h.c.). \tag{7.14}$$

This result is termed linear because it has been assumed that $|b_{\mathbf{k}+\mathbf{q}}|^2 \ll 1$, but the perturbation expression already used made this assumption. The result for $\delta\rho$ contains only the terms in the perturbation, but these are the terms that cause the local density to deviate from its average, or background, value. Only the positive momentum change has been used so far, so that the contribution from the negative momentum change still must be added to (7.14) (and is represented by the $h.c.$ term). Incorporating this term expands (7.14) to

$$\delta\rho = -e\sum_{\mathbf{k}} f(\mathbf{k})\left[\frac{1}{E(\mathbf{k}+\mathbf{q}) - E(\mathbf{k}) + \hbar\omega - i\hbar\alpha} + \frac{1}{E(\mathbf{k}-\mathbf{q}) - E(\mathbf{k}) - \hbar\omega + i\hbar\alpha}\right]eU_0 + c.c. \tag{7.15}$$

With a change of momentum $\mathbf{k} - \mathbf{q} \to \mathbf{k}$ in the second term in the square brackets, we can rewrite this as

$$\delta\rho = -e^2U_0\sum_{\mathbf{k}}\left[\frac{f(\mathbf{k}) - f(\mathbf{k}+\mathbf{q})}{E(\mathbf{k}+\mathbf{q}) - E(\mathbf{k}) + \hbar\omega - i\hbar\alpha}\right] + c.c. \tag{7.16}$$

This fluctuation in charge density produces, in turn, a fluctuation in the potential itself. Poisson's equation for the potential is just

$$\nabla^2\Phi = -\frac{\delta\rho}{\varepsilon_0}, \quad \text{or} \quad \Phi(\mathbf{q},\omega) = \frac{\delta\rho}{q^2\varepsilon_0}e^{i\mathbf{q}\cdot\mathbf{r}} + c.c. \tag{7.17}$$

Hence, one obtains the resulting potential of the perturbation to be just

$$\Phi = -\frac{e^2 U_0}{q^2 \varepsilon_0} \sum_{\mathbf{k}} \left[\frac{f(\mathbf{k}) - f(\mathbf{k}+\mathbf{q})}{E(\mathbf{k}+\mathbf{q}) - E(\mathbf{k}) + \hbar\omega - i\hbar\alpha} \right]. \tag{7.18}$$

The total potential δU (which gives rise to U_0) is the one that produces the perturbation in the charge density. But this total potential is composed of the applied potential V plus the response term Φ, which is the self-consistent solution (7.18), and this allows us to solve for U_0 through

$$U_0 = V + \Phi, \quad \text{or} \quad U_0 = \frac{\varepsilon_0 V}{\varepsilon(\mathbf{q},\omega)}, \tag{7.19}$$

and the *dielectric function* is now

$$\varepsilon(\mathbf{q},\omega) = \varepsilon_0 + \frac{e^2}{q^2} \sum_{\mathbf{k}} \left[\frac{f(\mathbf{k}) - f(\mathbf{k}+\mathbf{q})}{E(\mathbf{k}+\mathbf{q}) - E(\mathbf{k}) + \hbar\omega - i\hbar\alpha} \right]. \tag{7.20}$$

This last expression, termed the *Lindhard* dielectric function, includes the free space permittivity ε_0 as the initial term. In the following sections, a number of important limits of this function will be examined.

7.1.2 The optical dielectric constant

The sum over the carrier distribution in (7.20) is over *all* electrons, free carriers as well as valence electrons. Before proceeding to treat the free carriers, it is desirable to separate out the portion of the summation representing the valence electrons. These excitations can only be observed with a very high frequency source in which electrons are excited across the band gap. Hence, these excitations are primarily those involving a shift of k by a reciprocal lattice vector. The two sums inherent in (7.20) are separated, and in the second we replace k + q by k, which essentially undoes the simplification achieved in going from (7.15) to (7.16). This gives

$$\begin{aligned}\varepsilon(\mathbf{q},\omega) &= \varepsilon_0 + \frac{e^2}{q^2} \sum_{\mathbf{k}} f(\mathbf{k}) \left[\frac{1}{E(\mathbf{k}+\mathbf{q}) - E(\mathbf{k}) + \hbar\omega - i\hbar\alpha} - \frac{1}{E(\mathbf{k}) - E(\mathbf{k}-\mathbf{q}) + \hbar\omega - i\hbar\alpha} \right] \\ &= \varepsilon_0 + \frac{e^2}{q^2} \sum_{\mathbf{k}} f(\mathbf{k}) \left[\frac{2E(\mathbf{k}) - E(\mathbf{k}-\mathbf{q}) - E(\mathbf{k}+\mathbf{q})}{[E(\mathbf{k}+\mathbf{q}) - E(\mathbf{k}) + \hbar\omega - i\hbar\alpha][E(\mathbf{k}) - E(\mathbf{k}-\mathbf{q}) + \hbar\omega - i\hbar\alpha]} \right].\end{aligned} \tag{7.21}$$

In this expression, interest is focused on the role of the full valence band, so that the k vector summation runs over the full band. But the only states available lie vertically above a particular value of k, and this is separated by the effective optical gap. To reach these, k must span into the second

Brillouin zone, which is shifted back in the $E(\mathbf{k} \pm \mathbf{q})$ by introducing a reciprocal lattice vector G. Then the denominator terms are essentially just an average $\pm E_G$. It will be assumed that $\alpha = 0$, and

$$E(\mathbf{k} \pm \mathbf{G} \pm \mathbf{q}) - E(\mathbf{k}) \equiv E_G \gg \hbar\omega, \tag{7.22}$$

where this gap is an *effective optical gap*. Further, to leading order,

$$\begin{aligned} numerator &= 2E(\mathbf{k}) - E(\mathbf{k}+\mathbf{q}) - E(\mathbf{k}-\mathbf{q}) \\ &\approx -q^2 \frac{\partial^2 E}{\partial k^2} + \ldots \approx -\frac{\hbar^2 q^2}{m_d}, \end{aligned} \tag{7.23}$$

where we have used a Taylor series expansion of the energy and introduced the density of states effective mass m_d. With these approximations, plus the assumption that the summation of the distribution function over momentum just gives the total density of valence electrons N, (7.21) now becomes

$$\begin{aligned} \varepsilon(\mathbf{q},\omega) &= \varepsilon_0 + \frac{e^2}{q^2}\frac{\hbar^2 q^2}{m_d}\sum_{\mathbf{k}} f(\mathbf{k})\frac{1}{E_G^2} = \varepsilon_0 + \frac{Ne^2\hbar^2}{m_d E_G^2} \\ &= \varepsilon_0\left[1 + \left(\frac{\hbar\omega_P}{E_G}\right)^2\right] \equiv \varepsilon(\infty). \end{aligned} \tag{7.24}$$

In this expression, we have introduced the *high-frequency permittivity* $\varepsilon(\infty)$. In the last line, the valence plasma frequency

$$\omega_P = \sqrt{\frac{Ne^2}{m_d \varepsilon_0}} \tag{7.25}$$

also has been introduced (note that we use a capital "P" as the subscript for this quantity). The form of (7.25) is that due to Penn (1962), and clearly shows that the optical dielectric function is related to the properties of excitations across the average energy gap between valence and conduction bands. As noted above, the energy gap is an average over the Brillouin zone, and is not just the fundamental band gap E_{gap}. Furthermore, the result must be modified by Fermi level effects and by overlap with d-bands (both of which modify the actual electron density contributing to the plasma frequency).

Using the foregoing considerations, the dielectric function may be written in a relatively complete form as

Table 7.1 Some electronic optical properties

	$\hbar\omega_p$ (eV)	$\varepsilon_r(\infty)$	$\varepsilon_r(0)$
C	31.2		
Si	16.6	11.9	11.9
Ge	15.6	15.9	15.9
GaAs	15.6	11.1	13.1
GaP	16.5	8.5	10.2
GaSb	13.9	14.4	15.7
InAs	14.2	12.3	14.5
InP	14.8	9.6	12.4
InSb	12.7	15.7	17.9
AlAs	15.5	8.2	10.06
HgTe	12.8	6.9	14.9
CdTe	12.7	7.2	10.2

$$\varepsilon(\mathbf{q},\omega) = \varepsilon(\infty) + [\varepsilon(0) - \varepsilon(\infty)]\frac{\omega_{TO}^2}{\omega_{TO}^2 - \omega^2} + \frac{e^2}{q^2}\sum_{\mathbf{k}}\left[\frac{f(\mathbf{k}) - f(\mathbf{k}+\mathbf{q})}{E(\mathbf{k}+\mathbf{q}) - E(\mathbf{k}) + \hbar\omega - i\hbar\alpha}\right]. \tag{7.26}$$

In this expression, we have added (7.8), which represents the lattice contribution to the dielectric function in polar materials. Of course, this second term vanishes in non-polar materials such as Si or Ge. In the last term, *the sum now runs only over the free carriers* – electrons or holes, since the valence contribution is included in $\varepsilon(\infty)$. In Table 7.1, the valence plasma frequency, and the relative dielectric constants for a number of semiconductors are listed.

7.1.3 The plasmon–pole approximation

A similar approximation can be used in the free carrier term when we assume that the frequency is large compared to the energy exchange in the denominator of the last term of (7.26). In the previous section, we found an effect due to the valence plasmons; here we will find similar plasma oscillations from the free carriers. Transport of carriers is affected significantly by the interaction between electrons and plasmons, and between the plasmons and the phonons. We will pursue an approach similar to that of the last section, concentrating on the free carrier plasmons, and this approximation has been termed the ***plasmon–pole approximation***. Here, we rewrite the last term of (7.26) exactly as in (7.21), so that the numerator can be written as (7.23). In the denominator, however, we will assume that $\hbar\omega \gg E(\mathbf{k} \pm \mathbf{q}) - E(\mathbf{k})$, α so that the last term can be written as

$$\varepsilon_{fc}(\mathbf{q},\omega) = -\frac{e^2}{q^2}\frac{\hbar^2 q^2}{m_d}\sum_{\mathbf{k}} f(\mathbf{k})\frac{1}{(\hbar\omega)^2} = -\frac{ne^2}{m_d\omega^2} = -\varepsilon(\infty)\frac{\omega_p^2}{\omega^2}. \tag{7.27}$$

The free carrier plasma frequency is given by

$$\omega_p^2 = \frac{ne^2}{m_d\varepsilon(\infty)}. \tag{7.28}$$

(Note here that a lower case "p" is used for the free carrier plasma frequency. While this is confusing, it is common notation.) The dielectric function can now be written as

$$\varepsilon(\mathbf{q},\omega) = \varepsilon(\infty)\left[1 - \frac{\omega_p^2}{\omega^2}\right] + [\varepsilon(0) - \varepsilon(\infty)]\frac{\omega_{TO}^2}{\omega_{TO}^2 - \omega^2}. \tag{7.29}$$

Most people are familiar with Coulombic scattering of electrons from impurities and from other electrons. They are not so familiar with the scattering of individual electrons by the collective oscillations of the electron gas. The latter is represented by the plasmons, and this scattering process has only recently been treated in transport calculations to any great extent (Mooradian and Wright, 1966; Kim *et al.*, 1978; Hollis *et al.*, 1983; Lugli and Ferry, 1985a). The plasma oscillations are longitudinal mode oscillations, by which is meant that there is a longitudinal charge vibration along the direction of propagation (q direction). In this regard, it is almost the same as the polar mode of the longitudinal optical phonon. As a consequence, these two longitudinal oscillations can actually couple to each other, providing for hybrid modes, which are combined oscillations of the lattice and the electron, with characteristics of each. This is discussed later in this chapter by treating the total dielectric function (in the small q limit, however). Here, we want to examine further this coupling.

The total dielectric function, which includes both the contributions of the electrons and the lattice, in the long-wavelength limit is just (7.29). In treating the role of the dielectric function on transport, it must be remembered that the quantity $1/\varepsilon$ always appears in the scattering rates, in potentials, in screening, and so on. Thus the relative singularities that must be examined are those of the inverse dielectric function, which are the zeros of (7.29). To be fully correct, it would also be necessary to include the imaginary parts that arise due to decay of the electronic states by scattering as a fully self-consistent calculation. This is a difficult problem and is beyond the treatment considered here, but we will return to the results of such a calculation later in this chapter. A quite good estimate of the important effects can be obtained without resorting to the full, and complicated, treatment.

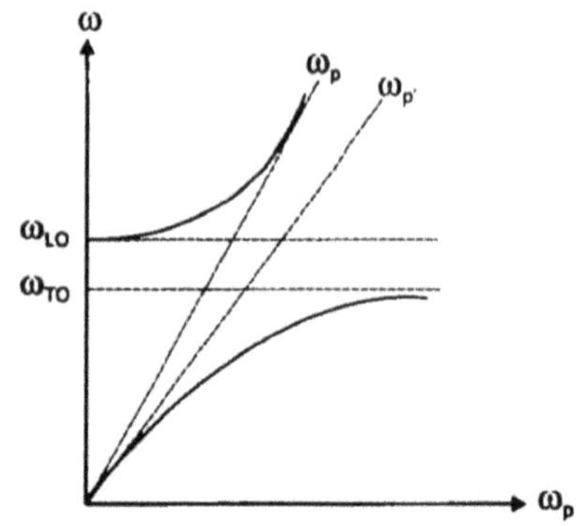

Figure 7.1 The dispersion curves for the plasmon–phonon coupling, showing the upper and lower hybrid frequencies, as the plasma frequency is varied.

The zeros of the dielectric function are found simply by setting the left-hand side of (7.29) to zero and solving for the relevant values of ω. In fact, this process gives an equation for ω as

$$\omega^4 - \omega^2(\omega_{LO}^2 + \omega_p^2) + \omega_p^2\omega_{TO}^2 = 0. \tag{7.30}$$

If $\omega_p \ll \omega_{TO}$, (7.30) can easily be found to give the limiting cases

$$\omega_u^2 = \omega_{LO}^2 + \omega_p^2, \quad \omega_l^2 = \frac{\varepsilon(\infty)}{\varepsilon(0)}\omega_p^2 = \omega_{p'}^2, \tag{7.31}$$

where the upper and lower "hybrid frequencies" have been introduced and the last term defines the modified plasmon frequency $\omega_{p'}$, which includes the static (low frequency) dielectric constant rather than the optical (high-frequency) dielectric constant. On the other hand, if the plasma frequency satisfies $\omega_p \gg \omega_{LO}$, we arrive at

$$\omega_u^2 = \omega_p^2 + \omega_{LO}^2 - \omega_{TO}^2 \sim \omega_p^2, \quad \omega_l^2 = \omega_{TO}^2. \tag{7.32}$$

Although the upper hybrid frequency remains the same functional root, this mode changes its character from phonon-like at low density (small ω_p) to more plasmon-like at high density. Conversely, the lower hybrid frequency changes its character from plasmon-like at low density to phonon-like at high density. In Figure 7.1, the dispersion diagram for these coupled modes is plotted as a function of the plasma frequency (the square root of the density). The critical frequency is obviously where $\omega_p = \omega_{TO}$, and for GaAs this occurs at a density of approximately 7×10^{17} cm^{-3}.

7.1.4 Static screening

In the above treatments, we have mostly considered the high-frequency behavior. We now turn to the low-frequency regime, where we can mostly ignore the terms in ω and α. The main task is once again to evaluate the summation inherent in the last term of (7.26). As mentioned, the low-frequency limit will be taken, so that the first two terms combine to just give $\varepsilon(0)$, the low-frequency dielectric function in the absence of free carriers. In the summation over the free carriers, the denominator is

$$E(\mathbf{k}+\mathbf{q}) - E(\mathbf{k}) \cong \mathbf{q}\cdot\frac{\partial E}{\partial \mathbf{k}} + \ldots. \tag{7.33}$$

In a similar expansion, the numerator can be written as

$$f(\mathbf{k}) - f(\mathbf{k}+\mathbf{q}) \cong -\mathbf{q}\cdot\frac{\partial f}{\partial \mathbf{k}} + \ldots = -\left[\mathbf{q}\cdot\frac{\partial E}{\partial \mathbf{k}}\right]\frac{\partial f}{\partial E} + \ldots. \tag{7.34}$$

For most transport in semiconductors, the distribution function is a non-degenerate Maxwellian, especially since the inter-electronic scattering serves to drive the system toward this distribution in equilibrium. Then, we can approximate the derivative as

$$\frac{\partial f}{\partial E} \approx -\frac{1}{k_B T} f(E). \tag{7.35}$$

As previously, the sum over the distribution function just gives the density (in this case, the free electron density n), so that (7.33)–(7.35) can be used in (7.26) to give

$$\varepsilon(\mathbf{q},\omega = 0) = \varepsilon(0) + \frac{ne^2}{q^2 k_B T} = \varepsilon(0)\left[1 + \frac{q_D^2}{q^2}\right], \tag{7.36}$$

where

$$q_D = \sqrt{\frac{ne^2}{\varepsilon(0) k_B T}} \tag{7.37}$$

is the *Debye screening wave vector*. We have used this form in previous chapters for the impurity and polar optical phonon scattering, and now see that this is a static screening approximation.

This also is the form found in the introduction of this chapter, where a simpler classical argument was used. The Debye screened potential is quite often used in the literature to describe the radial variation of the impurity potential. In truth, a rather dramatic approximation has been made. Here, it has been assumed that $q \ll k$ in the Taylor expansion use for both the numerator and the denominator. For a free-electron concentration of 10^{17} cm^{-3}, the Debye wave vector $q_D = 7.47 \times 10^5$ cm^{-1} in GaAs at 300 K. This value corresponds to a screening energy of about 3.2 meV. The inequality is actually satisfied for all but the slowest electrons. Consequently, Debye screening is almost universally used. To improve this, we need to consider what has been left out so far, which is the dynamic character of the dielectric response, and we turn to this now.

7.1.5 Momentum-dependent screening

We now want to remove the approximation that $q \ll k$. While the low-frequency approximation will continue to be used, the desire now is to evaluate the summation over the free carriers without limits on the range of q. To proceed, the dielectric function is rewritten as

$$\varepsilon(\mathbf{q},0) = \varepsilon(0) + \frac{e^2}{q^2}\sum_{\mathbf{k}} f(\mathbf{k})\left[\frac{1}{E(\mathbf{k}+\mathbf{q}) - E(\mathbf{k})} - \frac{1}{E(\mathbf{k}) - E(\mathbf{k}-\mathbf{q})}\right]. \tag{7.38}$$

The energy denominators can generally be expanded as

$$E(\mathbf{k} \pm \mathbf{q}) - E(\mathbf{k}) = \frac{\hbar^2 q^2}{2m} \pm \frac{\hbar^2 kq}{m}\cos\vartheta, \tag{7.39}$$

where the mass is the appropriate effective mass, and the summation over the vector **k** may be transformed into an integration as done in Chapter 3. In three dimensions, (7.38) becomes

$$\varepsilon(\mathbf{q},0) = \varepsilon(0) + \frac{2e^2 m}{q^2\hbar^2}\int_0^\infty\int_0^\pi \frac{k^2 dk \sin\vartheta d\vartheta}{2\pi^2} f(k)\left[\frac{1}{q^2 + 2kq\cos\vartheta} + \frac{1}{q^2 - 2kq\cos\vartheta}\right]$$

$$= \varepsilon(0) + \frac{e^2 m}{\pi^2 q^3 \hbar^2}\int_0^\infty f(k) k \ln\left|\frac{k+2q}{k-2q}\right| dk. \tag{7.40}$$

The form of the argument of the logarithm arises from the integration, in that the magnitude sign is required to assure that the argument is positive definite. To proceed further, the following normalized variables are now introduced:

$$\xi^2 = \frac{\hbar^2 q^2}{8mk_BT}, \quad x^2 = \frac{\hbar^2 k^2}{2mk_BT}. \tag{7.41}$$

It may be noted that the temperature here is that of the distribution function and represents the electron temperature, not that of the lattice. Although this has not been indicated by a subscript, it should not be confusing as the only temperature in the problem is that of the electrons. By incorporating the normalization factors on the Maxwellian distribution function, (7.40) is

$$\varepsilon(\mathbf{q},0) = \varepsilon(0)\left[1 + \frac{q_D^2}{q^2}F(\xi)\right], \tag{7.42}$$

where, for non-degenerate semiconductors,

$$F(\xi) = \frac{1}{\sqrt{\pi}\,\xi}\int_0^\infty x e^{-x^2} \ln\left|\frac{x+\xi}{x-\xi}\right| dx. \tag{7.43}$$

It should be noted that the density was introduced into (7.42) to achieve the Debye screening length from the normalizing integral over the non-degenerate Maxwellian distribution function, which just provides a numerical factor. In general, $F(\xi)$ may readily be computed numerically. However, Hall (1962) has shown that $F(\xi)$ can be rewritten as

$$F(\xi) = \frac{e^{-\xi^2}}{\xi}\int_0^\xi e^{t^2}\, dt, \tag{7.44}$$

which is related to an error function of imaginary argument. Actually, except for the $1/\xi$ prefactor, (7.44) is also known as Dawson's integral (Abramowitz and Stegun, 1964) and is a tabulated function. This function is also closely related to the *plasma dispersion function.*

In the case of degenerate semiconductors, the form of the function $F(\xi)$ is somewhat more complicated. Equation (7.40) is still correct, but the introduction of the density follows from the use of the full Fermi–Dirac distribution, as

$$n = \sum_{\mathbf{k}} f(\mathbf{k}) = \frac{1}{2\pi}\int_0^\infty f(k)k^2 dk = \frac{2N_c}{\sqrt{\pi}}\int_0^\infty f(x)x^2 dx = N_c F_{1/2}(\mu), \tag{7.45}$$

where $\mu = E_F/k_B T$, and N_c is the effective density of states, given by

$$N_c = \frac{1}{4}\left(\frac{2mk_BT}{\pi\hbar^2}\right)^{\frac{3}{2}}. \tag{7.46}$$

With these definitions, (7.43) is replaced with

$$F(\xi) \to F(\xi,\mu) = \frac{1}{\sqrt{\pi}\,\xi F_{1/2}(\mu)}\int_0^\infty \ln\left|\frac{x+\xi}{x-\xi}\right| \frac{xdx}{1+\exp(x^2-\mu)}. \tag{7.47}$$

If μ is large and negative, the nondegenerate condition (7.43) is recovered.

The behavior of $F(\xi,\mu)$ is generally not very dramatic. As $q \to 0$ ($\xi \to 0$), $F(\xi,\mu) \to 1$ for the non-degenerate case, and the usual Debye screened behavior is obtained. On the other hand, as $q \to \infty$, $F(\xi,\mu) \to 0$, and the screening is broken up completely. Thus, for high momentum transfer in the scattering process, the scattering potential is completely de-screened. Although screening is not eliminated in the materials of interest, the strength of the scattering can vary by some 20 to 30 percent, depending on the momentum transfer, and the de-screening must be considered for completeness. The function $F(\xi,\mu)$ is plotted in Figure 7.2 with the reduced Fermi energy as a parameter. It is clear that higher densities lead to a reduction in the classical concept of screening, but the wave vector dependence is also greatly reduced.

7.1.6 High-frequency dynamic screening

For a further development, we consider a number of factors. First, the difference of the Fermi Dirac functions for small $q \to 0$ give essentially the derivative of the function, which is equivalent to a delta function at the Fermi energy (at sufficiently low temperature). But, for this term the angular average vanishes. In reality, we wish to keep the frequency variables in the dielectric function, so that we have two different approaches to consider. In one, we expand the term in f_{k+q} and expand the energy differences, keeping only the lowest-order terms in q. The expansion of the energy around the momentum k yields a function of the angle between the two vectors, which is involved in the averaging process of the d-dimensional integration. In the second approximation, we use the expansion already found in (7.21). In the former case, we arrive at the derivative of the distribution, which is

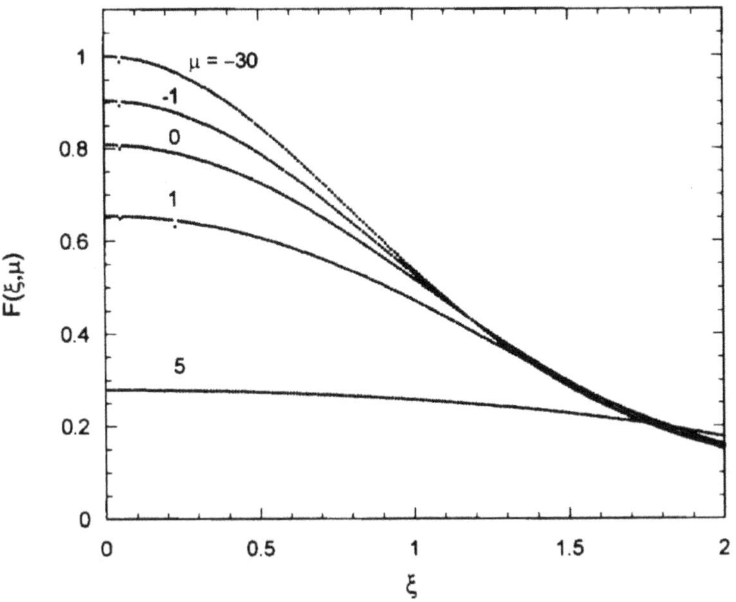

Figure 7.2 The momentum dependent screening function $F(\xi,\mu)$ is shown for the case of a degenerate Fermi–Dirac distribution function. The quantity μ is the reduced Fermi energy. The upper curve, $\mu = -30$, corresponds to a non-degenerate distribution.

useful for non-degenerate materials. However, in the case of degenerate semiconductors, we must use the second approach. Let us examine how these differ.

We begin with the form (7.26), and expand the numerator of the summation just as was done for (7.34). On the other hand, the denominator is expanded using the relationship

$$E(\mathbf{k}+\mathbf{q}) - E(\mathbf{k}) = \mathbf{q}\cdot\frac{\partial E}{\partial \mathbf{k}} + \ldots = \hbar\mathbf{q}\cdot\mathbf{v} + \ldots. \tag{7.48}$$

This allows us to write the dielectric function as

$$\varepsilon(\mathbf{q},\omega) = \varepsilon(0) + \frac{e^2}{q^2}\sum_{\mathbf{k}}\frac{\partial f(\mathbf{k})}{\partial E}\frac{-\hbar\mathbf{q}\cdot\mathbf{v}}{\hbar\mathbf{q}\cdot\mathbf{v} + \hbar\omega - i\hbar/\tau}. \tag{7.49}$$

In this expression, we have replaced α with $1/\tau$, recognizing the importance of scattering processes on the result. Equation (7.49) can now be rewritten for a non-degenerate distribution using (7.35) as

$$\varepsilon(\mathbf{q},\omega) = \varepsilon(0) + \frac{e^2}{q^2 k_B T} \sum_{\mathbf{k}} f(\mathbf{k}) \frac{\mathbf{q}\cdot\mathbf{v}}{\mathbf{q}\cdot\mathbf{v} + \omega - i/\tau}. \qquad (7.50)$$

The sum can be converted to an integral in spherical coordinates (three dimensions), so that the dot product involves the polar angle. Integration over this angle is complicated. However, we can expand the fraction as

$$F = \frac{\mathbf{q}\cdot\mathbf{v}}{\mathbf{q}\cdot\mathbf{v} + \omega - i/\tau} = 1 - \frac{\omega - i/\tau}{\mathbf{q}\cdot\mathbf{v} + \omega - i/\tau}$$

$$= 1 - \left[1 - \frac{\mathbf{q}\cdot\mathbf{v}}{\omega - i/\tau} + \frac{(\mathbf{q}\cdot\mathbf{v})^2}{(\omega - i/\tau)^2} + \dots\right]. \qquad (7.51)$$

In carrying out the angular average, the second term in the square brackets will always vanish. If we are in three dimensions, then the integral involves a $\sin\theta d\theta$ factor which couples with the $\cos\theta$ of this term to yield zero. A similar effect occurs in two and one dimension. On the other hand, the third term will give a factor of $1/d$, where d is the dimensionality. Hence, we may carry out the angular average, noting that the rest of the integration over momentum gives the density, to rewrite (7.50) as

$$\varepsilon(\mathbf{q},\omega) = \varepsilon(0) + \frac{e^2}{q^2 k_B T} \sum_{\mathbf{k}} f(\mathbf{k}) \left\{1 - \left[1 + \frac{q^2 D/\tau}{(\omega - i/\tau)^2}\right]\right\}$$

$$= \varepsilon(0) + \frac{ne^2}{q^2 k_B T}\left[1 - \frac{(\omega - i/\tau)^2}{(\omega - i/\tau)^2 - q^2 D/\tau}\right]$$

$$= \varepsilon(0) - \frac{ne^2}{k_B T} \frac{D/\tau}{(\omega - i/\tau)^2 - q^2 D/\tau}. \qquad (7.52)$$

Here, we have used the generalization of (3.25) to identify the diffusion constant as $D = v^2\tau/d$, although it must be identified with an average velocity in bringing this quantity outside of the integration (the summation over $\mathbf{k}$). We note that if $\omega \to 0$, we recover the Debye screening result (7.36), while if $q \to 0$, $\omega \gg i/\tau$, we replace D/τ with $k_B T/m^*$ and recover the free-carrier plasma result (7.27). Hence, (7.52) satisfies the requisite limits obtained previously.

A somewhat different approach must be followed for degenerate material, since we cannot relate simply the derivative of the distribution function to the distribution itself as was done in (7.50). Instead, we begin with the expansion of (7.21), and use our approximations of (7.23) and (7.48) to write

$$\varepsilon(\mathbf{q},\omega) = \varepsilon_0 - \frac{\hbar^2 e^2}{m}\sum_{\mathbf{k}} f(\mathbf{k}) \frac{1}{[\hbar\mathbf{q}\cdot\mathbf{v} + \hbar\omega - i\hbar/\tau]^2}$$

$$= \varepsilon_0 + \frac{e^2\tau^2}{m}\sum_{\mathbf{k}} f(\mathbf{k}) \frac{1}{[1 + i\omega\tau + i\mathbf{q}\cdot\mathbf{v}\tau]^2}. \tag{7.53}$$

The fraction term can now be written, by expanding and performing the angular integration, as

$$F = \frac{1}{(1 + i\omega\tau + i\mathbf{q}\cdot\mathbf{v}\tau)^2}$$

$$\approx \frac{1}{(1+i\omega\tau)^2}\left[1 - 2i\frac{\mathbf{q}\cdot\mathbf{v}\tau}{(1+i\omega\tau)} - 3\frac{(\mathbf{q}\cdot\mathbf{v}\tau)^2}{(1+i\omega\tau)^2} - \dots\right]$$

$$\approx \frac{1}{(1+i\omega\tau)^2 + Dq^2\tau}. \tag{7.54}$$

Here, we have used the same arguments on the angular integration as for (7.51), so that the dielectric function becomes

$$\varepsilon(\mathbf{q},\omega) = \varepsilon(0) + \frac{ne^2}{m}\frac{\tau^2}{(1+i\omega\tau)^2 + Dq^2\tau}. \tag{7.55}$$

Again, this satisfies the proper limits.

The difference in the two approaches lies in the portion of the spectrum in which the effects will occur. First, we note that in each case, the Coulomb interaction that appears as a product with the Lindhard potential is that appropriate for a three-dimensional system. We will address the changes appropriate for reduced dimensions in the next section. Here, however, we note that (7.52) may be rewritten, using (7.37) as

$$\varepsilon(\mathbf{q},\omega) = \varepsilon(0) - \frac{ne^2}{k_B T}\frac{D/\tau}{(\omega - i/\tau)^2 - q^2 D/\tau}$$

$$= \varepsilon(0)\left[1 + \frac{q_D^2 D\tau}{(1+i\omega\tau)^2 + Dq^2\tau}\right], \tag{7.56}$$

which should be compared with the rewritten form of (7.55), using (7.28),

$$\varepsilon(\mathbf{q},\omega) = \varepsilon(0) + \frac{ne^2}{m}\frac{\tau^2}{(1+i\omega\tau)^2 + q^2D\tau}$$

$$= \varepsilon(0)\left[1 + \frac{\omega_p^2\tau^2}{(1+i\omega\tau)^2 + Dq^2\tau}\right]. \tag{7.57}$$

These two forms are essentially the same, once it is recognized that there is a close relationship between D and k_BT in the non-degenerate case (the Einstein relationship). Thus, (7.56) is the proper form for non-degenerate material, while (7.57) is the proper form for degenerate material.

7.2 Screening in low-dimensional situations

The approach that was followed above was focused upon three-dimensional systems. This dimensionality appeared in at least two different places. First, the Fourier transform of the induced and applied potentials assumed a three-dimensional transform (e.g., q was a three-dimensional variable). Secondly, the wave function was taken to be a plane wave in three dimensions. The corrections for these two assumptions, in the case of a reduced dimensionality, are actually closely related. Let us consider, as an example, the case of a quasi-two-dimensional system arising in the inversion layer of an Si–SiO_2 or GaAs–AlGaAs interface. In the first case, the Fourier transform is changed to a two-dimensional variation in the plane of the interface and an integration over the z-component (normal to the interface) accounting for the localized wave functions in this direction. In the second case, the correction accounts for momentum plane waves in two dimensions and the localized wave function in the z-direction normal to the interface. While we will pursue the corrections for the two-dimensional case, the results are easily extended to lower dimensionality.

We basically start by rewriting the wave function that appears in (7.11) as a plane wave in two dimensions coupled to a z-direction envelope function as

$$|\mathbf{k}\rangle \to \varphi_n(z)e^{i\mathbf{k}\cdot\mathbf{r}} \equiv |\mathbf{k},n\rangle, \tag{7.58}$$

except that k is now a two-dimensional vector in the plane of the interface. The function $\varphi_n(z)$ is the envelope function in the z-direction. Similarly, the perturbed function appearing in (7.11) is now written as

$$\psi(\mathbf{k},z) = |\mathbf{k},n\rangle + b_{\mathbf{k+q},n'}|\mathbf{k}+\mathbf{q},n'\rangle + b_{\mathbf{k-q},n'}|\mathbf{k}-\mathbf{q},n'\rangle. \tag{7.59}$$

The heart of the change lies in the two-dimensional Fourier transform of the potential, as

$$V_{nn'}(\mathbf{q}) = \langle \mathbf{k},n|V(\mathbf{r},z)|\mathbf{k}+\mathbf{q},n'\rangle$$

$$= \int d^2\mathbf{r} \int_{-\infty}^{\infty} dz \varphi^*_{n'}(z)\varphi_n(z)V(\mathbf{r},z)e^{i\mathbf{q}\cdot\mathbf{r}}e^{-qz}. \tag{7.60}$$

Here, $\mathbf{r}$ is also a two-dimensional vector. Within the linear response and random-phase approximation that we have been using, the z-variation of the potential is ignored as being small. Normally, the z-integration yields a δ-function on the indices, but the e^{-qz} term spoils this orthogonality, even when the z-variation of the potential is ignored. The result is

$$V_{nn'}(\mathbf{q}) = \frac{e^2}{2\varepsilon(0)q}\zeta_{nn'}, \quad \zeta_{nn'} = \int_{-\infty}^{\infty} \varphi^*_{n'}(z)\varphi_n(z)e^{-qz}dz. \tag{7.61}$$

However, when this term is included in the summation appearing in the Lindhard potential, one has to account for contributions both to the local potential and to the distribution function from the envelope functions in the z-direction. Hence, in this case, the dielectric function becomes

$$\varepsilon_{nn'}(\mathbf{q},\omega) = \varepsilon(0)\left[\delta_{nn'} + \frac{e^2}{2\varepsilon(0)q}\sum_{mm'} F^{nn'}_{mm'}(q)L_{mm'}(\mathbf{q},\omega)\right], \tag{7.62}$$

where

$$F^{nn'}_{mm'}(q) = \int_{-\infty}^{\infty} dz \int_{-\infty}^{\infty} dz' \varphi^*_{m'}(z)\varphi_m(z)\varphi^*_{n'}(z')\varphi_n(z')e^{-q|z-z'|} \tag{7.63}$$

and

$$L_{mm'}(\mathbf{q},\omega) = \sum_{\mathbf{k}}\left[\frac{f_m(\mathbf{k}) - f_{m'}(\mathbf{k}+\mathbf{q})}{E_{m'}(\mathbf{k}+\mathbf{q}) - E_m(\mathbf{k}) + \hbar\omega - i\hbar\alpha}\right]. \tag{7.64}$$

Finally, the two-dimensional density is given by

$$n_s = 2\sum_{\mathbf{k},n} f_n(\mathbf{k}), \quad f_n(\mathbf{k}) = \left[1 + \exp\left(\frac{E - E_n - E_F}{k_B T}\right)\right]^{-1}, \tag{7.65}$$

where E_n is the subband energy of the nth subband.

In the special case of a quasi-two-dimensional system in the quantum limit with only one subband occupied, then (7.62) reduces to

$$\varepsilon_{00}(\mathbf{q},\omega) = \varepsilon(0)\left[1 + \frac{e^2}{2\varepsilon(0)q} F_0(q) L_{00}(\mathbf{q},\omega)\right]. \tag{7.66}$$

To evaluate F_0, we need to use some assumed form for the envelope function. A common choice is the variational wave function

$$\varphi_0(z) = \sqrt{\frac{b^3}{2}}\, z e^{-bz/2}, \tag{7.67}$$

this form factor, and that for the potential itself, are

$$F_0 = \frac{b^6}{(b^2 + q^2)^3}, \quad \zeta_0 = \frac{b^3}{(b+q)^3}. \tag{7.68}$$

In general $b > q$, and is sufficiently larger that $F_0 \sim \zeta_0 \sim 1$. In this situation, the major change to the previous sections is only the Fourier transform of the potential (7.61). With this change, the results of the previous sections can be used in the low-dimensional case as well.

7.3 Electron–electron scattering

The study of the interacting electron gas, and the resulting electron–electron scattering process, has a long history. From the earliest days, there has been interest in computing the self-interactions that arise from the Coulomb force between individual pairs of electrons, as opposed to the interaction between the electrons and the impurities that was of primary interest in the previous chapters. Moreover, we discussed in Section 4.6 how the inter-electronic scattering would create a Maxwellian distribution function, so that carriers with energy above the average would lose energy to those with energy below the average. This of course is an over-simplification, but the importance of electron–electron scattering was made at that point. In this section, we want to go through several versions of the scattering rate for this process, under different conditions.

There are two different methods to find the scattering rate for electron–electron scattering. In the first, we proceed as in previous chapters, and use the Fermi golden rule from perturbation theory to determine the scattering rate. This is the procedure we will begin with for several cases. There is, however, another approach in which the self-energy of the interacting gas is determined and the imaginary part of this is equated to the scattering rate. We will also illustrate this approach, particularly for the low-dimensional situations of interest. We begin with the general three-dimensional electron–electron scattering process.

7.3.1 Electron–electron scattering by energetic carriers

Here, we will treat the Coulomb scattering potential in the static screening approximation, given in Section 7.1.4 above. This means that the scattering matrix element is given simply by

$$M(\mathbf{q}) = \frac{e^2}{\varepsilon_s(q^2 + q_D^2)}, \tag{7.69}$$

where ε_s is the static dielectric function $\varepsilon(0)$ and q_D is the Debye screening wave vector. Following Takanaka *et al.* (1979), the scattering rate can be written as

$$\Gamma_{ee}(k) = \frac{2\pi}{\hbar} \sum_{\mathbf{k}_2,\mathbf{k}',\mathbf{k}_2'} \left[\frac{e^2}{\varepsilon_s(q^2 + q_D^2)}\right]^2 \delta_{\mathbf{k}+\mathbf{k}_2,\mathbf{k}'+\mathbf{k}_2'} f(k_2)\delta(E_{\mathbf{k}'} + E_{\mathbf{k}_2'} - E_{\mathbf{k}} - E_{\mathbf{k}_2}). \tag{7.70}$$

Here, $\mathbf{k}$ and $\mathbf{k}_2$ are the momentum wave vectors of the initial particle and the particle with which it interacts, respectively. The wave vectors $\mathbf{k}'$ and $\mathbf{k}_2'$ are the corresponding quantities after the scattering event (the final states). Further, it is assumed that the distribution is non-degenerate so that $f(k'), f(k_2') \ll 1$ can be used.

The summation over $\mathbf{k}_2'$ is accomplished using the δ-function, so that momentum conservation is enforced through

$$\mathbf{k} + \mathbf{k}_2 = \mathbf{k}' + \mathbf{k}_2'. \tag{7.71}$$

We now introduce two new wave vectors through

$$\mathbf{g} = \mathbf{k}_2 - \mathbf{k}, \quad \mathbf{g}' = \mathbf{k}_2' - \mathbf{k}' = \mathbf{k} + \mathbf{k}_2 - 2\mathbf{k}'. \tag{7.72}$$

With these definitions, the argument of the energy δ-function can be written as

$$\begin{aligned} E_{\mathbf{k}'} + E_{\mathbf{k}_2'} - E_{\mathbf{k}} - E_{\mathbf{k}_2} &= E_{\mathbf{g}'} - E_{\mathbf{g}} + \frac{\hbar^2}{m^*}(\mathbf{k}' - \mathbf{k}) \cdot (\mathbf{k}_2 - \mathbf{k}') \\ &= \tfrac{1}{2}(E_{\mathbf{g}'} - E_{\mathbf{g}}). \end{aligned} \tag{7.73}$$

In addition, we note that

$$\mathbf{k}' = \mathbf{k}_2 - \tfrac{1}{2}(\mathbf{g} + \mathbf{g}'), \tag{7.74}$$

so that the summation over $\mathbf{k}'$ can be taken into a summation over $\mathbf{g}'$ (there is a factor of 8 in the new denominator to account for the factor of $\frac{1}{2}$ in each of the three directions). We can then rewrite (7.70) as

$$\Gamma_{ee}(k) = \frac{\pi}{4\hbar}\sum_{\mathbf{k}_2,\mathbf{g}'}\left[\frac{e^2}{\varepsilon_s(\frac{1}{4}|\mathbf{g}-\mathbf{g}'|^2 + q_D^2)}\right]^2 f(k_2)\delta(\tfrac{1}{2}(E_g - E_{g'})). \qquad (7.75)$$

To proceed further, we introduce another change of variables, through

$$\mathbf{u} = \mathbf{g}' - \mathbf{g}. \qquad (7.76)$$

The summation over $\mathbf{g}'$ can then be carried out in the following manner:

$$\sum_{\mathbf{g}'} \to \frac{1}{4\pi^2}\int u^2 du \int_0^{\pi} \sin\vartheta d\vartheta \frac{e^4}{\varepsilon_s^2[u^2/4 + q_D^2]^2}\delta\left(\frac{\hbar^2}{4m^*}(u^2 + 2ug\cos\vartheta)\right)$$

$$= \frac{e^4 m^*}{2\pi^2\hbar^2\varepsilon_s^2 g}\int_0^{2g}\frac{udu}{[u^2/4 + q_D^2]^2} = \frac{e^4 m^*}{2\pi^2\hbar^2\varepsilon_s^2 q_D^2}\left(\frac{g}{g^2 + q_D^2}\right). \qquad (7.77)$$

This can be used in (7.75) to give

$$\Gamma_{ee}(k) = \frac{m^* e^4}{8\pi\hbar^3\varepsilon_s^2 q_D^2}\sum_{\mathbf{k}_2} f(\mathbf{k}_2)\left(\frac{|\mathbf{k}_2 - \mathbf{k}|}{|\mathbf{k}_2 - \mathbf{k}|^2 + q_D^2}\right). \qquad (7.78)$$

At this point, some assumptions have to be made about the distribution function that appears in the above equation. Our interest is in the case of non-equilibrium situations, where the distribution function is not that of equilibrium. A proper treatment would solve for this from e.g. a Monte Carlo procedure, in which the electron–electron scattering was explicitly included. Here, however, we will illustrate the result with a Maxwellian at an electron temperature T_e. Since $\mathbf{k}_2$ is a dummy variable which will disappear from the integral, we can shift this coordinate so that the new integration variable is $\xi = \mathbf{k}_2 - \mathbf{k}$. The integration can then be rewritten as

$$\Gamma_{ee}(k) = \frac{m^* e^4}{8\pi\hbar^3\varepsilon_s^2 q_D^2}\int_0^{\infty}\frac{\xi^2 d\xi}{2\pi^2}\int_0^{\pi}\sin\vartheta d\vartheta f(\xi + \mathbf{k})\left(\frac{\xi}{\xi^2 + q_D^2}\right)$$

$$= \frac{nm^* e^4}{8\sqrt{\pi}\hbar^3\varepsilon_s^2 q_D^2 k}e^{-x^2}\int_0^{\infty} e^{-y^2}\frac{y^2}{y^2 + \lambda^2}\sinh(2yx)dy \qquad (7.79)$$

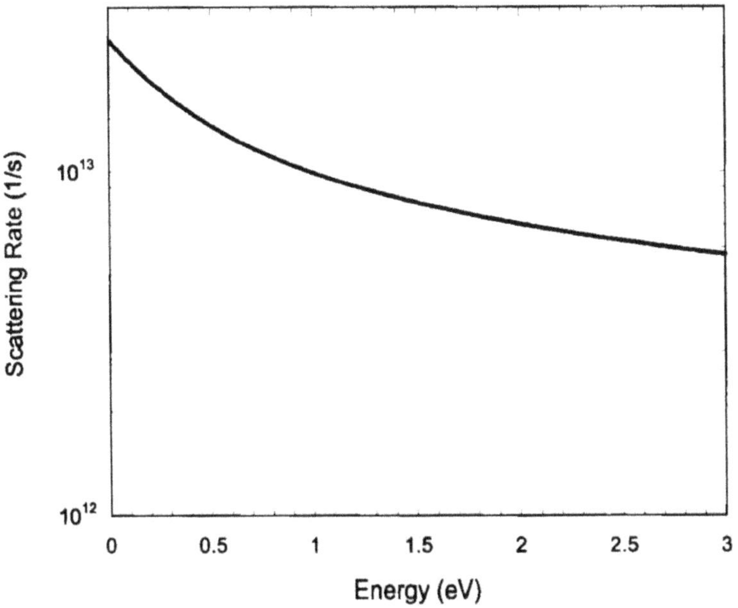

Figure 7.3 The electron–electron scattering rate in Si for an assumed electron density of 10^{17} cm^{-3} and an electron temperature of 2500 K.

and we have introduced the reduced units

$$y^2 = \frac{\hbar^2\xi^2}{2m^*k_BT_e}, \quad x^2 = \frac{\hbar^2k^2}{2m^*k_BT_e}, \quad \lambda^2 = \frac{\hbar^2q_D^2}{2m^*k_BT_e}. \tag{7.80}$$

The result (7.79) can be expressed as a combination of error functions and exponential integrals, but such a result provides no new insight over the simpler integral expression given. The density appears as a result of the normalization of the distribution function, and represents the number of other electrons with which the particular electron can interact. In Figure 7.3, we plot the scattering rate (7.79) as a function of the energy $E(\boldsymbol{k})$ for an electron temperature of 2500 K and a density of 10^{17} cm^{-3}, which correspond to values found in modern MOSFETs. This total scattering rate includes both emission and absorption of energy (to the other electrons). We will separate these two effects in the next section.

7.3.2 Energy gain and loss

The treatment of the last few paragraphs did not distinguish between emission and absorption of energy by the incident electron. That is, we can not determine just how much energy is gained or lost by the incident electron during the interaction. It is possible, however, to reformulate the above treatment in a manner in which this can be done explicitly. The key is to decompose the energy-conserving δ-function in (7.70) as

$$\delta(E_{\mathbf{k}'} + E_{\mathbf{k}_2'} - E_{\mathbf{k}} - E_{\mathbf{k}_2}) = \hbar\int d\omega\,\delta(E_{\mathbf{k}} - E_{\mathbf{k}'} - \hbar\omega)\delta(E_{\mathbf{k}_2} - E_{\mathbf{k}_2'} + \hbar\omega). \tag{7.81}$$

In this form, integration over the frequency (the energy exchange) produces the single δ-function that appears in (7.70). Hence, by leaving this integration until the end, the energy gained or lost by the incident electron can be studied. As previously, the integration over $\mathbf{k}_2'$ introduces the momentum conservation that is represented by (7.71), and the integration over $\mathbf{k}'$ can be changed into an integration over either $\mathbf{q} = \mathbf{k} - \mathbf{k}'$ or $\mathbf{g} = \mathbf{k}_2' - \mathbf{k}'$. However, the integration over $\mathbf{k}_2$ now involves the frequency ω, as well as the other variables, so that this integration cannot be deferred until the end as was done above. In other words, the argument of the second δ-function in (7.81) involves

$$E_{\mathbf{k}_2'} - E_{\mathbf{k}_2} - \hbar\omega = E_{\mathbf{k}-\mathbf{k}'+\mathbf{k}_2} - E_{\mathbf{k}_2} - \hbar\omega$$
$$= E_{\mathbf{q}} + \frac{\hbar^2 q k_2}{m^*}\cos\chi - \hbar\omega. \tag{7.82}$$

Here, χ is the angle between $\mathbf{k}_2$ and $\mathbf{q}$. It now makes more sense to integrate over $\mathbf{k}_2$ prior to integrating over $\mathbf{q}$ (or $\mathbf{g}$). This integration, as can be seen from (7.70), involves the summation over $\mathbf{k}_2$, the distribution function itself, and the second δ-function on the right hand side of (7.81), which has the expansion (7.82). The integration involves an angle term that uses the argument of the δ-function to set the limits on the integral over the magnitude of k_2, just as in the case of phonon emission in Chapter 4. Thus, we can write this integration as

$$I = \sum_{\mathbf{k}_2} f(k_2)\delta\left(E_{\mathbf{q}} + \frac{\hbar^2 q k_2}{m^*}\cos\chi - \hbar\omega\right)$$
$$= \frac{1}{4\pi^2}\int f(k_2)k_2^2\,dk_2\int_0^{\pi}\delta\left(E_{\mathbf{q}} + \frac{\hbar^2 q k_2}{m^*}\cos\chi - \hbar\omega\right)\sin\chi\,d\chi$$
$$= \frac{m^*}{4\pi^2\hbar^2 q}\int_{k_0}^{\infty} f(k_2)k_2\,dk_2. \tag{7.83}$$

Here,

$$k_0 = \frac{m^*}{\hbar^2 q}|E_q - \hbar\omega| \tag{7.84}$$

is the limit imposed by the integration over the δ-function. If we now assume, as previously, that we have a Maxwellian distribution function at temperature T_e, then (7.83) can be integrated to yield

$$I = \frac{n}{2\hbar q}\left(\frac{m^*}{2\pi k_B T_e}\right)^{\frac{1}{2}} \exp\left[-\frac{\hbar^2}{8m^* k_B T_e}\left(q - \frac{2m^*\omega}{\hbar q}\right)^2\right]. \tag{7.85}$$

The result (7.85) can now be used in the remaining summation over k′, which becomes an integration in the same manner. This integration involves the first δ-function on the right-hand side of (7.81), which will be evaluated with the angular integration and also will lead to a set of limits on the final q integration. Here, the limits are more directly related to those obtained in optical phonon scattering, as the energy $\hbar\omega$ plays a similar role to the phonon energy. Thus, we may now write (7.70) as

$$\begin{aligned}
\Gamma_{ee}(k,\omega) &= \frac{2\pi}{\hbar}\sum_{\mathbf{k}'}\left[\frac{e^2}{\varepsilon_s(q^2 + q_D^2)}\right]^2 I\delta\left(E_q - \frac{\hbar^2 kq}{m^*}\cos\vartheta + \hbar\omega\right) \\
&= \frac{2\pi}{\hbar}\sum_{\mathbf{q}}\left[\frac{e^2}{\varepsilon_s(q^2 + q_D^2)}\right]^2 I\delta\left(E_q - \frac{\hbar^2 kq}{m^*}\cos\vartheta + \hbar\omega\right) \\
&= \frac{e^4}{2\pi\varepsilon_s^2\hbar}\int\frac{q^2 I dq}{(q^2 + q_D^2)^2}\int_0^\pi \delta\left(E_q - \frac{\hbar^2 kq}{m^*}\cos\vartheta + \hbar\omega\right)\sin\vartheta d\vartheta \\
&= \frac{m^* e^4}{2\pi\varepsilon_s^2\hbar^3 k}\int_{q_-}^{q_+}\frac{Iqdq}{(q^2 + q_D^2)^2}.
\end{aligned} \tag{7.86}$$

The limits on the integration are given as

$$k - \sqrt{k^2 - \frac{2m^*\omega}{\hbar}} = q_- < q < q_+ = k + \sqrt{k^2 - \frac{2m^*\omega}{\hbar}}. \tag{7.87}$$

We can now introduce (7.83) and the integration over ω to give the final result

$$\Gamma_{ee}^{em}(k) = \frac{nm^*e^4}{4\pi\varepsilon_s^2\hbar^3 k}\left(\frac{m^*}{2\pi k_B T_e}\right)^{\frac{1}{2}} \int_0^{E(k)/\hbar} d\omega \int_{q_-}^{q_+} \frac{dq}{(q^2+q_D^2)^2} \exp\left[-\frac{\hbar^2}{8m^* k_B T_e}\left(q - \frac{2m^*\omega}{\hbar q}\right)^2\right]. \quad (7.88)$$

The result (7.88) is for the case of energy loss by the incident (primary) electron. An equivalent expression can be found for the case of energy gain by the incident electron. In this latter case, the limit on the frequency integral is ∞, rather than the energy of the carrier (an incident electron cannot lose more than the energy that it possesses, but there is no limit on the amount of energy it can gain, in principle). In addition, the limits on both the q-integration and the k_2 are also changed because of the change of sign of $\hbar\omega$ in the δ-functions. The change of the limits of the q integration are straightforward, but those of the k_2 integration change the functional form of the result, primarily by changing the sign within the exponential. The result is

$$\Gamma_{ee}^{ab}(k) = \frac{nm^*e^4}{4\pi\varepsilon_s^2\hbar^3 k}\left(\frac{m^*}{2\pi k_B T_e}\right)^{\frac{1}{2}} \int_0^{\infty} d\omega \int_{q_-}^{q_+} \frac{dq}{(q^2+q_D^2)^2} \exp\left[-\frac{\hbar^2}{8m^* k_B T_e}\left(q + \frac{2m^*\omega}{\hbar q}\right)^2\right], \quad (7.89)$$

with the limits

$$\sqrt{k^2 + \frac{2m^*\omega}{\hbar}} - k = q_- < q < q_+ = k + \sqrt{k^2 + \frac{2m^*\omega}{\hbar}}. \quad (7.90)$$

In Figure 7.4, we plot the scattering rates for emission and absorption as a function of the energy of the primary electron. The conditions here are the same as for Figure 7.3. It is clear that, at very low energies, absorption processes dominate the scattering, while emission dominates at higher energies, just as expected. The sum of the two processes yields the rates shown earlier in Figure 7.3. In Figure 7.5, the relative scattering rates for the loss of a particular amount of energy are shown for primary energies of 0.5, 1.0, and 2.0 eV. The integration of each of these curves over frequency gives the total scattering rates shown in Figure 7.4.

Goodnick and Lugli (1988) studied non-equilibrium transport in a quantum well and incorporated electron–electron scattering explicitly into the calculation. They used the quasi-two-dimensional version of (7.79), and monitored the energy exchanged in each of the collisions during the simulation. They found that, indeed, the energy exchange was in general quite

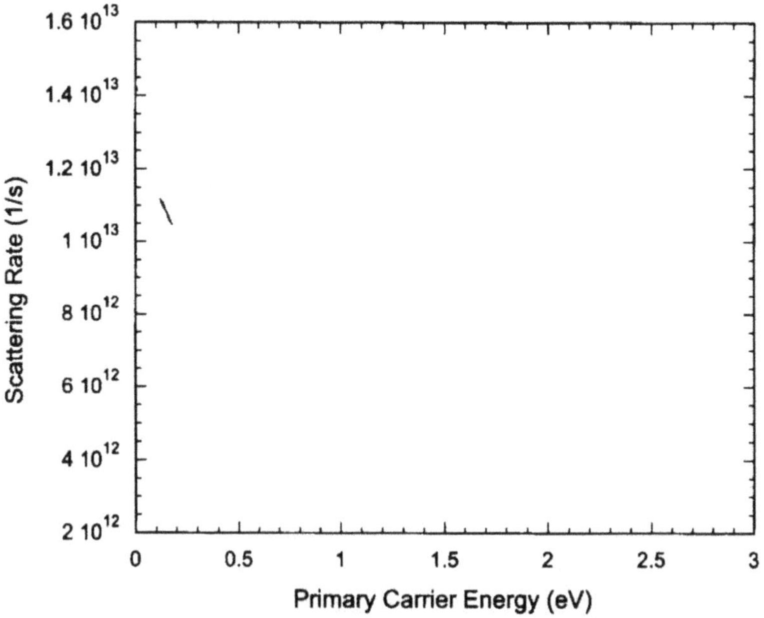

Figure 7.4 The scattering rates for emission (solid curve) and absorption (dashed curve) of energy by the primary electron. The sum of these two give the results of Figure 7.3.

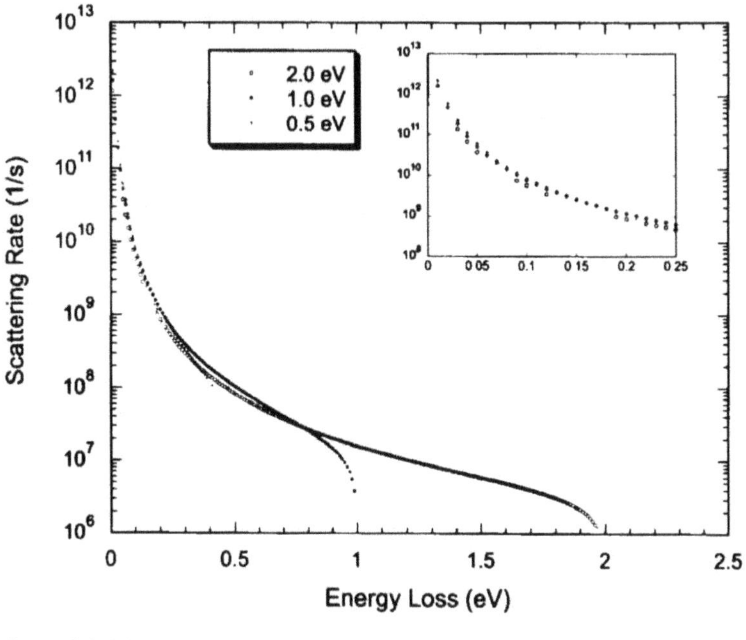

Figure 7.5 The relative scattering rate for the emission of energy by a carrier of fixed energy (the curves are for 0.5, 1.0, and 2.0 eV) of the primary electron. The inset is an expanded view of the region at low energy loss.

small, but non-zero in most cases. The interaction was especially effective in transferring energy between different subbands, and would thus thermalize a laser excited electron–hole plasma (into an effective electron temperature) within a picosecond.

7.3.3 *Plasmon scattering*

Both electron–plasmon and electron–(polar)–phonon scattering are caused by the perturbing potential that arises from the polarization of the charge oscillation with the vibrating lattice or electron system. In both cases, the strength of the scattering, as in the case of the free carrier Coulomb interaction, depends upon the dielectric function. The zeros at ω_p and, in the case of the polar phonons, at ω_0 correspond to the excited modes of the two hybrid frequencies. As the density is increased, ω_p moves up toward ω_{TO}, but never passes it, so that the basic two-mode behavior is not modified. The distinction as to how much of the total behavior to assign to each of the modes is rather arbitrary, but can be related to the inverse of the dielectric function. As mentioned earlier, one approach to calculating the scattering strength is from the self-energy, and this incorporates the inverse of the dielectric function in a normal manner. This is the approach we shall follow in this section.

Scattering from the (screened) potential of the other electrons, and from the collective plasmon modes, are both part of the total electron–electron interaction among the free carriers. Generally, it is not possible to consider the full Coulomb interaction beyond the lowest order of perturbation theory because of the long range of the potential associated with this interaction. As discussed above, the dielectric function has singularities at the plasmon frequency and at zero frequency, corresponding to the plasmon modes and the single-particle scattering, respectively. Formally, one can split the summation over q that appears in a Fourier transform of the potential into a short-range (in real space, which means large q) part, for which $q > q_c$, and a long-range (or small q) part, for which $q < q_c$, where q_c is a cut-off wave vector defining this split. It was shown some time ago that the short-range part of the potential corresponds to the screened Coulomb interaction for single-particle scattering (Madelung, 1978), which was discussed above. The long-range part, on the other hand, is responsible for scattering by the collective oscillations of the electron gas, which describe the motion of the electrons in the field produced by their own Coulomb potential, the *plasmons*. In fact, these are just the modes we discussed in the plasmon–pole approximation to the dielectric function in Section 7.1.3. These collective oscillations are bosons, so their scattering rate can be calculated in much the same way as phonons – the distribution function that we include is that of the plasmons rather than the free electrons. The only difference from the formal approach of the phonons is that there is now a maximum

q (= q_C) that can be involved in the scattering, and this cutoff is essentially the Debye wave vector.

To proceed, we will calculate the inverse of the dielectric function, and use the poles of this function to define the scattering strengths. The form that we want is a variant of the plasmon-pole approximation, in which we will *not* assume that the frequency ω is large. In fact, we will assume the external frequency is small, and ignore it. Thus, we return to (7.27) to write the carrier contribution to the dielectric function, but we recover the denominator terms from (7.21) (while ignoring the frequency), descriptively as

$$\varepsilon_{fc}(\mathbf{q},\omega) = -\frac{e^2}{q^2}\frac{\hbar^2 q^2}{m_d}\sum_{\mathbf{k}} f(\mathbf{k})\frac{1}{(E_{\mathbf{k+q}} - E_{\mathbf{k}} - i\hbar\alpha)(E_{\mathbf{k}} - E_{\mathbf{k-q}} - i\hbar\alpha)}$$
$$= -\frac{e^2}{m_d}\sum_{\mathbf{k}} f(\mathbf{k})\frac{1}{(\mathbf{q}\cdot\mathbf{v} - i\alpha)^2}. \tag{7.91}$$

The angle in the denominator will be ignored for now, but will be evaluated later after we have restored the full form of the energy functions. The dielectric function is now

$$\varepsilon(q) = \varepsilon(0)\left[1 - \frac{\omega_p^2}{(\mathbf{q}\cdot\mathbf{v} - i\alpha)^2}\right]. \tag{7.92}$$

The scattering strength is given by

$$\frac{1}{\pi}\mathrm{Im}\left(\frac{1}{\varepsilon(q)}\right) = \frac{1}{\pi\varepsilon(0)}\frac{(\mathbf{q}\cdot\mathbf{v} - i\alpha)^2}{(\mathbf{q}\cdot\mathbf{v} - i\alpha)^2 - \omega_p^2}$$
$$\cong \frac{1}{\pi\varepsilon(0)}\frac{(\mathbf{q}\cdot\mathbf{v})^2}{(\mathbf{q}\cdot\mathbf{v} - i\alpha + \omega_p)(\mathbf{q}\cdot\mathbf{v} - i\alpha - \omega_p)}$$
$$\approx \frac{\omega_p}{2\varepsilon(0)}[\delta(\mathbf{q}\cdot\mathbf{v} + \omega_p) + \delta(\mathbf{q}\cdot\mathbf{v} - \omega_p)]. \tag{7.93}$$

In this last expression we have used the relationship

$$\lim_{\alpha\to 0}\frac{1}{x - i\alpha} = P\frac{1}{x} + i\pi\delta(x), \tag{7.94}$$

where P denotes the principal part. The two δ-functions correspond to the absorption and emission of a plasmon, respectively. Hence, we can now write the scattering rates, after restoring the full arguments of the δ-functions, as

$$\Gamma_{e\text{-}pl}(k) = \frac{2\pi}{\hbar}\frac{1}{4\pi^2}\int q^2 dq \frac{\hbar\omega_p e^2}{2\varepsilon(0)q^2}$$
$$\times \int_0^{\pi} \sin\vartheta d\vartheta [N_q \delta(E_{\mathbf{k+q}} - E_{\mathbf{k}} + \hbar\omega_p)$$
$$+ (N_q + 1)\delta(E_{\mathbf{k-q}} - E_{\mathbf{k}} - \hbar\omega_p)]. \quad (7.95)$$

As in other cases, the integration over the angle involves the arguments of the delta functions, and sets the limits on the q-integration. Here, however, only the lower limit is set in this fashion, as the upper limit is set by the cutoff wave vector. In addition, the Bose–Einstein distribution

$$N_q = \left[\exp\left(\frac{\hbar\omega_p}{k_B T_e}\right) - 1\right]^{-1} \quad (7.96)$$

is a function of the *carrier* temperature rather than the lattice temperature. Hence, in non-equilibrium situations, the plasmon distribution varies with the same carrier temperature as the electron distribution. With these caveats, the integration can be performed, and

$$\Gamma_{e\text{-}pl}(k) = \frac{m^* e^2 \omega_p}{4\pi\varepsilon(0)\hbar^2 k}\left[N_q \int_{q_-^e}^{q_D} \frac{dq}{q} + (N_q + 1)\int_{q_-^e}^{q_D} \frac{dq}{q}\right]$$
$$= \frac{m^* e^2 \omega_p}{4\pi\varepsilon(0)\hbar^2 k}\left[N_q \ln \frac{q_D/k}{\sqrt{1 + \hbar\omega_p/E_k} - 1}\right.$$
$$\left. + (N_q + 1)\ln \frac{q_D/k}{1 - \sqrt{1 - \hbar\omega_p/E_k}}\right]. \quad (7.97)$$

As was mentioned, one important attribute of electron–plasmon scattering is that the Bose–Einstein distribution for plasmons does not remain in equilibrium for hot carriers. Studies of transport in high electric fields, carried out by ensemble Monte Carlo techniques, show that the plasmon temperature comes into local equilibrium with the same electron temperature that describes the free-carrier distribution in the high electric field, as it should (Lugli and Ferry, 1985b). This again demonstrates a certain consistency of approach, since the two distributions describe properties of the same electrons.

7.3.4 *Scattering in a quasi-two-dimensional system*

Electrons that are in excited states (or states with energy above the average energy) will on average lose energy to the overall electron gas, as discussed above. This is true regardless of the dimensionality of the semiconductor. In the previous paragraphs, the loss to the plasmon modes was discussed. In two dimensions, the plasma frequency is not constant, but is a function of the wave vector, hence approaches zero at $q = 0$. In the present section, we want to compute the scattering rate in this quasi-two-dimensional system. We will assume, however, that the carriers are in the lowest subband, and that the wave function factors (7.68) are unity and can be ignored. We will find that, contrary to the last section, the inverse of the dielectric function does not give the δ-function found there, so will have to directly incorporate this energy conserving function. A second factor that will become important is to limit the *lower* range of the integration over q (in a sense, this is using a long range cutoff), just as is done for inelastic phonon scattering. Here, however, we will limit this value to the inverse of the mean free path, or $1/v\tau = 1/\sqrt{2D\tau}$ in two dimensions, predicated on the fact that the scattering will break up any process that would emit a plasmon with lower momentum value (or that the coherence can not be maintained for more than a diffusion length).

We can now formulate the process of energy loss by an energetic electron in a quasi-two-dimensional electron gas. We treat a nearly free electron, and not the case of a strongly impurity-damped process popular in mesoscopic physics. Our treatment will be applicable to that of high mobility carriers in e.g. a heterostructure with little impurity scattering, even though we will find essentially the same result as that of the disordered system. Following the approach of the previous section, we can write the scattering rate from the self-energy as

$$\frac{1}{\tau_{ee}} = \frac{2\pi}{\hbar}\sum_{q}\int_{-\infty}^{\infty}\frac{d\omega}{2\pi}\coth\left(\frac{\hbar\omega}{2k_BT}\right)\left|\mathrm{Im}\left\{\frac{V(q)}{\varepsilon(q,\omega)}\right\}\right|\delta\left(\omega - \frac{E_{\mathbf{k}+\mathbf{q}} - E_{\mathbf{k}}}{\hbar}\right). \tag{7.98}$$

Here, we ignore the details of the Fermi factor that arises from the possibility of the final states being full, and we have combined the emission and absorption terms through

$$N_q + (N_q + 1) = \frac{1}{e^x - 1} + \frac{e^x}{e^x - 1} = \frac{e^{x/2} + e^{-x/2}}{e^{x/2} - e^{-x/2}} = \coth(x/2). \tag{7.99}$$

In two dimensions, the collective excitations (the plasmons) have a frequency that goes to zero as q goes to zero, and our basic interest lies in small frequency (small energy) exchange. Hence, we will use the approximation that

$$\coth(x/2) \sim \frac{2}{x}. \tag{7.100}$$

For the dielectric function, we take the form (7.57), and use the two-dimensional plasma frequency $\omega_p^2|_{2d} = n_s e^2 q/\varepsilon_s 2m^*$ to give

$$\varepsilon(q,\omega) = \varepsilon(0) + \frac{n_s e^2 q}{2m_d}\frac{\tau^2}{(1-i\omega\tau)^2 + q^2 D\tau}$$

$$= \varepsilon(0)\left[1 + \frac{\chi_2 q\tau^2}{(1-i\omega\tau)^2 + q^2 D\tau}\right]. \tag{7.101}$$

Here $\chi_2 = n_s e^2/2m^*\varepsilon_s$, and we recognize that we would define the square of a plasma frequency as $\omega_p^2 = \chi_2 q$, as discussed above. The difference here lies in the use of the bare potential as the two-dimensional form $e^2/2\varepsilon_s q$ in the screening function. In the low frequency limit, we can write the imaginary part of the potential times the inverse dielectric function as

$$-\frac{e^2}{2\varepsilon_s q}\frac{2\omega}{Dq^2 + \chi_2 q\tau}, \tag{7.102}$$

The integral over the frequency can now be written as

$$\int_{-\infty}^{\infty} \frac{d\omega}{2\pi}\frac{2k_B T}{\hbar\omega}\frac{e^2}{\varepsilon_s q}\frac{\omega}{Dq^2+\chi_2 q\tau}\delta\left(\omega - \frac{E_{k+q}-E_k}{\hbar}\right) = \frac{e^2 k_B T}{\pi\hbar\varepsilon_s q(Dq^2+\chi_2 q\tau)}. \tag{7.103}$$

Now, we can write the scattering rate as

$$\frac{1}{\tau_{ee}} = \frac{2\pi}{\hbar}\sum_q \frac{e^2 k_B T}{\pi\hbar\varepsilon_s q(Dq^2 + \omega_p^2\tau)} = \frac{e^2 k_B T}{\pi\hbar^2\varepsilon_s}\int_{q_{min}}^{\infty}\frac{dq}{Dq^2 + \chi_2 q\tau}$$

$$\sim \frac{e^2 k_B T}{\pi\hbar^2\varepsilon_s\chi_2\tau}\ln\left(1 + \chi_2\tau\sqrt{\frac{2\tau}{D}}\right). \tag{7.104}$$

Here, we have cut off the integration at a lower value for q, as discussed above. This form has a characteristic variation of $\tau_{ee} \sim 1/T$, which is typical for a two-dimensional system. We can estimate the value of τ_{ee} for a GaAs/AlGaAs heterostructure at low temperature. We assume that the density is 4×10^{11} cm^{-2}, and that the mobility is 10^6 cm^2/Vs, typical of a high mobility structure. Then, $\tau \sim 3.8 \times 10^{-11}$ s, and $\chi_2 \sim 8.5 \times 10^{18}$ cm/s^2. This gives the value $\tau_{ee} \sim 4.4 \times 10^{-10}$ s at 1 K. This value is comparable to actual measured values in such semiconductor structures at low temperatures. In diffusive systems, where the mobility is quite low, other forms need to be used, particularly to account for the interactions between the impurities and the electrons which can occur simultaneously. These are reviewed in Altshuler and Aronov (1985) and Fukuyama (1985). From these approaches, the same temperature dependence has been obtained by Altshuler *et al.* (1982), Fukuyama and Abrahams (1983), and Giuliani and Quinn (1982), although the latter appears only after a correction to determine the proper temperature dependence (Ferry and Goodnick, 1997).

7.3.5 Scattering in a quasi-one-dimensional system

We now turn to the case of a quantum wire. Although we think of these as one-dimensional systems, in most cases they are narrow two-dimensional systems, with the Fermi energy being determined by the quasi-two-dimensional electron gas to which the wires are attached. In this case, the one-dimensional density is more properly given by $n_1 = n_2 W$, where W is the wire width. Nevertheless, the treatment here assumes a quasi-one-dimensional wire, for which the Coulomb interaction is approximately

$$U = \frac{e^2}{4\pi\varepsilon_s \ln(1 + q_0^2/q^2)}. \tag{7.105}$$

Here, we will continue to ignore the wave function correction terms for the lateral dimensions by the assumption that the scattering remains within a single subband. The logarithmic factor in (7.105) will cancel from the resulting development, so will also be ignored in the following. If we now utilize this form of the Coulomb interaction in the dielectric function (7.57), we can write this as

$$\varepsilon(q,\omega) = \varepsilon(0)\left[1 + \frac{\chi_1 q^2 \tau^2}{(1 + i\omega\tau)^2 + Dq^2\tau}\right], \tag{7.106}$$

where $\chi_1 = n_1 e^2/4\pi m_d \varepsilon_s$. We can now determine the imaginary part of the inverse dielectric function as

$$\frac{e^2}{2\pi\varepsilon_s} \frac{\omega}{q^2(D + \chi_1\tau)}. \tag{7.107}$$

This can now be used in (7.98).

In evaluating (7.98), it will prove more convenient to carry out the integration over q prior to that over ω. The former will entail the δ-function, and we will need to invoke some cutoffs on the ω integration. In fact, we are interested in frequencies that lie between $1/\tau_{ee}$ and $1/\tau$. We will take these as the lower and upper cutoffs, respectively, when we need to utilize such in the evaluation of the integrals. We can now write the q-integration as

$$\sum_q \left| \mathrm{Im}\left(\frac{U}{\varepsilon}\right) \right| \delta\left(\omega - \frac{E_{k+q} - E_k}{\hbar}\right) = \sum_{\pm} \int_0^{\infty} dq \frac{e^2 \omega}{2\pi\varepsilon_s q^2 (D + \chi_1 \tau)} \delta\left(\omega - \frac{\hbar q^2}{2m^*} \mp \frac{\hbar k q}{m^*}\right). \tag{7.108}$$

The summation that remains is over forward- and back-scattering. In fact, the dominant contribution to the phase breaking is by back-scattering through plasmon emission, so that $q \gg k$, and the integration yields

$$\sum_q \left| \mathrm{Im}\left(\frac{U}{\varepsilon}\right) \right| \delta\left(\omega - \frac{E_{k+q} - E_k}{\hbar}\right) = \frac{m^* e^2}{2\pi\hbar\varepsilon_s (D + \chi_1 \tau)} \left(\frac{\hbar}{2m^*}\right)^{\frac{3}{2}} \frac{1}{\sqrt{\omega}}. \tag{7.109}$$

We can now use this in the remainder of (7.98) to give

$$\begin{aligned} \frac{1}{\tau_{ee}} &= \frac{e^2}{2\hbar\varepsilon_s (D + \chi_1 \tau)} \sqrt{\frac{\hbar}{2m^*}} \int_{-\infty}^{\infty} \frac{d\omega}{2\pi} \frac{2k_B T}{\hbar\omega} \frac{1}{\sqrt{\omega}} \\ &= \frac{e^2 k_B T}{\pi\hbar^{\frac{3}{2}} \sqrt{2m^*}\, \varepsilon_s (D + \chi_1 \tau)} \tau_{ee}^{\frac{1}{2}}. \end{aligned} \tag{7.110}$$

The lower cutoff frequency has been used in this last form, and we can now solve for the scattering rate as

$$\frac{1}{\tau_{ee}} = \left[\frac{e^2 k_B T}{\pi\hbar^{\frac{3}{2}} \sqrt{2m^*}\, \varepsilon_s (D + \chi_1 \tau)}\right]^{\frac{2}{3}}. \tag{7.111}$$

We now find that the carrier–carrier relaxation time decays as $T^{-\frac{2}{3}}$, a result also found for disordered systems (Altshuler *et al.*, 1982; Golubev and Zaikin, 1998), although we have not accounted for the disorder in this treatment. Using the same parameters as the previous section, and a wire width of 6 nm, we find that $\tau_{ee} \sim 6.9 \times 10^{-11}$ s at 1 K.

At lower temperatures, the hyperbolic cotangent function will actually saturate at a value of unity. In this low limit, the integral (7.110) becomes

$$\begin{aligned} \frac{1}{\tau_{ee}} &= \frac{e^2}{2\hbar\varepsilon_s (D + \chi_1 \tau)} \sqrt{\frac{\hbar}{2m^*}} \int_{-\infty}^{\infty} \frac{d\omega}{2\pi} \frac{1}{\sqrt{\omega}} \\ &= \frac{e^2}{2\pi\hbar^{\frac{1}{2}} \sqrt{2m^* \tau}\, \varepsilon_s (D + \chi_1 \tau)}. \end{aligned} \tag{7.112}$$

This latter form is independent of temperature, a result found for some semiconductor wires (Ikoma *et al.*, 1992) as well as for some metallic wires (Mohanty *et al.*, 1997), although these are most likely disordered in most cases. Saturated behavior also has been predicted for disordered wires by Golubev and Zaikin (1998). Using the values for the high electron mobility heterostructure, the saturated value of $\tau_{ee} \sim 9.3 \times 10^{-10}$ s.

7.4 Molecular dynamics

The treatment of the electron–electron interaction in semiconductors through the dielectric function has always involved the need to approximate the full frequency- and wave-vector-dependent dielectric function in order to obtain useful results. One way to avoid having to make these approximations is to invoke a numerical simulation of the carriers and their transport, such as with an ensemble Monte Carlo technique. This method can be expanded to incorporate the full electron–electron interaction, without approximations, in *real space*, as opposed to the momentum-space scattering and perturbation approaches used above. Here the forces between the individual particles are calculated from direct Coulomb forces between each pair of carriers.

Consider an electron distribution in which normal scattering and transport in a high electric field is treated through an ensemble Monte Carlo calculation. Inter-electronic Coulomb interaction is retained as a real-space potential, just as is the case with any self-consistent device potential, and the effect of inter-particle potential on the motion of the electrons is computed through a molecular dynamics procedure. Here, the local force on each electron due to the electric field and the repulsion of all other electrons is calculated for each time step of the Monte Carlo process (Jacoboni, 1976; Lugli and Ferry, 1986). Coulomb interaction between electrons is now treated in real space, whereas phonon scattering processes are treated in momentum space. Such a procedure has the advantage that no approximations to the range of the scattering wave vector in the Coulomb interaction must be made. Moreover, there is no need to separate the inter-particle force into direct Coulomb and plasmon terms. However, only a finite number of particles can be treated, so that a small cell of real space is considered and this cell is assumed to be replicated throughout the entire crystal. The cell is the simulation volume, which contains the number of particles being treated by the ensemble Monte Carlo procedure. The size of this volume is given by the ratio of the number of particles considered (N_0) to the simulated carrier density n, as $V = n/N_0$. Care must be taken that the simulated volume and the number of particles are sufficiently large that artifacts from periodic replication of this volume do not appear in the calculation results.

The Coulomb force is considered only through the shortest connecting vector between each pair of carriers, so that two primitive cells, of equal

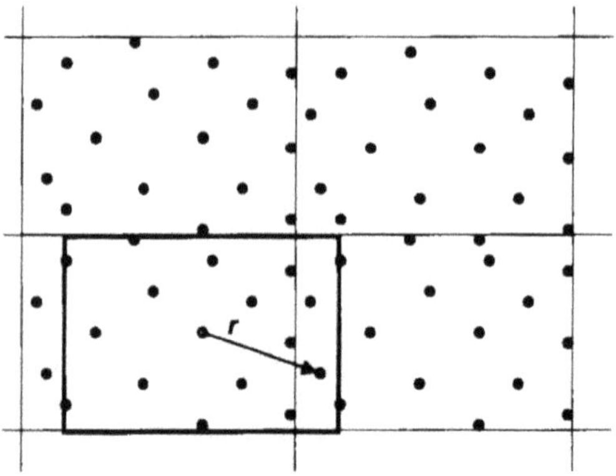

Figure 7.6 Various boxes used in the molecular dynamics simulation are illustrated here. They are discussed in the text.

volume, must be considered. The second volume is an equal-sized cell which is centered on each carrier. Thus, when the inter-particle forces are being calculated at each time step, each of the other simulated particles interacts with the particle of interest through the shortest distance between them (so that the latter may occur in one of the replications of the first volume). This is shown in Figure 7.6. Since the simulated volume may not correspond to each electron interacting through the shortest distance with the carrier of interest, replicas of the electrons that lie in the second cell must be used to compute the force through the well-known form for the electric field

$$\mathbf{F}_i = -\frac{e}{4\pi\varepsilon_s}\sum_{j\neq i}\frac{1}{r_{ji}^2}\mathbf{a}_{ij}. \tag{7.113}$$

Here, r_{ij} is the distance to the interacting particle j (from the reference particle i) and $\mathbf{a}_{ij}$ is a unit vector pointing *to the interacting particle*, with both particles assumed to have the same charge.

The two boxes are explained best by reference to Figure 7.6. In the figure, the replicated rectangular cells are the basic computational volume repeated over a so-called lattice. There are four of these basic unit cells shown in Figure 7.6, and these actually form a "superlattice" defined by the vectors $\mathbf{L} = L_x\mathbf{a}_x + L_y\mathbf{a}_y + L_z\mathbf{a}_z$. Each of these cells contains the basic number of particles N_0 used in the ensemble Monte Carlo technique; in this case, 16 carriers are illustrated in each basic cell in the figure (typically, several thousand are used). The "on-site" particle from which the vector originates in Figure 7.6 is the one at which the field sum is being computed, and its local box is the one with the heavy border. A typical vector r is indicated and shows how a particle from the next "zone" is used for the calculation, as it is closer to the on-site particle of interest than its image within the same zone as the "on-site" particle. The reason for using this centered volume can be explained simply. The desire is to sum only over the actual N_0 electrons used in the summation. However, if this is done for the test-site particle in Figure 7.6, the other $N_0 - 1$ are not uniformly distributed around this particle, and the force will have a net average value pointing to the lower left corner. This average has nothing to do with the true force and is an artifact of the test particle being off-center in the cell. Use of a centered cell avoids this problem. The appearance of the Ewald sums can also be illustrated. The total potential arising from the inter-particle potential can be written as

$$\Phi = \frac{e}{4\pi\varepsilon_s}\frac{1}{2}\sum_{i\neq j,j}\frac{1}{r_{ij}}. \tag{7.114}$$

The additional factor of 2 arises from double counting each pair-wise force in the summation over both i and j. The summation in (7.114) runs over all the particles in the solid, both those within the basic cell *and those that lie in all replicas of the basic cell.* Consider once again the test particle shown in Figure 7.6, along with the indicated single vector **r** indicated. This term and its replicas contribute to (7.114) the terms

$$\sum_{cells}\frac{1}{r} \rightarrow \frac{1}{r} + \sum_{\mathbf{L}>0}\frac{1}{|\mathbf{r}+\mathbf{L}|} \sim \frac{1}{r} + \sum_{\mathbf{L}>0}\frac{1}{L} - r\sum_{\mathbf{L}>0}\frac{1}{L^2} + r^2\sum_{\mathbf{L}>0}\frac{1}{L^3} + \ldots. \tag{7.115}$$

The denominator has been expanded since, in general, $L > \sqrt{2}r$ when the centered cell is used for the direct field calculation. The summations over the lattice vectors are known as Ewald sums (Ewald, 1921), since they were worked out (actually in reciprocal space) for X-ray scattering some years ago. The method of evaluating these inter-particle forces and incorporating the Ewald sum terms to account for the longer-range replications of the principal cell have been studied for some time in connection with molecular dynamics calculations in other fields (Potter, 1973; Brush *et al.*, 1964; Adams and Dubey, 1987). If (7.115) is multiplied by L, the expansion is in terms of the normalized distance L/r. Then, the resulting sums are independent of r. Rather, they are sums over the defining vectors of the superlattice and become constants that are well known for each choice of superlattice for the boxes. With the expansion (7.115), the potential of (7.114) and the field of (7.113) now need only be calculated over the actual number of particles used in the basic cell, provided that the particle-centered box is used in computing these sums. What these Ewald-sum terms contribute to the field (7.113) are correction terms that compensate for the fact that a finite simulation volume introduces certain Fourier periodicities that can upset the force calculation's accuracy, as well as a limiting of the number of particles available for screening the interaction. For example, the leading correction to the potential is a term linear in r rather than inversely proportional to r. If the number of particles is small, say less than 100, these corrections are large. However, if the number of particles used is large, the corrections are small. This is translated to the casual observation that the primitive cell volume should have an edge that is significantly larger than the screening length, and this translates into a large number of particles if finite-size effects are to be avoided.

The use of the molecular dynamics allows for a more exact inclusion of the electron–electron interaction within the ensemble Monte Carlo. This, in turn, allows for a more exact computation of the distribution function in high electric fields, or other places in which this function is far from equilibrium. On the other hand, it greatly magnifies the required computational time necessary to simulate such behavior. Typically, the molecular dynamics forces need to be updated with a time step smaller than 1 fs, whereas the scattering times are of the order of 0.1–5 ps. Hence, the required computational resources are significantly greater when the molecular dynamics approach is utilized, and its use is usually reserved for those cases in which significant effects on the distribution function are expected from the electron–electron interaction.

PROBLEMS

1. Using the two-dimensional plasma frequency for free carriers from the equation below (7.101) in the dielectric function (7.57), calculate the dispersion curves for the plasmon and phonon excitations for several values of wave vector in the range up to 10^8 cm^{-1}.
2. Compute the equivalent to the Debye screening wave-vector for the situation in which the distribution function is a Fermi–Dirac.
3. Assume that the screening reduction function $F(q)$ is given by

$$F(q) = \frac{1}{1 + (q/q_D)^2}.$$

What is the form of the screened interaction in real space? Assume both a three-dimensional Coulomb interaction as well as a two-dimensional Coulomb interaction.

REFERENCES

Abramowitz, M., and Stegun, I. A., 1964, *Handbook of Mathematical Functions* (Washington, D. C.: Government Printing Office).
Adams, D. J., and Dubey, G. S., 1987, *J. Comp. Phys.*, 72, 156.

Altshuler, B. L., and Aronov, A. G., 1985, "Electron–electron interaction in disordered conductors," in *Electron–Electron Interactions in Disordered Systems*, Ed. By Efros, A. L., and Pollak, M. (Amsterdam: North-Holland) 1–154.

Altshuler, B. L., Aronov, A. G., and Khmelnitskii, 1982, *J. Phys. C*, 15, 7367.

Brush, S. C., Salikin, H. L., and Teller, E., 1964, *J. Chem. Phys.*, 45, 2101.

Ewald, P. P., 1921, *Ann. Phys.*, 64, 253.

Ferry, D. K., and Goodnick, S. M., 1997, *Transport in Nanostructures* (Cambridge: Cambridge University Press) 403.

Fukuyama, H., 1985, "Interaction effects in the weakly localized regime of two- and three-dimensional disordered systems," in *Electron–Electron Interactions in Disordered Systems*, Ed. By Efros, A. L., and Pollak, M. (Amsterdam: North-Holland) 155–230.

Fukuyama, H., and Abrahams, E., 1983, *Phys. Rev. B*, 27, 5976.

Giuliani, G. F., and Quinn, J. J., 1982, *Phys. Rev. B*, 26, 4421.

Golubev, D. S., and Zaikin, A. D., 1998, *Phys. Rev. Lett.*, 81, 1074.

Goodnick, S. M., and Lugli, P., 1988, *Phys. Rev. B*, 37, 2578.

Hall, G. L., 1962, *J. Chem. Phys. Sol.*, 23, 1147.

Hollis, M. A., Palmeteer, S. C., Eastman, L. F., Dandekar, H. V., and Smith, P. M., 1983, *IEEE Trans. Electron Dev.*, 4, 440.

Ikoma, T., Odagiri, T., and Hirakawa, K., 1992, in *Quantum Effect Physics, Electronics, and Applications*, Ed. by Ismail, K., Ikoma, T., and Smith, H. I., *IOP Conf. Series*, 127, 346.

Jacoboni, C., 1976, *in Proceeding of the 13th International Conference on the Physics of Semiconductors* (Rome: Marves) 1195.

Kim, M. E., Das, A., and Senturia, S. D., 1978, *Phys. Rev.*, B18, 6890.

Lugli, P., and Ferry, D. K., 1985a, *IEEE Electron Dev. Lett.*, 6, 25.

Lugli, P., and Ferry, D. K., 1985b, *Appl. Phys. Lett.*, 46, 594.

Lugli, P., and Ferry, D. K., 1986, *Phys. Rev. Lett.*, 56, 1295.

Madelung, O., 1978, *Introduction to Solid State Theory* (Berlin: Springer-Verlag) pp. 104–9.

Mooradian, A., and Wright, G. B., 1966, *Phys. Rev.*, 16, 999.

Mohanty, P., Jariwala, E. M. Q., and Webb, R. A., 1997, *Phys. Rev. Lett.*, 78, 3366.

Penn, D., 1962, *Phys. Rev.*, 128, 2093.

Potter, D., 1973, *Computational Physics* (London: Wiley).

Takanaka, N., Inoue, M., and Inuishi, Y., 1979, *J. Phys. Soc. Jpn.*, 47, 861.

Ziman, J., 1964, *Principles of the Theory of Solids* (Cambridge: Cambridge University Press) Ch. 5.

Index